Parkinson's Disease in India

From Clinic to Bench

Editors
Madhuri Behari
S P Gorthi

MANIPAL
UNIVERSAL PRESS

First published 2026
by CRC Press
4 Park Square, Milton Park, Abingdon, Oxon, OX14 4RN

and by CRC Press
2385 NW Executive Center Drive, Suite 320, Boca Raton FL 33431

CRC Press is an imprint of Informa UK Limited

British Library Cataloguing-in-Publication Data
A catalogue record for this book is available from the British Library

ISBN: 9781032826486 (hbk)
ISBN: 9781032826493 (pbk)
ISBN: 9781003505488 (ebk)

DOI: 10.1201/9781003505488

Typeset in Garamond
by Manipal Universal Press

MANIPAL
UNIVERSAL PRESS

This book is dedicated to thousands of Parkinson's disease patients who have suffered the brunt of disease silently and cooperated in collecting data for studies which form the foundation of this work.

"to my parents Tara and Krishna Bihari"

Editors

Madhuri Behari
Formerly, Professor and Head of Department, Neurology Department, All India Institute of Medical Sciences, New Delhi

Currently, Professor and Chairperson, Flt. Lt. Rajan Dhall Fortis Hospital, Vasant Kunj, New Delhi

SP Gorthi, MBBS, MD, DM
Brigadier, Specialist, Medical and Neurologist, R&R Hospital, New Delhi

Currently, Professor and Head of Neurology, Kasturba Medical College, Manipal

Contributors

Sumit Singh, MBBS, MD, DM
Additional Professor, Neurology Department, AIIMS

Currently, Chairperson, Artemis Hospital

Ravi Yadav, MD, DM
Additional Professor
Neurology Department
National Institute of Mental Health and Allied Sciences, Bagaluru

Somdutta Ray, MBBS, DM
Senior Resident, NIMHANS
Bangaluru

Sunil Singhla, MBBS, MD
Senior Resident, Neurology Department, Artemis Hospital, Gurugram, Haryana

Roopam Singh, MBBS, DNB
Resident, Neurology Department
Artemis Horpital, Gurugram, Haryana

Soaham Dilip Desai, MBBS, MD, DM
Professor and Head, Depaartment of Neurology

Professor & Head, Department of Neurology

Pramukhswami Medical College, Privilege Centre, Shree Krishna Hospital, Karamsad, Anand, Gujarat, India-388325
soahamdd@charutarhealth.org

Shree Krishna Hospital, Karamsad, Anand, Gujarat, India-388325

Devangi Desai, MBBS, MD
Associate Professor in Medicine, Pramukh Swami Medical College and Shree Krishna Hospital , Charutar Arogya Mandal, Karamsad, Anand, Gujarat, India

Siddharth Singh, MBBS
Junior Resident, Sri Krishna Hospital, Karamsad, Anand
siddharthsinghkmc@hotmail.com

Deepti Vibha, MBBS, MD, DM
Additional Professor
All India Institute of Medical Sciences, New Delhi

Muthukani Sankaranarayanan
MBBS, MD, DM
Post Doctoral Epilepsy Fellow,
Amrita Advance Centre for Epilepsy,
Amrita Institute of Medical Sciences,
Kochi - 682041 India

RK Agrawal, MTech, PhD
Proessor and Dean
School of Computer & Systems Sciences, JawaharLal Nehru University, New Delhi- 110 067

Bharti Rana, BSc, MCA, MTech, PhD
Assistant Professor, Hansraj College, University of Delhi, Delhi, INDIA
bhartirana.it@gmail.com

Akanksha Juneja, B Tech, Mtech, Phd
Assistant Professor
National Institute Technology Delhi

S Senthil Kumaran, MSc PhD
Professor, Department of NMR
All India Institute of Medical Sciences, New Delhi
senthilssk@yahoo.com

Nirendra K Rai, MBBS, MD, DM
Associate Professor of Neurology
All India Institute of Medical Sciences Bhopal

Mohit Saxena, BSc, MSC, MPhil, PhD
Assistant Research Fellow
2545, Overlook Road, Cleveland Heights, (OH) USA
Email: mohitdelhi@yahoo.com

Vaishna Narang, BSc, MSc, PhD
Ex. Professor and Head
School of Linguist and Lnguages, J.N.U., New Delhi

Laxmi Koirala, BSc, MSc, PhD
Post doctoral Fellow, JC Self research Institute, Greenwood Genetic Center, Greenwood, SC 29646, USA

Vipin Ola, MBBS, MD, DM
Consultant Neurologist
Apex group of Hospitals
Jaipur

Vibhor Pardasani
Senior consultant
Bombay Hospital
Mumbai

Rajeev Agarwal, PhD
Physiotherapist and In-charge, Neurosciences Centre, All India Institute of Medical Sciences, Ansari Nagar, New Delhi

Vikas Agarwal, MBBS, MD, DM
Senior Consultant Neurophysician
Yashoda Hospital, Secunderabad
Telangana

Dr Aparajita Chatterjee
MBBS, MD
Registrar, Neurology, Kasturba Hospital, Manipal -576104.
tinkschat@gmail.com

Arvind, MBBS
Junior Resident
Kasturba Medical College
Manipal
Bangaluru

Thelma BK, BSc, MSc, PhD
Professor
Human Genetics Department
University of Delhi
South Campus
Delhi

Abhishek Pathak
Assistant Professor
Department of Neurology
Institute of medical Sciences
Banaras Hindu University
Varanasi
Uttar Pradesh

Vijay Chandran
Assistant Professor
Kasturba Medical College
Manipal
Bangaluru

Surya Prakash Pandey
BSC, MSC, PhD
Post doctoral Fellowship
Department of Zoology
Banaras Hindu University
Varanasi, UP

Chaturbhuj Rathore, MBBS, MD, DM

Professor & Head of Department of Neurology, SBKS Medical Institute Research Centre, Sumandeep Vidyapeeth, Vadodara, Gujrat

crathore@rediffmail.com

Vinay Goyal, MBBS, MD, DM

Professor of Neurology, All India Institute of Medical Sciences, New Delhi

drvinaygoyal@gmail.com

Ashash Sharma, MBBS, MD, DM
Associate professor
Dept of Neurology
DRPG Medical College Hospital
Kangra, (HP)-176001

Ashwwani Uttam
Consultantaa Neurology,
Regency Hospital,
Sarvodaya Nagar,
Kanpur.

Geetika Bajpai, MBBS, MD, DM
Fomerly Assistant Consultant
Neurology Department
Fortis Medical Research Institute
Gurugram

Bhawna Kaul, MBBS, MD, DM
Formerly Assistant Professor
Neurology Department
Vardhman Institute of Medical Sciences and Safdarjang Hospital
New Delhi

Jasmine Parihar, MBBS, MD
Senior Resident, All, India Institute of Medical Sciences, New Delhi

Sachin Sureshbabu
Sr Consultant Neurology
Aster MIMS Hospital
Kozhikode Kerala

Preface

It was in 1989, that I thought of concentrating on movement disorders and Parkinson's disease as my research and clinical interest. I like challenges and back in those days epilepsy and stroke were two common neurological disorders that were lucrative in terms of patient satisfaction and treatment satisfaction. Specific treatments were available for these disorders and hence most neurologists preferred to concentrate their attention to these two areas in neurology. To take up challenges and do something different has always been my way of working. So, when faced with the question of taking up a sub-specialty in neurology, I chose movement disorders and Parkinson's disease. My journey has been rewarding. As they say when something is started in true earnest, people keep on joining and thus the caravan is made. This is exactly what happened in my case; many of my colleagues joined me, many of my students preferred to take up topics in movement disorders and Parkinson's disease for their research in Movement during their Disorders course. I also looked after these patients with full dedication, trying to understand their problems, addressing them, educating them, inviting them to participate in research, which they participated with full enthusiasm.

The journey of looking after the persons with Parkinson's disease, spanning about 38 years, understanding and studying their problems, educating them about the disease and new research in the area has been very satisfying and rewarding. It has given me an opportunity to collect so much material about Parkinson's disease that when I approached my current students and former students to collate our findings in a book form, all of them confirmed and agreed enthusiastically to contribute. The book is a culmination of the affection of my students towards their teacher, my patients towards their doctor.

Usually, most research is done in laboratory and the findings are them translated in clinical use. However, since I am a clinical neurologist, we identified clinical problems first and tried to find the reason and explanations in the laboratory. This explains the name of our book "*Parkinson's Disease in India - From Clinics to Bench*". This book intends to fill a gap between a text book and reference book for researchers on Parkinson's disease in Indian patients and will provide much needed information on status of research in India on Parkinson's disease. It is not a text book. It can best be termed as a reference book for scholars and neurologists, physicians aspiring to know and gain in-depth knowledge about Parkinson's disease.

I hope the readers find something of interest to them about Parkinson's disease in India in this.

Dr Madhuri Behari

Contents

Chapter 1

How Indian Parkinson's Disease Patients are Different: Demographic Profile

Madhuri Behari

Demographic and epidemiological data on Parkinson's disease (PD) patients in India are not available in literature. PD is predominantly a disease of elderly, usually starting after age of 60 years, however, in a small percent it starts in people before the age of 40 years, when it is known as young onset PD (YOPD). No large-scale data exist for natural history of PD. PD is a relentlessly progressive neurodegenerative movement disorder which is estimated to affect approximately 1% of the population, older than 65 years of age.[1] The diagnostic criteria of disease are bradykinesia, resting tremor, rigidity, and postural instability[2] which are essentially motor symptoms; non-motor symptoms in the form of autonomic, cognitive, neuro-psychiatric disturbances, and sensory symptoms are present in large majority of patients, occurring before motor symptoms and profoundly affecting quality of life of PD patients. Majority of motor symptoms of PD result from severe and selective loss of dopaminergic (DA) neurons in the substantia nigra pars compacta (SNc). Additionally, there is loss of other neurotransmitters as well with the involvement of adrenergic, serotonergic cholinergic neuronal loss which results in non-motor symptoms of the disease.[3] The prevalence rate also varies across the globe. Several prevalence studies in India have shown age adjusted prevalence rate to vary from 33-53 per 1,00,000 population. Crude prevalence rate (CPR) varies from 68 to 74/1,00,000 except for Parsi community in which it is 148/1,00,000 (CPR 328.3/100,000).[4-7] In Orientals (Chinese and Japanese), it ranges from 99.5 to 130 (CPR 118-328/1,00,000), [8-10] and 194/1,00,000 in Caucasians (CPR 1400/1,00,000).[11] On the other hand in Africa, the CPR is very low at 10/1,00,000[12] (Figure 1). It appears density of melanin in skin may in some way be an indicator of density in substantia nigra; and those with higher number of melanized neurons in substantia nigra have higher amount of dopamine, thus being protected from PD. However, a study by Muthane et al. refuted this hypothesis and observed that Indians with lower incidence of PD indeed have lower number melanized neurons in substantia nigra,[13] though afterwards lots of questions were raised regarding the authenticity of this study.

DOI: 10.1201/9781003505488-1

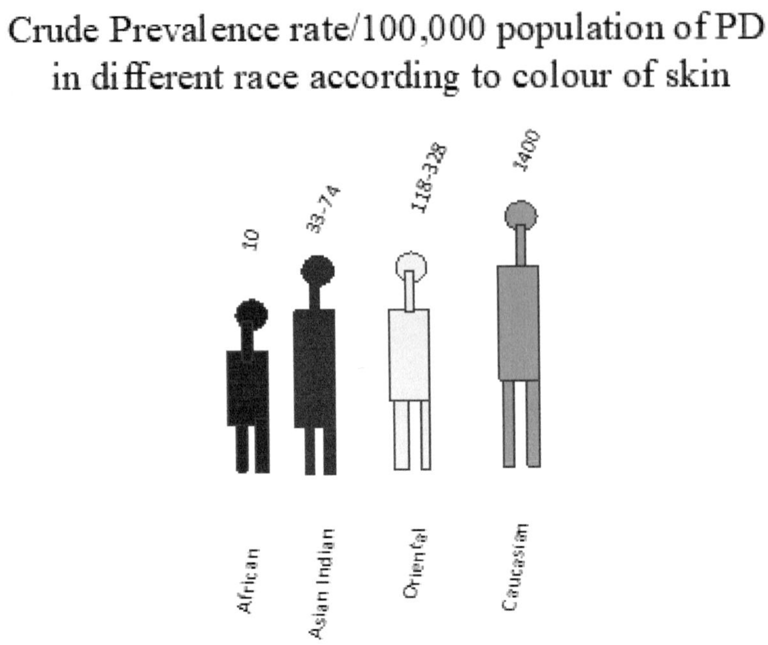

Figure 1: *Association of skin colour and prevalence of Parkinson's disease*

Darkest skin colour as seen in Black Africans have lowest crude prevalence rate followed by browns (Asian Indians) followed by yellow skinned (orientals) and highest in caucations.

Age

Though PD is considered a disease of elderly, it usually starts in the 6th decade, however, in a small minority symptoms starts in early age (between 20-40 years) when it is called as Young onset PD (YOPD). In a still smaller minority they start before 20 years of age. This subset is termed as Juvenile Parkinsonism. The mean age of the patients in our cohort consisting of 759 PD patients, was 56.87 ± 11.64 yrs. (range 20 to 85 years) and mean age of onset of symptoms was 50.36 ± 11.84 yrs. (range 16-83 years), which is about a decade younger than reported in other countries [14,15]. The age of onset was similar in both genders ($p=0.12$) in our cohort. In a study from Cambridge, UK, the mean age of cohort was 64.4 (SD 9.3) years.[14] In a large cohort from China consisting of 1,119 PD patients, the mean age of patients was 61.3 ± 10.3 years.[15] It is seen that generally younger patients usually consult tertiary care specialty canters such as ours. This is similar to what is reported from other tertiary care centers in our country (Trivandrum, Bangalore). In my personal experience of treating PD patients in a corporate hospital over the last two years, I have observed mean age of presentation as 60.84 years in a cohort of 118 PD patients (unpublished data).

Sex

Most studies show an excess of males among PD patients. The average standardized ratios of female to male are 1:3.5 for prevalence studies and 1:3.4 for incidence studies.[16] In our cohort, males were 76.3% (female to male ratio was 1:3.21). These figures are comparable suggesting male predominance. In the Indian context, more men than women seek medical care for comparably severe medical illness;

hence this may not be true reflection of gender ratio in India. In a study from China involving 1,119 PD patients, men formed 51% of the cohort (ratio 1:1.06), almost equal among the sexes.[15] In a cohort of 120 PD patients from Cambridge, UK with Hoehn and Yahr stages I and III, there were 43 women and 77 men (ration 1:1.8).[14]

Though seen commonly in men; women seem to be protected during active sexual life due to sex hormones. The difference in clinical features between the sexes is part of another chapter authored by Yadav et al. in this book.

Asymmetry of symptoms

At onset, the disease is usually asymmetrical, starting from one side of the body. Asymmetrical onset is also a supportive feature of UK PD brain bank clinical diagnostic criteria. In our cohort, the onset was from right side in 55.3% of PD patients. We found the onset from upper limb earlier than lower limb in 75.2% cases. We also observed males have onset of symptoms in upper limbs more commonly (77.5%, p=0.05) as compared to women. Asymmetry of symptoms not only occurs at onset but persists well into as the advanced stage of disease. This is also a supportive feature in the UK PD brain bank clinical diagnostic criteria. In our study, 89% subjects had asymmetrical onset with onset on right side 55.3%. This is similar to a previous study.[16] Majority of the study population was right handed and thus when dominant side is affected patients notice it earlier. Involvement of dominant side as the first side affected has been observed in other study as well.[16] The reason for earlier involvement of dominant side is not available in literature. It has been postulated that association between initial symptoms starting from dominant side and hand preference might be due to an attention bias for the dominant hand,[11] seems unlikely since distribution of symptoms persists while the disease progresses. However, as handedness reflects brain asymmetry, it is not clear whether such asymmetry reflects lateralization or progression of Parkinson symptoms.[17, 18]

Akinetic-Rigid vs. Tremor Dominant

It has been observed that some PD patients have tremor as the dominant major disabling symptoms, while in others akinesia-rigidity is a major problem. Based on this, it is customary to divide patients as tremor dominant or akinetic – rigid sub-types of PD. Several clinical features have been attributed to tremor dominant or akinetic/ postural instability gait disorder (PIGD) sub-type of PD. One of them is prognosis. There are reports stating that patients with tremor-dominant PD have a relatively slower disease progression and have better prognosis. This idea was first proposed by Hoehn and Yahr in 1967,[19] who found a greater proportion of death and disability in PD patients with non-tremor onset mode, at least during the first 10 years of the disease compared to patients with tremor-dominant sub-type of PD. Patients with PD with PIGD subtype had higher annual increase in symptom severity[20] and shorter survival rate.[21, 23] Furthermore, a post-mortem study showed that patients with tremor-dominant PD progressed more slowly to Hoehn and Yahr (H&Y) stage IV than patients with PIGD type of PD, using the sub-typing scheme suggested by earlier authors.[15, 24] Finally, patients with tremor-dominant PD have been shown to have better cognitive performance than non-tremor-dominant PD[15,25,26] and are less likely to develop dementia.[27,28] Categorization of PD patients in to tremor dominant vs. akinetic rigidity dominant sub-types was obtained by dividing patient's "total tremor score" by their "non-tremor score". The tremor score was derived from the sum of items 16 and 20–26 on the Unified Parkinson's Disease Rating Scale (UPDRS) divided by 8 (the number of items included) and represented the degree of tremor reported in the activities of daily living section of the UPDRS, along with objective tremor at rest and with action, determined on physical examination. The non-tremor score was derived from the sum of items 5, 7, 12–15, 18, 19, and 27–44 on the UPDRS divided by 26 (the number of items included). This measure assesses speech, swallowing, ability to turn in bed, falls, freezing, and walking from the activities of daily living section as well as speech, facial expression, rigidity, bradykinesia, ability to stand, posture, gait, and postural stability determined by motor examination on the UPDRS.[15] In our study, for tremor dominance we took all scores of tremor items on UPDRS-III (item 20 - on 5 body parts + item 21- on 2 body parts). The total tremor score was divided by 7 to give "unified tremor score". To obtain akinetic rigidity score, we took all the scores on bradykinesia and rigidity items on UPDRS-III (item 22 - 4 body parts + item 23 - 2 body parts + item 24 - 2 body parts + item 25-2 body parts + item 26 - 2 body parts + item 27-1 body part + and item 31- 1 body part). The sum of the total score was then divided by 14 to give a "unified akinetic–rigidity score". The subtype of PD was decided on the basis of whichever score was higher i.e. "unified tremor score" or "unified akinetic rigidity score". In our cohort of 759 patients, tremor dominant PD formed 66.93% (508 patients) and akinetic rigid sub-type was 33.069% (251 patients). At enrolment, duration of disease (49.87 vs. 35.11 months), tremor as first symptom (75.00% vs. 75.69%), gait disturbance (69.88% vs. 64.44%), disturbance of balance (9.84 vs. 9.16%), falls (4.33% vs. 3.18%), voice problem (10.23% vs. 7.17%), dystonia (9.44% vs. 11.55%), wearing off (23.62% vs. 28.28%), on–off phenomenon (16.73% vs. 15.93%), motor blocks (24.01% vs. 18.72%), Schwabb & England score [(mean ± SD) 76.10 ±17.88 vs. 77.89 ±19.61] and H&Y stage (2.84 vs. 2.43) were not significantly different in the two groups i.e. tremor dominant vs akinetic – rigid type. On the non-motor front too, hallucinations (4.92% vs. 3.58%), depression (9.05% vs. 10.35%), thought disorder (esp. delusions) (0.78% vs. 1.195%), sleep disturbance (14.17% vs. 9.96%), and autonomic symptoms (9.87% vs. 12.01%) were also not significantly different in the two groups. However, dyskinesia (17.71 % vs. 9.56%), first symptom starting from dominant side (85.79% vs. 67.84%), upper extremity involvement before lower extremity (65.79% vs. 57.84%), higher levodopa equivalent daily dose (LEDD) (mean ± SD) (848.96 ± 598.93 vs. 673.42 ± 565.15) were significantly more in tremor dominant PD patients. It was also noted that dyskinesia occurred significantly earlier in tremor dominant PD patients (88.78 ± 25.62 vs. 96.24 ± 26.12) months after onset of motor symptoms. On the other hand, patients with akinetic rigid dominant PD patients had significantly faster rate of progression of disease (4.12 vs 3.67) points on UPDRS motor score per year, suggesting more aggressive course in akinetic rigid type as compared to tremor subtype. At onset of symptoms, tremor was the dominant symptom in 67.5% among our cohort. Males were more likely to have tremor at onset (69.94% vs. 59.44% $p=0.01$) as compared to women. This finding is comparable with DATATOP cohort,[29] in which males predominantly had tremor dominant subtype of PD as prominent gender difference.

Males generally had a higher incidence of disabling tremor at rest and early onset of dyskinesia. On the other hand, in another study,[30] patients with tremor dominant sub-type formed 23.42% of all PD patients. However, this was a small study with only 98 participants and that could have resulted in this bias. The authors of this study also reported significantly slower progression in tremor predominant disease. They also noted poor performance on Mini-Mental State Examination (MMSE) scores in tremor predominant patients.

Among the PD patients treated at our centre, akinetic rigid sub-type had faster rate of disease progression and early progression to involve the second side of body (bilateral) and higher incidence of depression. These results were comparable with previous study of DATATOP cohort.[29] In this study too, although there was no difference in duration of symptoms, the akinetic rigid sub-type had greater motor disability, suggesting more aggressive course in this group as compared with tremor dominant sub-type.

Clinical features vs. side of the body first involved

Some authors have also tried to examine the effect of disease starting from left or right side. They noted neuropsychological difference when the disease started from right side vs. left side.[31] We, however, did not find any such difference.

Rate of progression

As regards to progression of the disease, there are no fool proof methods to assess it. Dopamine receptor uptake scans are expensive and it is not feasible to do in any study routinely. We devised a method based on clinical examination; though it stands to criticism as it has not been tested objectively, nor has inter rater reliability established. The slopes (i.e. the annual rates of decline) of progression in the UPDRS-III scores, is what we measured by dividing UPDR-III score by the number of years the patient reported to have had symptoms, to give a rough idea about the rate of worsening per year. Alternately, in patients who were on follow up, we had yearly UPDRS motor scores recorded, and the difference in scores on two consecutive years gave figure of worsening per year. Mean rate of progression of PD was 3.67±1.22 per year in tremor dominant PD sub-type as compared to 4.12 ± 1.2/ year in akinetic rigid sub-type, which was faster in akinetic-rigid type ($p=0.04$). Overall mean time for disease to become bilateral was 30.78 ± 27.0 months (with very wide standard deviation). Overall mean rate of progression was 4.02 ± 1.32/ year.

Motor complication

Gait disturbance is a major cause of disability. Problems of gait consist of shuffling, dragging of feet, diminished arm swing, propulsion sometimes leading to falls and difficulty in turning. Gait disturbance is also one of the leading causes of poor quality of life. Gait disturbance was observed in 68.11% patients in the cohort. It started after a mean interval of 57.73 ± 33.51 (range 36-240) months after onset of motor symptoms. Falls were reported by 4% of PD patients. Mean duration of illness, before falls were reported was 96.60 ± 14.12 (range 48-264) months. Other problems like dyskinesia, dystonia, and motor blocks were reported by 114 (15%), 77 (10.14%), and 169 (22.3%) patients respectively. Whereas, dyskinesia occurred after a mean interval of 95.78 ± 36.39 (range 84-216) months, dystonia was noted after mean interval of 84.06 ± 20.10 (range 60-192) months, and motor blocks took on an average of 106.63 ± 47.78 (range 72-248) months after the motor symptoms were observed. Mean duration of levodopa therapy to onset of dyskinesia was 88.32 (range 36-216) months and mean duration of levodopa therapy before dystonia was observed was 37.75 ± 13.25

months. Phenomenon of wearing off was noted in 191 (25.2%) of patients, which started 70.94 ± 29.47 (range 60-276) months after onset of motor symptoms. Similarly, on – off phenomenon was seen in 16.5% of patients which developed 93.58 ± 36.38 (range 48-200) months after onset of symptoms. Voice disturbance was noted in 20.55% of patients.

We did not find any difference in motor fluctuations apart from dyskinesia (significantly higher in tremor predominant) and non-motor features apart from depression (significantly more in akinetic rigid) between akinetic-rigid vs. tremor dominant type of PD. Low incidence of dementia in our subjects may be because we used MMSE as a tool to diagnose dementia and detailed neuropsychological tests were not done.

The initial symptom of tremor predominant PD was tremor (75.00%), but patients with akinetic rigid dominant PD too had tremor as the initial symptom in 75.67%. We also found that the mean levodopa equivalent daily dose (LEDD) was higher in tremor predominant PD patients. The reason for this could be the presence of disabling tremor in these patients which might have necessitated increase in drug dosage to control it, subsequently leading to higher incidence of dyskinesia; which was indeed observed.

Treatment and dose of dopaminergic drugs

The LEDD was calculated as described earlier.[32] Mean LEDD was 790.78 (range 15-2300) mg. In a report on Italian patients almost similar dose was used.[33] LEDD was higher in YOPD (872.96, $p=0.04$) and tremor dominant cases (848.96, $p<0.0005$) as compared to later onset PD (LOPD) and akinetic rigidity sub-type of PD. In a recent study consisting of 2247 PD patients from Glasgow, UK, LEDD was 295 and 926 in LOPD and YOPD cases, whose duration of symptoms was 1.3 (±0.9) and 10.2 (± 6.7) years, respectively.[34] For similar duration of disease, our Indian patients took a much smaller LEDD. We observed that most of our patients had very good initial response to dopamine replacement therapy being in the range of 70-100%.

Sleep benefit in motor symptoms

In questionnaire-based studies on sleep benefit, 30–55% of thc patients reported to have experienced the phenomenon of benefit in motor symptoms after sleep.[35-37] However, objective examination (UPDRS-III) did not corroborate it.[38] Another study on subjective assessment of two tasks in PD patients reporting sleep benefit and those not reporting sleep benefit showed that the subjective experience of sleep benefit is not paralleled by an actual improvement in motor functioning. Sleep benefit therefore appears to be a subjective phenomenon and not a Parkinson-specific reduction in symptoms.[39]

YOPD vs LOPD

In our study on Indian PD patients seen at a tertiary care centre, 23.58% (179 out of 759 patients) of PD cohort had their disease onset before or at 40 years of age and formed the YOPD group. The rest, whom we termed as LOPD had some characteristics different from YOPD. Most YOPD had tremor predominant disease. Duration of disease was significantly higher in YOPD (88.97 ± 58.21 vs. 77.04 ±

58.58 months) and also the LEDD (872.96 ± 584.93 vs. 765.42 ± 594.15) respectively. YOPD patients also had higher incidence of gait disturbance (75.97% vs. 65.68%), falls (6.14% vs. 3.27%), wearing off (31.28% vs. 23.27%), motor blocks (28.33% vs. 20.34%), depression (13.4% vs. 8.27%), and sleep disturbance (20.11% vs. 10.51%); all statistically significant. On the other hand, other clinical features did not show any significant difference between the two groups. These clinical features are onset of symptoms from right side (56.42% vs. 55%), time taken for the disease to become bilateral after onset (68.97 ± 58.21 vs. 67.4 ± 58.30) months, Schwabb & England score (78.64 ±19.80 vs. 77.02 ± 18.01 %), H&Y stage (2.77 ± 0.82 vs. 2.45 ±0.31), rate of progression (3.69 ±2.12 vs. 3.48 ±1.2) per year. Other motor complications such as imbalance, dystonia, dyskinesia, change in voice, on-off phenomenon, thought disorder, hallucinations, and autonomic disturbance were also not significantly different in the two groups. Although in literature,[40, 41] the rate of progression has been reported to be higher in older subjects (>70) we got different results, since we divided our cohort in two groups as follows: Equal to or less than 40 years for YOPD and more than 40 years for LOPD. This may account for different rates of progression in our series. Also, YOPD patients presented to us after having symptoms for longer period (mean duration of disease 88.97 ± 58.21 months) and had higher requirement of daily dopaminergic drugs as measured by mean LEDD as compared to LOPD. Higher frequency of hallucinations and falls are unique findings in our cohort and has not been described in earlier study.[36] Depression before onset of motor symptoms of PD was not prominent in this group. We could not attribute any reason for this difference, but it may be possible that with the advent of newer treatment strategies, longevity is seen in the cohort and so also the prevalence of depression which happens with chronic diseases. Other motor and non-motor complications such as motor blocks, wearing off, and sleep disorders were also more often seen in YOPD group, which is again possibly due to longer duration of disease in these patients. [42] However, we did not find any difference as regards to prevalence of dementia and dystonia. We observed that YOPD experience a higher disabling disease than LOPD patients do, when measured by UPDRS.

Non- motor symptoms

Non-motor symptoms (NMS) was subject of study in another different cohort consisting of 74 patients with PD with mean ± age and age of onset of symptoms were 59 ± 11.56 (range 33-85 years) and 51.7±11.66 (range 22-78) years respectively. All the patients in this study had one or more NMS. Common NMS were sleep disturbance (83.78%), urinary disturbance (83.43%), mood and cognition and attention and memory or both (81.08%), GIT symptoms (75%), CVS/falls (72.97%), sexual disturbance (51.35%), and perceptual disturbance and hallucinations (40.54%) according to NMSS (Singhla et al., unpublished data). Common NMS reported by Zhang and colleagues were constipation in 28% patients, (of which 17.4% were pre-motor symptoms and 10.6% post-motor symptoms), lowered sense of smell or taste was 39.8% (total), 21.3% (pre-motor symptom), and 18.5% (post- motor symptom) respectively. Similar figures for hallucinations, depression, REM sleep behavior disorder and pain were 5.6%, 4.1%, 1.4% (total, pre-motor, post-motor); 30.4%, 26.8% and 3.6% (total, pre-motor, post-motor); 18.5%, 11.6% and 6.9% (total, pre-motor, post-motor); and 18.4%, 14.3%, 4.1% (total, pre-motor, post-motor), respectively in a cohort consisting of 1,119 PD patients from China.[15]

Genetic factors

Recently, genetic factors are being recognized in the pathogenesis of PD. Though true monogenic Mandelian forms are rare, polygenic inheritance is now seen more commonly. Family history suggesting autosomal dominant or autosomal recessive inheritance is also identified rarely in PD. A positive family history is observed in about 21% of unselected PD patients in areas of India, where consanguinity is not common (Wani et al; unpublished data). Two most commonly tested autosomal dominant Parkinson disease (ADPD) genes namely Syn-nuclein Alpha (*SNCA)* and (Leucine Rich Kinase2) *LRRK2* are negligible or almost non-existent in our cohort. Mutations in *SNCA, LRRK2, PARKIN, PINK1,* and *DJ1* genes account for <30% of all familial PD cases tested to date in the country. Two novel causal genes namely *RIC3* and *PODXL* have been identified in an ADPD and autosomal recessive juvenile parkinsonism (ARJP) family each and a novel mutation in *SYNJ1* has recently been discovered in an ARJP family.[43]

In our cohort, genetic factors and identified genes are also given in another chapter authored by Thelma et al. in this book.

Conclusion

To conclude, PD patients who attended tertiary care facility are generally younger, mean age of 56 years, males generally dominate with a ratio of about 1: 3.2, family history is present in 21%. YOPD patients i.e. those in whom symptoms start at or before 40 years of age form 23.6%. The disease is asymmetrical at onset in 89.9% and remained asymmetrical in advanced stage as well. In 53%, it started from dominant side. On an average it took 30.78 months to become bilateral. A subset of tremor dominant PD is identified separate from akinetic rigid dominant sub-type, with more fulminant disease in akinetic rigid form in which the yearly slope of decline in UPDRS (part 3) is 4.12 as compared to tremor dominant PD sub-type in which it was 4.02/year. In patients with tremor dominant PD progression to become bilateral was also slower as compared to akinetic rigid sub-type. In them it took 65.79 months. Frequency of motor blocks was common in akinetic rigid type; whereas dyskinesias are more common in tremor predominant disease. Mean LEDD that was used was 790.78 (range 15-2300) mg. LEDD was higher in YOPD (872.96, p=0.04) and in tremor dominant cases (848.96, p<0.0005) as compared to LOPD and akinetic rigidity sub-type of PD. YOPD cases generally had longer duration of illness, required higher LEDD, with higher rates of motor complications like motor blocks, gait disturbance, and wearing off. YOPD also had higher prevalence of sleep disorders and depression.

References

1. Lang AE, Lozano AM. Parkinson's disease. First of two parts. N Engl J Med. 1998; 339(15);1044-53.
2. Hughes AJ, Daniel SE, Kilford L, Lees AJ. Accuracy of clinical diagnosis of idiopathic Parkinson's disease: A clinico-pathological study of 100 cases. J Neurol, Neurosurg Psychiatry. 1992; 55;181–4.
3. Braak H, Ghebremedhin E, Rüb U, Bratzke H, Del Tredici K. Stages in the development of Parkinson's disease-related pathology. Cell Tissue Res. 2004; 318(1);121-34.
4. Saha SP, Bhattacharya S, Das SK, Maity B, Roy T, Raut DK. Epidemiological study of neurological disorders in a rural population of Eastern India. J Indian Med Assoc. 2003;101;299 –300, 302–304.

5. Gourie-Devi M, Gururaj G, Satishchandra P, Subbakrishna DK. Prevalence of neurological disorders in Bangalore, India: A community-based study with a comparison between urban and rural areas. Neuroepidemiology. 2004;23;261–268.
6. Das SK, Biswas A, Roy T, et al. A random sample survey for prevalence of major neurological disorders in Kolkata. Indian J Med Res. 2006;124;163–172.
7. Bharucha NE, Bharucha EP, Bharucha AE, Bhise AV, Schoenberg BS. Prevalence of Parkinson's disease in the Parsi community of Bombay, India. Arch Neurol. 1988;45;1321–1323.
8. Wang SJ, Fuh JL, Teng EL, et al. A door-to-door survey of Parkinson's disease in a Chinese population in Kinmen [published erratum appears in Arch Neurol 1996;53:508]. Arch Neurol. 1996;53;66 –71.
9. Chen RC, Chang SF, Su CL, et al. Prevalence, incidence, and mortality of PD: A door-to-door survey in Ilan County, Taiwan. Neurology. 2001;57;1679 –1686.
10. Kusumi M, Nakashima K, Harada H, Nakayama H, Takahashi K. Epidemiology of Parkinson's disease in Yonago City, Japan: Comparison with a study carried out 12 years ago. Neuroepidemiology. 1996;15;201–207.
11. de Rijk MC, Breteler MM, Graveland GA, et al. Prevalence of Parkinson's disease in the elderly: The Rotterdam Study. Neurology. 1995;45;2143–2146.
12. Osuntokun BO, Adeuja AO, Schoenberg BS, et al. Neurological disorders in Nigerian Africans: A community-based study. Acta Neurol Scand. 1987;75;13–21.
13. Muthane U, Yasha TC, Shankar SK. Low numbers and no loss of melanized nigral neurons with increasing age in normal human brains from India. Ann Neurol. 1998;43(3);283-7.
14. Lewis SJ, Foltynie T, Blackwell AD, Robbins TW, Owen AM, Barker RA. Heterogeneity of Parkinson's disease in the early clinical stages using a data driven approach. J Neurol Neurosurg Psychiatry. 2005;76(3);343-8.
15. Zhang Z, Roman GC. Worldwide occurrence of Parkinson's disease: An update review. Neuroepidemiology. 1993;12;195–208.
16. Zhang T, Shu-yang Yu, Guo P, et al.. Non-motor symptoms in patients with Parkinson disease A cross-sectional observational study. Medicine. 2016;95;50.
17. Yust-Katz S, Tesler D, Treves TA, Melamed E, Djaldetti R. Handedness as a predictor of side of onset of Parkinson's disease. Parkinsonism Relat Disord. 2008;14(8);633-5.
18. van der Hoorn A, Bartels AL, Leenders KL, de Jong BM. Handedness and dominant side of symptoms in Parkinson's disease. Parkinsonism Relat Disord. 2011;17(1);58-60.
19. Uitti RJ, Baba Y, Whaley NR, Wszolek ZK, Putzke JD. Parkinson disease: handedness predicts asymmetry. Neurology. 2005;64;1925-30.
20. Hoehn MM, Yahr MD. Parkinsonism: onset, progression and mortality. Neurology. 1967;17(5);427-42.
21. Jankovic J, Kapadia AS. Functional decline in Parkinson disease. Arch Neurol. 2001;58(10);1611-5.
22. Lo RY, Tanner CM, Albers KB, et al. Clinical features in early Parkinson disease and survival. Arch Neurol. 2009;66(11);1353-8.
23. Forsaa EB, Larsen JP, Wentzel-Larsen T, Alves G. What predicts mortality in Parkinson disease?: a prospective population-based long-term study. Neurology. 2010;75(14);1270-6.
24. Rajput A, Vilariño-Güell C, Rajput ML, et al. Alpha-synuclein polymorphisms are associated with Parkinson's disease in a Saskatchewan population. Mov Disord. 2009;24(16);2411-4.

25. Rajput AH, Pahwa R, Pahwa P, Rajput A. Prognostic significance of the onset mode in parkinsonism. Neurology. 1993;43(4);829-30.
26. Vakil E1, Herishanu-Naaman S. Declarative and procedural learning in Parkinson's disease patients having tremor or bradykinesia as the predominant symptom. Cortex. 1998;34(4);611-20.
27. Burn DJ, Rowan EN, Allan LM, Molloy S, O'Brien JT, McKeith IG. Motor subtype and cognitive decline in Parkinson's disease, Parkinson's disease with dementia, and dementia with Lewy bodies. J Neurol Neurosurg Psychiatry. 2006;77(5);585-9.
28. Aarsland D, Andersen K, Larsen JP, Lolk A, Kragh-Sørensen P. Prevalence and characteristics of dementia in Parkinson disease: an 8-year prospective study. Arch Neurol. 2003;60(3);387-92.
29. Williams-Gray CH, Foltynie T, Brayne CE, Robbins TW, Barker RA. Evolution of cognitive dysfunction in an incident Parkinson's disease cohort. Brain. 2007;130(7);1787-98.
30. Jankovic J, McDermott M, Carter J, et al. Variable expression of Parkinson's disease: a base-line analysis of the DATATOP cohort. The Parkinson Study Group. Neurology 1990;40;1529–1534.
31. Liu ZL, Wang LM, Tao EX. Study of the clinical heterogeneity of movement disorder in Parkinson disease. Nan Fang Yi Ke Da Xue Xue Bao. 2008;28(3);447-8.
32. Adwani S, Yadav R, Kumar K, Chandra SR, Pal PK. Neuropsychological profile in early Parkinson's disease: Comparison between patients with right side onset versus left side onset of motor symptoms. Ann Indian Acad Neurol. 2016;19(1);74-8.
33. Tomlinson CL, Stowe R, Patel S, Rick C, Gray R, Clarke CE. Systematic review of levodopa dose equivalency reporting in Parkinson's disease. Mov Disord. 2010;25(15);2649–53.
34. Cova I, Di Battista ME, Vanacore N, et al. Validation of the Italian version of the Non Motor Symptoms Scale for Parkinson's disease. Parkinsonism and Related Disorders 34. 2017;38-42.
35. Malek N, Swallow DMA, Grosset KA, Lawton MA, Marrinan SL, Lehn AC, Bresner C et al on behalf of the PRoBaND clinical consortium. Tracking Parkinson's: Study Design and Baseline Patient Data. Journal of Parkinson's Disease 5. 2015;947–959.
36. Diamond SG, Markham CH, Hoehn MM, McDowell FH, Muenter MD. Effect of age at onset on progression and mortality in Parkinson's disease. Neurology. 1989;39 (9);1187-90.
37. Merello M, Hughes A, Colosimo C, Hoffman M, Starkstein S, Leiguarda R. Sleep benefit in Parkinson's disease. Mov Disord. 1997;12;506–8.4.
38. van Gilst MM, Bloem BR, Overeem S. Prospective assessment of subjective sleep benefit in Parkinson's disease. BMC Neurol. 2015;15;2.5.
39. Currie LJ, Bennett JP Jr, Harrison MB, Trugman JM, Wooten GF. Clinical correlates of sleep benefit in Parkinson's disease. Neurology. 1997;48;1115–7.25.
40. Högl BE, Gomez-Arevalo G, Garcia S, Scipioni O, Rubio M, Blanco M, Gershanik OS. A clinical, pharmacologic, and polysomnographic study of sleep benefit in Parkinson's disease. Neurology. 1998;50;1332–9.
41. van Gilst MM, van Mierlo P, Bloem BR., Overeem S, Quantitative Motor Performance and Sleep Benefit in Parkinson Disease SLEEP. 2015;38(10);1567-1573.
42. Ferguson LW, Rajput ML, Muhajarine N, Shah SM, Rajput A. Clinical features at first visit and rapid disease progression in Parkinson's disease. Parkinsonism Relat Disord. 2008;14(5);431-5.
43. Sudhaman S, Prasad K, Behari M, Muthane UB, Juyal RC, Thelma BK. Discovery of a frameshift mutation in podocalyxin-like (PODXL) gene, coding for a neural adhesion molecule, as causal for autosomal-recessive juvenile Parkinsonism. Med Genet. 2016;53(7):450-6. doi: 10.1136/jmedgenet-2015-103459.

Chapter 2

Parkinson's Disease and Gender Differences

Somdatta Ray, Ravi Yadav

Parkinson's disease (PD) is a progressive neuro-degenerative disorder, second only to Alzheimer's disease. According to western literature it affects 0.3% of the population.[1] The prevalence of PD in India shows marked topographic variation, ranging from 14.1 per 1,00,000 in rural Kashmir[2] to 328.3 per 1,00,000 in Mumbai.[3] Epidemiological data from Europe, Asia, and North America have shown that PD is more common in men.[4-7] PD has slight preponderance in males in China,[8] however in Japan; the disease is more common in females.[9] In India, the data shows slightly higher prevalence in males from Kolkata[10] and higher prevalence in females.[11] In a large meta-analysis, it was suggested that the prevalence in men is two times higher as compared to women.[12] There has been evidence from the animal models, clinical studies, and trials regarding the role of male and female sex hormones on the various aspects of dopaminergic pathways and PD. This chapter reviews the current evidence, highlighting the gender differences on clinical features of PD and laboratory evidence in animal models' of PD.

Clinical research

Sexual dimorphisms have been noted in the basal ganglia, with younger women having a higher baseline number of dopaminergic neurons.[13] Animal experiments have revealed that there is an upgradation of genes responsible for signal transduction and neuronal maturation in females, while genes causing oxidative phosphorylation were found in males.[14]

Parkinson's disease is most commonly seen in post-menopausal women, indicating that the pathogenesis of PD may be associated with estrogen level reduction.[15] Risk of PD increases with shorter estrogen exposure,[16] longer length of endogenous estrogen exposure or a shorter fertile period, correlated with onset of PD at an older age with less severe impairment of motor function in many studies.[17] Similarly, higher parity and higher cumulative length of pregnancies are shown to be associated with higher and longer duration of estrogen exposure that not only is protective, but also delays the onset of clinical manifestation in women who will eventually develop PD.[18] Bilateral oophorectomy also

DOI: 10.1201/9781003505488-2

increased the risk of PD.[19] Postmenopausal administration of hormonal replacement therapy has been linked with lower risk of PD.[20]

Table 1: *The location and distribution of the major estrogen receptors in the brain*

Regions with predominantly ER alpha in rats	Regions with predominantly ER beta in rats (35)
1. Amygdala areas 2. Lateral septal nucleus 3. Media septal nucleus 4. Bed nuclei of stria terminalis, anterior, and posterior part 5. Thalamus 6. Median pre-optic nucleus 7. Posterior ventricular area of hypothalamus 8. Area postrema 9. Medial zone of hypothalamus 10. Nucleus ambiguous 11. Substantia nigra 12. Periaquductal 13. Locus ceorulus	• Deep nuclei of cerebellum • Interpeduncular nucleus • Medial mammillary nucleus • Anterior hypothalamic nucleus • Premamillary nucleus • Anterodorsal pre-optic nucleus • Medial parvicellular part of hypothalamus • Supra-optic nucleus • Substantia innominate • Anterior amygdaloid area • Isocortex

Reports showing effects of estrogen on women with PD have not shown any consensus. Earlier, reports claim estrogen as an antidopaminergic agent.[21] A study had shown that treatment with estrogen in women with hysterectomy may actually increase the risk of developing PD, thus opining that although endogenous estrogen may be protective in PD, exogenous estrogenous therapy might accentuate the risk of PD.[22] This increased risk of PD was seen only when hysterectomized women were treated with estrogen, indicating that unopposed estrogen has effects dissimilar to that of combined therapy. This study also found that if menopause occurred at an earlier age, there was reduction of risk of PD by about 50%.[23] However, a randomized controlled study demonstrated more "on time" after administering estrogen in women with PD. [24]

Testosterone deficiency was found in men with PD, which was similar or higher than in general population.[25] Exposure to certain pesticides like paraquat that have estrogen blocking or mimicking effect is associated with increased chances of PD.[26] Head trauma predisposes men to develop PD as they are more likely to participate in contact sports and be involved in road traffic accidents.[27]

The impact of emotional stress on brain structure and function has shown major differences in men and women that could be an additional risk in increasing risk of PD in males. Functional magnetic resonance demonstrates sex differences in response to stress. When stressful events accumulate, they increase the likelihood of developing PD.[28] Moreover; inability to place oneself against the negativity of stress additionally worsens life in PD.

Sex differences have been observed in the clinical profile of PD. Women develop PD two years later than men.[29,30] Clinical manifestations are less severe in women in early stages of PD compared to men.[29,31] Rapid eye movement behaviour disorder and rigidity were found to be more commonly seen in men whereas depression associated with PD and dyskinesias were more frequently seen in women.[32] Women are also more likely to present with tremor,[29] impaired visuospatial cognition,[32] nervousness, depression, and constipation.[33] Women with PD had better speech articulation and fine motor control.[34] Facial emotional recognition and verbal fluency is impaired among men[32] and they also suffered more from urinary symptoms like dribbling as well as excessive daytime sleepiness.[33]

Laboratory evidence in animal models

Animal experiments indicate sex difference in PD to be contributed mainly by estrogen. It has been shown that ovariectomy in rats and primates result in dopaminergic neuron loss that could be reversed with exogenous estrogen administration.[35,36] Studies have demonstrated that female mice are more resistant to striatal dopaminergic neuron loss after MPTP (-methyl-4-phenyl-1,2,3,6 tetrahydropyridine) and methamphetamine administration[37, 38] whereas only male mice can be rendered parkisonian after administration of these toxins. Data on the effect of estrogen on the toxin induced Parkinsonism models have demonstrated both protective and non-protective impact of estrogen on substantia nigra neurons. However, it was observed that the protective effects of estrogen could be demonstrated only when estrogen was administered prior to toxin exposure,[39] suggesting that physiological levels of estrogen during lesioning, whether attained through endogenous or exogenous administration, is protective in partially injured female nigrostriatal dopaminergic system.[40] Estrogen has been shown to alter the release or metabolism of dopamine but does not affect striatonigral dopamine loss.[41] 17 beta estradiol, with its highest receptor affinity is most neuroprotective.

It is observed that 17 beta estradiol exerts its neuroprotective effects in multiple ways. Firstly, it acts on MAPK/ERK and P13/AKT pathways causing down regulation of pro apoptotic proteins like GSK3 and BAD, thus protecting against glutamate, H_2O_2 and beta amyloid.[39] There is an increased expression of neurotrophic factors like glial cell line derived neurotophic factor mediated by estrogen receptors and this is an another mechanism of neuroprotection. In addition, estrogen causes increased expression of superoxide dismutase, glutathione peroxidase, and glutaredoxin.[42] Estrogen also exerts protective effects on mitochondrial membrane potential, prevents free radical production and ATP depletion increases glucose utilization by increasing production of glycolytic enzyme pathways. Estrogen also increases mitochondrial DNA expression. Estrogen also prevents formation of Lewy body in idiopathic PD.[43] Interestingly, estrogen has also been found to have an inhibitory function on dopamine pathways by causing prolactin release which suppresses the dopaminergic system.[44]

Progesterone prevented striatal dopamine loss in gonadectomized male and female models.[45] Hence, progesterone can also be considered as a neuroprotective agent.

Testosterone appears to have detrimental effect when tested in laboratory. When male mice, whether gonadectomized or not are treated with testosterone, striatal dopaminergic loss worsened with mild to moderate lesions, but not in extensive lesions.[46] Data suggests that nigro, striatal dopaminergic depletion in male rats is mediated by testosterone, after aromatase enzyme located in certain areas of the brain converts it into estradiol.[47]

However, sex hormones are not the sole cause for sex differences in Parkinsonism; dopaminergic neurons grown in culture with identical hormonal milieu exhibit sex differences.[45] The effect could be due to the genes located on the Y chromosome itself or due to incomplete silencing of X chromosome genes in females with XX chromosome.[48] The studies on the effects of SRY region of Y chromosome on dopamine metabolism highlight its association with increased monoamine oxidase, DOPA decarboxylase, and Tyrosine hydroxylase. [49] SRY also plays a role in the development of dopamine neurons.[50]

Summary

There is some scientific evidence to suggest that estrogen has some neuroprotective action on dopaminergic neurons. Additionally, clinical observations suggest women PD patients have slightly different phenotypic features in the form of higher tremor dominance, more dyskinesia and less rapid progression to dementia. Hence to conclude, clinical and experimental studies suggest that female sex hormones have important neuroprotective role in PD and also in determining age of onset and to some extent a benign phenotype.

References

1. Dexter DT, Jenner P. Parkinson disease: from pathology to molecular disease mechanisms. Free Radic Biol Med. 2013;62;132-44.
2. Razdan S, Kaul RL, Motta A, Kaul S, Bhatt RK. Prevalence and pattern of major neurological disorders in rural Kashmir (India) in 1986. Neuroepidemiology. 1994;13(3);113–9.
3. Bharucha EP, Bharucha NE. Epidemiological study of Parkinson's disease in Parsis in India. Adv Neurol. 1993;60;352–4.
4. Muangpaisan W, Hori H, Brayne C. Systematic review of the prevalence and incidence of Parkinson's disease in Asia. J Epidemiol. 2009;19(6);281–93.
5. Twelves D, Perkins KS, Counsell C. Systematic review of incidence studies of Parkinson's disease. Mov Disord. Jan 2003;18(1);19–31.
6. Mayeux R, Marder K, Cote LJ, Denaro J, Hemenegildo N, Mejia H, et al. The frequency of idiopathic Parkinson's disease by age, ethnic group, and sex in northern Manhattan, 1988–1993. Am J Epidemiol. 1995;142(8);820–7.
7. Bower JH, Maraganore DM, McDonnell SK, Rocca WA. Incidence and distribution of Parkinsonism in Olmsted County, Minnesota, 1976–1990. Neurology. 1999;52(6);1214–20.
8. Baldereschi M, Di Carlo A, Rocca WA, et al. Parkinson's disease and parkinsonism in a longitudinal study: Two-fold higher incidence in men. ILSA Working Group. Italian Longitudinal Study on Aging. Neurology. 2000;55(9);1358–63.
9. Kimura H, Kurimura M, Wada M, et al. Female preponderance of Parkinson's disease in Japan. Neuroepidemiology. 2002;21(6);292–6.
10. Das SK, Misra AK, Ray BK, et al. Epidemiology of Parkinson disease in the city of Kolkata, India: A community-based study. Neurology. 2010;75(15);1362–9.
11. EP Bharucha, NE Bharucha Epidemiological study of Parkinson's disease in Parsis in India. Adv Neurol. 1993;352–354.

12. Yu L, Kuo Y, Cherng CG, Chen HH, Hsu CH. Ovarian hormones do not attenuate methamphetamine-induced dopaminergic neurotoxicity in mice gonadectomized at 4 weeks postpartum. Neuroendocrinology. 2002;75(5);282-7.

13. Bernheimer H, Birkmayer W, Hornykiewicz O, Jellinger K, Seitelberger F. Brain dopamine and the syndromes of Parkinson and Huntington Clinical, morphological and neurochemical correlations. Journal of the neurological sciences. 1973;20(4);415-55.

14. Cantuti-Castelvetri I, Keller-McGandy C, Bouzou B, et al. Effects of gender on nigral gene expression and Parkinson disease. Neurobiology of disease. 2007;26(3);606-14.

15. Smith KM, Dahodwala N. Sex differences in Parkinson's disease and other movement disorders. Experimental neurology. 2014;259;44-56.

16. Ragonese P, D'amelio M, Salemi G, et al. Risk of Parkinson disease in women Effect of reproductive characteristics. Neurology. 2004;62(11);2010-4.

17. Cereda E, Barichella M, Cassani E, Caccialanza R, Pezzoli G. Reproductive factors and clinical features of Parkinson's disease. Parkinsonism & related disorders. 2013;19(12);1094-9.

18. Yadav R, Shukla G, Goyal V, Singh S, Behari M. A case control study of women with Parkinson's disease and their fertility characteristics. J Neurol Sci. 2012;319(1);135-8.

19. Rocca WA, Bower JH, Maraganore DM, et al. Increased risk of Parkinsonism in women who underwent oophorectomy before menopause. Neurology. 2008;70(3);200-9.

20. Currie LJ, Harrison MB, Trugman JM, Bennett JP, Wooten GF. Postmenopausal estrogen use affects risk for Parkinson disease. Archives of neurology. 2004;61(6);886-8.

21. Quinn NP, Marsden CD. Menstrual-related fluctuations in Parkinson's disease. Movement Disorders. 1986;1(1);85-7.

22. Garcia-Segura LM, Azcoitia I, DonCarlos LL. Neuroprotection by estradiol. Prog Neurobiol. 2001;63(1);29–60.

23. Popat RA, Van Den Eeden SK, Tanner CM, et al. Effect of reproductive factors and postmenopausal hormone use on the risk of Parkinson disease. Neurology. 2005;65(3);383-90.

24. Tsang KL, Ho SL, Lo SK. Estrogen improves motor disability in parkinsonian postmenopausal women with motor fluctuations. Neurology. 2000;54(12);2292-8.

25. Okun MS, Fernandez HH, Rodriguez RL, et al. Testosterone therapy in men with Parkinson disease: results of the TEST-PD Study. Archives of neurology. 2006;63(5);729-35.

26. Brown TP, Rumsby PC, Capleton AC, Rushton L, Levy LS. Pesticides and Parkinson's disease-is there a link? Environmental health perspectives. 2006;156-64.

27. Lehman EJ, Hein MJ, Baron SL, Gersic CM. Neurodegenerative causes of death among retired National Football League players. Neurology. 2012;79(19);1970-4.

28. Clark AJ, Ritz B, Prescott E, Rod NH. Psychosocial risk factors, pre-motor symptoms and first-time hospitalization with Parkinson's disease: A prospective cohort study. European journal of neurology. 2013;20(8);1113-20.

29. Haaxma CA, Bloem BR, Borm GF, et al. Gender differences in Parkinson's disease. J Neurol, Neurosurg Psychiatry. 2007;78(8);819-24.

30. Alves G, Müller B, Herlofson K, et al. Incidence of Parkinson's disease in Norway: the Norwegian ParkWest study. J Neurol Neurosurg Psychiatry. 2009;80(8);851-7.

31. Shulman LM, Bhat V. Gender disparities in Parkinson's disease. Expert review of neurotherapeutics. 2006;6(3);407-16.

32. Miller IN, Cronin-Golomb A. Gender differences in Parkinson's disease: Clinical characteristics and cognition. Mov Disord. 2010;25(16);2695-703.

33. Martinez-Martin P, Pecurariu CF, Odin P, et al. Gender-related differences in the burden of non-motor symptoms in Parkinson's disease. J Neurol. 2012;259(8);1639-47.

34. Jennings PJ, Janowsky JS, Orwoll E. Estrogen and sequential movement. Behavioral neuroscience. 1998;112(1);154.

35. Laflamme N, Nappi RE, Drolet G, Labrie C, Rivest S. Expression and neuropeptidergic characterization of estrogen receptors (ERa and ERb) throughout the rat brain: anatomical evidence of distinct roles of each subtype. J Neurobiol. 1998;36(3);357-78.

36. Le Saux M, Morissette M, Di Paolo T. ERβ mediates the estradiol increase of D 2 receptors in rat striatum and nucleus accumbens. Neuropharmacology. 2006;50(4);451-7.

37. Dluzen DE, McDermott JL, Liu B. Estrogen as a neuroprotectant against MPTP-induced neurotoxicity in C57/B1 mice. Neurotoxicology and teratology. 1996;18(5);603-6.

38. Miller DB, Ali SF, O'callaghan, JP, Laws SC. The impact of gender and estrogen on striatal dopaminergic neurotoxicity. Annals of the New York Academy of Sciences. 1998;844(1);153-65.

39. Smith KM, Dahodwala N. Sex differences in Parkinson's disease and other movement disorders. Exp Neurol. 2014;259:44-56.

40. Gilles GE, Pienaar IS, Vohra S, Qamhawi Z. Sex differences in Parkinson's disease. Front Neuroendocrinol. 2014;35(3);370-84.

41. Campos FL, Cristovao AC, Rocha SM, Fonseca CP, Baltazar G. GDNF Contributes to Oestrogen-Mediated Protection of Midbrain Dopaminergic Neurons. J Neuroendocrinol. 2012;24(11);1386-97.

42. Wang J, Green PS, Simpkins JW. Estradiol protects against ATP depletion, mitochondrial membrane potential decline and the generation of reactive oxygen species induced by 3-nitroproprionic acid in SK-N-SH human neuroblastoma cells. J Neurochem. 2001;77(3);804-11.

43. Hirohata M, Ono K, Morinaga A, Ikeda T, Yamada M. Anti-aggregation and fibril-destabilizing effects of sex hormones on α-synuclein fibrils in vitro. Exp Neurol. 2009;217(2);434-9.

44. Dupont A, Di Paolo T, Gagne B, Barden N. Effects of chronic estrogen treatment on dopamine concentrations and turnover in discrete brain nuclei of ovariectomized rats. Neuroscience letters. 1981;22(1);69-74.

45. Reisert I, Pilgrim C. Sexual differentiation of monoaminergic neurons-genetic or epigenetic? Trends Neurosci. 1991;14(10);468-73.

46. Lewis C, Dluzen DE. Testosterone enhances dopamine depletion by methamphetamine in male, but not female, mice. Neuroscience letters. 2008;448(1);130-3.

47. Küppers E, Ivanova T, Karolczak M, Beyer C. Estrogen: a multifunctional messenger to nigrostriatal dopaminergic neurons. J Neurocytol. 2000;29(5-6);375-85.

48. Federman DD. The biology of human sex differences. N Engl Med. 2006;354(14);1507-14.

49. Tao Q, Fan X, Li T, Tang Y, Yang D, Le W. Gender segregation in gene expression and vulnerability to oxidative stress induced injury in ventral mesencephalic cultures of dopamine neurons. J Neurosci research. 2012;90(1);167-78.

50. Dewing P, Chiang CW, Sinchak K, et al. Direct regulation of adult brain function by the male-specific factor SRY. Current Biology. 2006;16(4);415-20.

Chapter 3

Etiopathogenesis of Parkinson's Disease

Vibhor Pardasani

The pathology

The pathologic hallmark of Parkinson's disease (PD) is apoptosis of dopaminergic neurons in the substantia nigra pars compacta (SNc) in midbrain with accumulation of cytoplasmic inclusion body with dense eosinophilic cores and pale haloes, "Lewy bodies".[1] Lewy bodies consist of α-synuclein associated with ubiquitin, lipids, and iron. In PD, presence of α-synuclein has also been described in autonomic nuclei, sympathetic ganglia, peripheral cutaneous nerves, salivary glands, enteric neurons, spinal cord, lower brainstem, other cortical structures like limbic cortex including amygdala and hippocampus, and neocortex.[2]

Loss of dopaminergic neurons in the SNc results in reduction of striatal dopamine, resulting in increase in the inhibitory output from the globus pallidus interna which in turn leads to the characteristic motor manifestations of PD namely hypokinesia, bradykinesia, and rigidity.

Risk factors

Among the several risk factors, common ones are advanced age and male sex. 1-methyl-4-phenyl-1,2,3,6-tetrahydropyridine (MPTP) is a mitochondrial pro-toxin which irreversibly destroys striatal dopaminergic neurons casing severe form of parkinsonism, which responds to levodopa, has long been known to occur among drug-addicts who have taken an MPTP-contaminated analog of meperidine.[3] The discovery and isolation of MPTP in the causation of Parkinsonism gradually led to the production of a very reliable animal model of PD which further helped in understanding the disease, its neurochemistry, pathology and treatment better.

Certain pesticides like paraquat, rotenone, etc. are known to produce the animal models of PD.[4] Although the role of pesticides in causation of human PD remains undefined, several studies in India, China and Canada have suggested that the risk of PD is higher in people involved in agricultural industry as well as people exposed to pesticides. It has also been shown that individuals who take food

DOI: 10.1201/9781003505488-3

rich in dietary antioxidants have a lower risk of PD. Caffeine drinkers and smokers are included in this category.[5, 6]

High-level manganese (Mn) exposure in occupational settings has been known to trigger a severe extra-pyramidal syndrome with features of parkinsonism and dystonia, which may be accompanied by cognitive dysfunction with or without behavioral disturbances.[7] This syndrome has a lot in common with PD, including response to levodopa.[8] However, pathologically it is different in that the dopaminergic neurons of the substantia nigra are preserved.[9, 10] In addition, there is scant evidence that manganese exposure contributes to idiopathic PD.

In a case control study done at AIIMS, New Delhi to investigate factors contributing to idiopathic PD, it was observed that of the 377 patients of PD, male gender, family history, and previous history of mood disorder had significant association with occurrence of PD, while presence of pet animals in the household and use of tobacco for more than 20 year duration were protective. In the population studied, exposure to insecticides, pesticides or herbicides had no effect on the occurrence of PD.[11]

Oxidative stress also plays some role in the pathogenesis of PD. In another study at AIIMS, activities of antioxidant enzymes viz., superoxide dismutase (SOD), catalase (CAT) glutathione peroxidase (G-Px), and glucose-6-phosphate dehydrogenase (G6PD) in RBCs of 115 of PD patients was compared with 37 age and gender matched controls. It was observed that levels of these enzymes were significantly lower in PD patients. The enzyme activities also correlated inversely with the Unified Parkinson's Disease Rating Scale (UPDRS) scores as well as the severity of disease measured by Hoehn and Yahr stage of the disease. [12]

Genetic factors

Approximately 10-35% of PD patients have positive family history of PD with at least one first-degree relative suffering from the same disease. First-degree relatives of PD patients are 2 to 7 times more likely to develop PD.[13] The first suggestions that genetic factors might be important in PD came with the reports of a large Greek-Italian family with dominantly inherited PD.

Over the years, several genetic mutations have shown association with occurrence of PD; prominent of among these are the ones that code for Parkin (PARK2 gene), PTEN-induced putative kinase 1 (PINK1 gene), α-synuclein (SNCA gene), and leucine-rich repeat kinase 2 (LRRK2 gene) protein.

The genetic component in PD becomes especially significant in patients who have early onset PD. In people who have onset of PD symptoms before the age 40 years are termed Early onset Parkinson's Disease (EOPD). This group accounts for roughly 3-5% of all PD patients.[14] Among this group, those in which symptoms start before 21 years are termed juvenile PD and Young onset PD (YOPD), with onset of symptoms between 21 and 40 years. Autosomal recessive form of inheritance is seen in those patients with EOPD, who have PD due to mutations in PARK 2, DJ1 (PARK7), and PINK1 (PARK 6) genes.[15]

The juvenile form of parkinsonism who have atypical clinical features generally have ATP13A2 (PARK9), DNAJC6 (PARK19), PLA2G6 (PARK14), FBXO7 (PARK15), SYNJ1 (PARK20), SPG11, and POLG mutations. [15, 16]

PARK 1: The first monogenetic form of PD was found to have associations with mutations in the α-synuclein gene (SNCA). Later, PARK 4, another inheritable Parkinson's was found to be associated with overexpression of α-synuclein by way of duplication and triplication. Indeed, SNCA gene mutation is a not a common cause of PD.

PARK 2: Mutations in the Parkin gene is associated with YOPD, which has very good response to levodopa with autosomal recessive form of inheritance. The Parkin gene encodes a protein which has a E3 ubiquitin-ligase activity. This plays a critical role in the process of ubiquitination i.e. covalent labelling of molecules with ubiquitin before they get directed towards degradation in lysosomes. Mutations in parkin gene impairs this process and causes neural cell death without accumulation of Lewy bodies. Parkin mutations are seen in more than 50% of all EOPD patients and even in some with onset after 40 years of age.[17]

PARK 5: In this form of PD, mutations occurs in ubiquitin carboxy terminal hydrolase L1 (UCHL1) gene, on 4p14 which encodes for the deubiquinating enzyme, found in patients with autosomal dominant PD again pointing towards impairment of cellular transport and protein destruction as the main pathophysiological mechanism in PD.[18] Mutations were found in the phosphatase and tensin homolog (PTEN) induced putative kinase 1 (PINK1) gene on chromosome 1p36 in some patients with autosomal recessive EOPD (PARK6). This gene encodes serine-threonine kinase, which is a mitochondrial enzyme and provides further support that oxidative stress is an important contributor to the pathogenesis of PD. [17, 18]

PARK 6 and 7: Both PARK 6 and PARK 7 have been found to be associated with a benign extremely slowly progressive levodopa responsive form of PD.[18] This form results in autosomal recessive form of PD, which localizes to the same chromosomal region in the DJ-1 gene.

PARK 8: This form of PD is the most common form of autosomal dominant PD with onset in late life. Its locus is mapped to the leucine-rich repeat kinase 2 (LRRK) gene at 12p11.23-q13.11.[17, 18] This is again a benign slowly progressive levodopa responsive type of PD, clinically indistinguishable from sporadic idiopathic PD. LRRK2 codes for a protein called dardarin, a cytoplasmic kinase that causes phosphorylation of α-synuclein and tau. Penetrance is variable in various ethnic groups.

Among the first genetic studies done in PD patients in India was the one conducted at AIIMS, New Delhi. [19] It was done to investigate the prevalence of G88C, G209A, and other mutations in exons 3 and 4 of the SNCA gene in PD patients attending our clinic. Of the 169 PD patients investigated, 18 were familial, 3 patients were juvenile PD, 48 patients were EOPD and the rest 100 patients were sporadic. In this study, we could find neither G88C nor G209A mutations in exons 3 and 4.[19]

We also evaluated for presence of Parkin mutations among both familial as well as sporadic PD patients.[20] Out of total of 259 patients including 173 YOPD; 51 late onset PD; and 35 familial PD, the mutation frequency in Parkin gene was 8.5%. We were able to identify 9 point mutations; among them 7 were novel mutations. These mutations were more common among familial PD patients (14.3), less so in YOPD (6.9%), and least in sporadic late onset (5.9%). A point of interest was that the overall frequency of point mutations and exon rearrangements were higher among south Indian PD patients, suggesting that there is an intrinsic sequence difference close to the Parkin gene or in the Parkin gene itself between north Indian and south Indian populations. In addition, among south Indian subjects, there were only two different mutations - Gln34Arg and Arg334Cys; while all the nine mutations were recorded in the north Indian PD patients. [20]

When we investigated the commonly reported mutations in LRRK2 gene among our patients, we did not find any of the common G2019S, R1441C, R1441G, R1441H, I2012T, and I2020T mutations in our patients except in one female YOPD patient, who showed a heterozygous G2019S mutation.[21] On investigating the contribution of vacuolar protein sorting homolog 35 (VPS35) gene and eukaryotic translation initiation factor 4 gamma 1 (EIF4G1) gene in our population, we did not observe the reported p.Asp620Asn in VPS35 and p.Arg1205His in EIF4G1 mutations or any novel mutations. [22]

On investigating for the DJ1 mutations in our patients, we found that one patient with asymmetric onset, slowly progressive Parkinsonism had the (c.313 A>T, p.Ile105Phe) mutation, while four patients with early onset levodopa-responsive symmetrical Parkinsonism had c.293 G>A, p.Arg98Gln mutation.[23]

On looking for genetic markers associated with dopamine receptor genes (DRD1, DRD2, DRD3, and DRD4 and dopamine transporter gene) in our patients, we found a 120 bp duplication marker of DRD4 gene to be significantly prevalent.[24]

On investigating the contribution of polymorphisms in genes from the drug-metabolizing enzymes and the oxidative stress pathway to susceptibility to PD among our patients, we found GSTO1-rs4925 and NQO1-rs1800566 to be significantly associated. These may suggest an inherent population-specific genetic predisposition which may be getting compounded by a rise in environmental toxins and pesticide use leading to rising incidence of YOPD in India.[25]

In an Indian family with autosomal-recessive juvenile Parkinsonism (ARJP), we tested PARK2, PINK1, and DJ1 mutations and found these to be absent.[26] However, we did find a novel homozygous frameshift mutation, c.89_90insGTCGCCCC in exon 1 of podocalyxin-like gene (PODXL, 7q32–33). We also found significant differences in neurite branching and length ($p<0.0001$) in PC12 cells with wild-type and mutant constructs.[26] This suggested that a neurodevelopmental pathway may also be involved in PD causation.

We were also able to identify a novel homozygous mutation (c.1376C > G, p.Arg459Pro) in SYNJ1 in another family with autosomal-recessive juvenile Parkinsonism.[27]

In another 14-member PD family with two affected cousins, we performed whole exome sequencing (WES) and found a heterozygous mutation c.169C>A, p.P57T in RIC3 acetylcholine receptor chaperone (11p15).[28] RIC3 is a known chaperone of neuronal nicotinic acetylcholine receptor subunit α-7 (CHRNA7). These findings provided credence to the role of cholinergic pathway in the etiopathogenesis of PD.

Conclusions

If we group together our current understanding on the etiopathogenesis of PD, we find that SNCA gene expression and α-synuclein aggregation is the common pathway leading to neurotoxicity and neurodegeneration. Lewy body formation is the result of this process. There are certain PD-susceptibility genes, such as LRRK2, GBA, and MAPT which influence this central pathway along with aging which determines the expression of gene further leading to clinical symptoms. Other environmental factors such as toxins and infections may act as triggers. Reactive oxygen species produced in the mitochondria as a result of environmental influences acting on genetic susceptibilities including pesticide and other toxin use, may also contribute to cellular injury.

References

1. Dickson DW, Braak H, Duda JE, et al. Neuropathological assessment of Parkinson's disease: Refining the diagnostic criteria. Lancet Neurol. 2009;8(12);1150–57.
2. Braak H, Del Tredici K. Nervous system pathology in sporadic Parkinson disease. Neurology. 2008;70(20);1916–25.
3. Langston JW, Ballard P, Tetrud JW, Irwin I. Chronic parkinsonism in humans due to a product of meperidine-analog synthesis. Science. 1983;219;979–80.
4. Dauer W, Przedborski S. Parkinson's disease: Mechanisms and models. Neuron. 2003;39(6);889–909.
5. Seidler A, Hellenbrand W, Robra BP, et al. Possible environmental, occupational and other etiologic factors for Parkinson's disease: A case control study in Germany. Neurology. 1996;46(5);1275–84.
6. De Michele G, Filla A, Volpe G, et al. Environmental and genetic factors in Parkinson's disease: a case control study in southern Italy. Mov Disord. 1996;11(1);17–23.
7. Wang JD, Huang CC, Hwang YH, Chiang JR, Lin JM, Chen JS. Manganese induced Parkinsonism: An outbreak due to an unrepaired ventilation control system in a ferromanganese smelter. Br J Ind Med. 1989;46(12);856–859.
8. Mena I, Marin O, Fuenzalida S, Cotzias GC. Chronic manganese poisoning: clinical picture and manganese turnover. Neurology. 1967;17(2);128–136.
9. Yamada M, Ohno S, Okayasu I, et al. Chronic manganese poisoning: a neuropathological study with determination of manganese distribution in the brain. Acta Neuropathol. 1986;70(3-4);273–278.
10. Guilarte TR, Chen MK, McGlothan JL, et al. Nigrostriatal dopamine system dysfunction and subtle motor deficits in manganese-exposed non-human primates. Exp Neurol. 2006;202(2);381–390.
11. Behari M, Srivastava AK, Das RR, Pandey RM. Risk factors of Parkinson's disease in Indian patients. J Neurol Sci. 2001;190(1-2);49-55.
12. Abraham S, Soundararajan CC, Vivekanandhan S, Behari M. Erythrocyte antioxidant enzymes in Parkinson's disease. Indian J Med Res. 2005;121(2);111-5.
13. Marder K, Levy G, Louis E, et al. Familial aggregation of early- and late-onset Parkinson's disease. Ann Neurol. 2003;54(4);507–13.

14. Schrag A, Schott JM. Epidemiological, clinical, and genetic characteristics of early-onset parkinsonism. Lancet Neurol 2006;4;355-363.

15. Vincenzo Bonifati. Autosomal recessive parkinsonism. Parkinsonism Relat Disord. 2012 Jan;18(1);S4-6.

16. Edvardson S, Cinnamon Y, Ta-Shma A, et al. A deleterious mutation in DNAJC6 encoding the neuronal-specific clathrin-uncoating co-chaperone auxilin, is associated with juvenile Parkinsonism. PLoS One. 2012;7(5);e36458.

17. Shulman JM, De Jager PL, Feany MB. Parkinson's disease: Genetics and pathogenesis. Annu Rev Pathol. 2011;6;193-222.

18. Trinh J, Farrer M. Advances in the genetics of Parkinson disease. Nat. Rev. Neurol. 2013;9(8);445–454.

19. Nagar S, Juyal RC, Chaudhary S, et al. Mutations in the alpha-synuclein gene in Parkinson's disease among Indians. Acta Neurol Scand. 2001;103(2);120-2.

20. Chaudhary S, Behari M, Dihana M, et al. Parkin mutations in familial and sporadic Parkinson's disease among Indians. Parkinsonism Relat Disord. 2006;12(4);239-45.

21. Punia S, Behari M, Govindappa ST, et al. Absence/rarity of commonly reported LRRK2 mutations in Indian Parkinson's disease patients. Neurosci Lett. 2006;409(2);83-8.

22. Sudhaman S, Behari M, Govindappa ST, Muthane UB, Juyal RC, Thelma BK. VPS35 and EIF4G1 mutations are rare in Parkinson's disease among Indians. Neurobiol Aging. 2013;34(10);2442.

23. Abbas MM, Govindappa ST, Sudhaman S, et al. Early Onset Parkinson's disease due to DJ1 mutations: An Indian study. Parkinsonism Relat Disord. 2016;32;20-24.

24. Juyal RC, Das M, Punia S, et al. Genetic susceptibility to Parkinson's disease among South and North Indians: I. Role of polymorphisms in dopamine receptor and transporter genes and association of DRD4 120-bp duplication marker. Neurogenetics. 2006;7(4);223-9.

25. Punia S, Das M, Behari M, et al. Leads from xenobiotic metabolism genes for Parkinson's disease among north Indians. Pharmacogenet Genomics. 2011;21(12);790-7.

26. Sudhaman S, Prasad K, Behari M, Muthane UB, Juyal RC, Thelma BK. Discovery of a frameshift mutation in podocalyxin-like (PODXL) gene, coding for a neural adhesion molecule, as causal for autosomal-recessive juvenile Parkinsonism. J Med Genet. 2016;53(7);450-6.

27. Kirola L, Behari M, Shishir C, Thelma BK. Identification of a novel homozygous mutation Arg459Pro in SYNJ1 gene of an Indian family with autosomal recessive juvenile Parkinsonism. Parkinsonism Relat Disord. 2016;31;124-128.

28. Sudhaman S, Muthane UB, Behari M, Govindappa ST, Juyal RC, Thelma BK. Evidence of mutations in RIC3 acetylcholine receptor chaperone as a novel cause of autosomal-dominant Parkinson's disease with non-motor phenotypes. J Med Genet. 2016;53(8);559-66.

Chapter 4

Genetics of Parkinson's Disease: Contributions from India

Laxmi Kirola, Madhuri Behari, BK Thelma

Unveiling the aetiology of Parkinson's disease (PD) has come a long way from the early belief on PD being caused by exogenous and endogenous toxins,[1,2] to gene-environment interactions,[3] to a probable role of a single gene and combinations thereof.[4] It has also moved subsequently from the hypothesis based candidate gene testing approach to hypothesis free genome-wide association studies (GWAS) and recently to next generation sequencing tools (NGS) for rare variant identification. Indeed, it may not be an exaggeration to say that discovery genomics stepped into a golden era between the end of the last and beginning of this century. Concomitant progress in functional genomics and systems biology approaches are providing some insights into the etiological mechanisms underlying diseases with promising leads for new therapeutics. This overview attempts to capture some of the excitement in contemporary medical genomics with specific reference to PD genetics which has witnessed notable progress and with particular focus on the novel findings in the ethnically and genetically distinct Indian population. Recent advances in the country including: i) putative causal gene discoveries, ii) consequent recognition of newer pathways underlying disease biology, iii) shared and unique mutation spectra across ethnic groups, iv) implications of these findings for predictive medicine, and v) the most recent excitement with induced pluripotent stem cells (iPSCs) combined with genome editing technology for in vitro disease modelling and personalized medicine are presented in brief.

PD has a worldwide distribution with a 1-2% overall prevalence but the epidemiological data for PD from different parts of India are rather few. A crude prevalence rate of 14 and 14.1/1,00,000 in rural Karnataka in South India[5] and in rural Kashmir in North India respectively have been reported.[6] Age adjusted prevalence among individuals above 50 years of age was 134/1,00,000; and a slightly higher prevalence of 192/1,00,000 was reported among closely bred Parsi community from Bombay.[7] Several causative environmental agents have been considered for the different rates of prevalence in different parts of the world[5], but emerging data suggests a significant genetic component contributing to the observed differences.

DOI: 10.1201/9781003505488-4

Early genetic findings

Most of the PD cases are sporadic in occurrence with increasing incidence with age. Depending on the age of onset, PD is broadly grouped into juvenile (onset below 20 yrs of age), young onset (20-40 yrs), and late onset (above 40 yrs). However, a small proportion of familial forms of PD, following an autosomal dominant or autosomal recessive mode of inheritance have been reported globally and also in India, suggesting a likely monogenic nature of disease etiology in them. Considerable amount of data from a range of genes/pathways implicated in PD are now available,[8] but selective degeneration of dopaminergic neurons in the substantia nigra pars compacta (SNpc) part of the mid brain is, as yet the most accepted pathogenic mechanism underlying dopamine deficiency disorders such as PD (Fig 1).

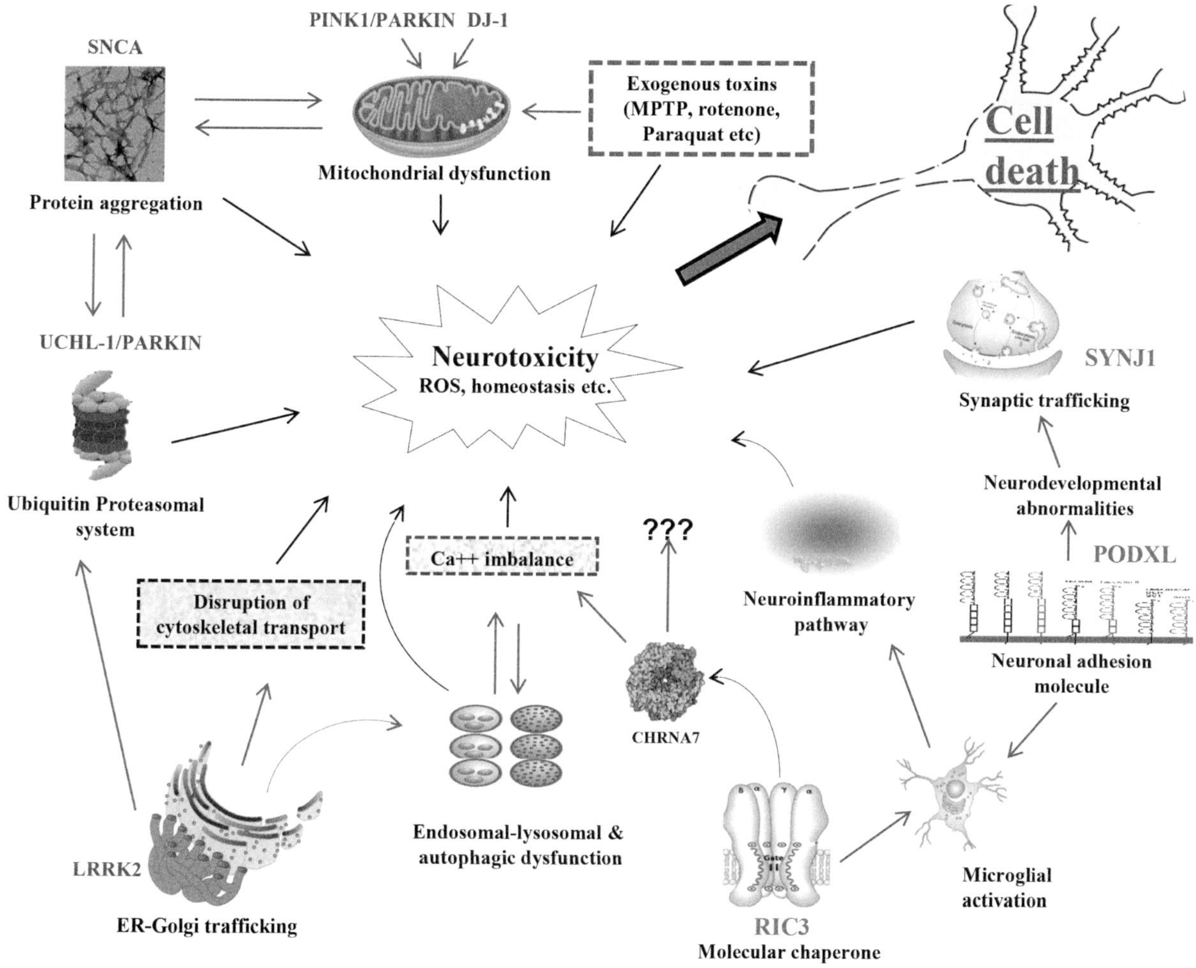

Figure 1

Early genetic studies were limited to sporadic, case-control based association testing of genetic variants in candidate genes. Such genes were largely prioritized based on available biochemical or pharmacological evidence and dopamine hypothesis being the most accepted with majority of the candidates from the dopaminergic pathway. With the parallel hypothesis of mitochondrial, oxidative free radicals, and neurotoxin involvement together with exogenous neurotoxins such as herbicides, pesticides, etc. in disease etiology, mitochondrial genes and those from the oxidative stress pathway and phase I and II drug metabolism pathways also became favourite candidates. Several reports of association of some of these genes with sporadic PD among Caucasians and a few other groups also emerged.[9] Similar studies were also carried out on a moderate PD patient resource from South and North India. Of a total of 12 candidate genes tested for their association with disease, most notable findings include: i) significant association of the 120bp duplication in the promoter of Dopamine receptor D4 gene (DRD4) in both South and North Indian PD cohorts[10]; ii) haplotypes of Dopamine β hydroxylase gene (DBH) markers with total and young onset PD cohort and a 3' UTR genotype in *DBH* with disease severity[11]; and iii) N-acetyl transferase 2 gene (*NAT2*) polymorphisms with slow acetylator phenotype among young and to a lesser extent in late onset PD among Indians.[12]Almost all these remain the first and only candidate gene based association studies in PD from the country. Around this time GWAS became possible and a few GWASs have been carried out on Caucasian populations. However, unlike in other complex traits such as type 2 diabetes or rheumatoid arthritis or cardiovascular disorders, GWAS findings are very limited in PD.[13,14] Functionally relevant candidate genes including *GBA, SMPD1, MAPT,* and *GCH1* were identified and the most notable finding has been the association of well known PD genes namely *SNCA* and *LRRK2* (discussed briefly below).

Genes and Mendelian forms of PD

Generation of the microsatellite map of the human genome[15] together with emerging recombinant DNA tools enabled notable linkage studies leading to identification of potential disease causal genes in single gene disorders including familial forms of PD. A landmark discovery of a single pathological mutation Ala53Thr in α-synuclein (*SNCA*) in a large Italian kindred with autosomal dominant PD (ADPD) using this approach,[16] marked the beginning of global success stories for monogenic forms of PD. Identification of mutations in *PARKIN* in a Japanese family with autosomal recessive juvenile Parkinsonism (ARJP) followed soon after.[17] Exploiting the linkage approach on informative families, more genes namely *LRRK2, PINK1, DJ-1*etc. were discovered for familial PD in quick succession.[18–20] This paved the way for mutation screening in PD families across ethnic groups. Though *SNCA* was the first and highly functionally relevant gene to be discovered in PD, except for six mutations namely Ala53Thr, Ala30Pro, Glu46Lys, His50Gln, Gly51Asp, and Ala53Glu in this gene reported from the index Italian family and others from German, Spanish, English/Welsh, French, and Finnish families,[21] no other mutations have been reported in this gene to date. However, copy number variations of *SNCA* have been subsequently implicated in etiology of PD in few studies.[8] Of note, of all the PD genes known to date, a range of mutations in *PARKIN* were among the most consistently observed across ethnic groups.[22,23] *LRRK2* was identified in a Basque family with autosomal dominant PD and three mutations namely G2019S, R1441C, and R1441G in this gene were most commonly reported across populations. However, the frequency of these mutations varied extensively from ~2% in Northern European and American populations to ~40% in Arab PD cohort. *DJ-1* and *PINK1* for autosomal recessive forms of PD were identified in families from Dutch and Italian origin[19,20] but negligible frequencies of mutations in other populations have been reported to date. At this juncture, advent of NGS in the last two decades marked a new phase in discovery genetics. Revisiting small to moderate sized PD families using this powerful tool has enabled identification of >16 genes to date[24] (Table 1). Screening for mutations in these genes is probably still underway in different populations. But of all those tested, mutation frequencies in *PARKIN* followed by *LRRK2* in sporadic and familial PD cases remain the highest to date.[24,25] It may be noteworthy to mention here that though the mutation frequencies in *SNCA* and *LRRK2* have been negligible, both these genes have been shown to be significantly associated in GWASs.[13,26,27] This clearly suggests that not the rare variants but common polymorphisms in these genes may be contributing substantially to sporadic PD aetiology. Most notable findings from the above-mentioned studies among familial PD account for only<30% of the families tested to date. The striking implication of this is that more disease causal genes need to be discovered.

Table 1. Genes identified to date from analysis of Parkinson's disease families

Gene, Locus	Gene name	Kind red, first identified (References)
Autosomal dominant		
PARK 1, 4q21-23	α-synuclein *(SNCA)*	Italian, Polymeropoulous et al., 1997
PARK 5, 4p14	*UCH L1*	German, Leroy et al., 1998
PARK 8, 12q12	*LRRK2*	Japanese, Zimprich et al., 2004
PARK11, 2q36-37	*GIGYF2*	Italian & French, Lautier at al., 2008 Bras et al., Meeus et al., 2009
PARK 13, 2p13	*HTRA2/OMI*	German, Strauss et al., 2005 Ross et al., Simon-Sanchez et al., 2008
PARK 17, 16q11.2	*VPS35*	Swiss, Vilariño-Güell et al., 2011
PARK18, 3q27.1	*EIF4G1*	French, Farrer et al., 2011
PARK21, 3q22.1	*DNAJC13*	Dutch/German/Russian, Vilarino-Guell et al., 2014
PARK22, 7p11.2	*CHCHD2*	Japanese, Funayama et al., 2015
Not assigned, 11p15.4	*RIC3*	Indian, Sudhaman et al., 2016b
Autosomal recessive		
PARK 2, 6q 25.2-27	*PARKIN*	Japanese, Kitada et al., 1998
PARK 6, 1p35-36	*PINK 1*	Sicilian, Valente et al., 2001
PARK 7, 1p36	*DJ-1*	Dutch, Bonifati et al., 2003
PARK 9, 1p36	*ATP13A2*	Jordanian, Ramirez et al., 2006
PARK 15, 22q12–q	*FBXO7?*	Iranian, Shojaee et al., 2008
PARK 14, 22q13.1	*PLA2G6?*	Indian/Pakistani, Paisan-Ruiz et al., 2009
PARK19, 3q22.1	*DNAJC6*	Saskatchewan, Farrer et al., 2013
PARK20, 21q22.11	*SYNJ1*	Iranian, Krebs et al., 2013; Quadri et al., 2013
Not assigned, 4q21.23	*COQ2*	Japanese, Mitsui et al., 2013
PARK23, 15q22.2	*VPS13C*	European, Deng et al., 2016
Not assigned,7q32-33	*PODXL*	Indian, Sudhaman et al., 2016a
PARK loci with unknown relevance		
PARK 3, 2p13	*? SPR*	Greek, Gasser et al., 1998
PARK10, 1p32	*?*	Iceland, Hicks et al., 2002
PARK 12, Xq21-q25	*?*	French, Pankratz et al., 2003
2q22-23	*NR4A2*	Mixed European, Le et al., 2003
PARK 16, 1q32	*? RAB7L1*	Japanese, Satake et al., 2009

Status of mutations in PD among Indians

In keeping pace with the global PD genetic findings mentioned above, mutation screening among familial PD as well as sporadic forms of PD has been carried out in our country, but to a very limited extent. PD patients/families from all regions of India have been investigated. A detailed (but not exhaustive) compilation of the findings from these studies including mutation spectra, sample sizes, and population sub-groups is presented in Table 2. It is noteworthy that mutations in the two most commonly tested ADPD genes namely *SNCA* and *LRRK2* and the more recently identified *VPS35* are negligible or almost non-existent in our cohort.[28–30] On the other hand, almost all research groups have identified novel and/or reported mutations in *PARKIN, DJ1,* and *PINK1* known for autosomal recessive forms of PD. This may not be unexpected considering that there is a comparatively higher rate of consanguinity in Indian populations compared to the west.[31] As for *SNCA* and *LRRK2*, the results to date may be interpreted with caution. While complete exonic regions of *SNCA* have been screened for mutations,[28] barring an ongoing study (unpublished) in the lab showing a comparatively higher frequency of duplications, no other studies to date have reported copy number variations (CNVs) in this gene. As for *LRRK2*, only three commonly reported mutations namely G2019S, R1441C, and R1441G but not all the 52 exons have been screened in the published reports. It is likely that allelic heterogeneity/population specific mutation(s) remains to be identified in this gene. Finally, mutation(s) in the large number of newly reported genes may be also minimal due to their near private status- that is being present in only the families that they have been originally identified in. In this context, whole exome sequencing (WES) of well powered and clinically well characterized Indian cohorts is highly warranted to obtain a reliable profile of the mutation spectra in the known PD genes in our genetically distinct population. Further, identification of novel causal genes which will be enabled by the WES approach will be very valuable to obtain additional insights into the pathogenic mechanisms underlying PD, which have remained obscure currently. These efforts would be of considerable translational value for predictive and preventive medicine as well as personalized treatment options in the near future.

Table 2. Genes/mutations identified in Parkinson's disease among Indians

Gene, Position First discovered	Mutations, Location	Status	Sample size		% Mutation Frequency$	Population* NI/SI/EI/ WI	References
			Cases (F/YO/LO)#	Controls			
Autosomal dominant							
SNCA, 4q21-22 (Alpha-synuclein) Polymeropoulos et al., 1997	-	-	169 (21+48+100)	-	-	NI & SI	Nagar et al., 2001
	-	-	140	201	-	SI	Padmaja et al.,2012
	-	-	100	100	-	SI	Kadakola et al., 2014
	Ala53Thr, ex.@ 4	Known	23	23	21.73@@	SI	Dhivya et al., 2016
LRRK2, 12q12 (Leucine-rich repeat kinase 2) Paisan-Ruiz et al., 2004	Gly2019Ser, ex. 41	Known	800	212	0.13	NI & SI	Punia et al., 2006
	-	-	150	170	-	EI	Sanyal et al., 2010
	-	-	186	100	-	SI	Vijayan et al., 2011
	-	-	140	201	-	SI	Vishwanathan et al., 2012
	IVS**34-1A>T IVS35+23T>A IVS47-9Tdel IVS48+52G>A	Novel Novel Novel Novel	300	446	-	EI	Sadhukhan et al., 2012
VPS35, 16q11.2 (Vacuolar protein sorting-associated protein 35) Vilariño-Güell et al., 2011	rs168745, synonymous, ex.15	Novel	320	100	-	NI & SI	Sudhaman et al., 2016
RIC3, 11p15.4 (RIC3 Acetylcholine Receptor Chaperone) Sudhaman et al., 2016b	Pro57Thr, ex. 2 Val168Leu, ex. 4	Novel Novel	288	186	-	SI NI	Sudhaman et al., 2016b

Gene, Position First discovered	Mutations, Location	Status	Sample size Cases (F/YO/LO)[#]	Controls	% Mutation Frequency[$]	Population[*] NI/SI/EI/WI	References
Autosomal recessive							
	Thr240Met, ex. 6 Hom. del. ex. 8, 9 Arg42his, ex.2 Ser167Asn, ex.4 Arg334cys, ex.9 Ala339ala, ex.9 Gly359asp, ex.9 Val380leu, ex.10 Glu444gln, ex.12 IVS2+25T>C IVS2+35G>A IVS9+20C>A IVS3-20T>C IVS7-35G>A	Known Known Novel Known Known Known Novel Known Novel Known Novel Novel Known Known	102	105	6.76	SI	Madegowda et al., 2005
PRKN, 6q25.2–q27 (Parkin RBR E3 Ubiquitin Protein Ligase) Kitada et al., 1998	Ala31Asn , ex. 2 Glu34Arg , ex. 2 Arg42Cys, ex. 2 Arg104Trp, ex. 3 Gly359Asp, ex. 9 Cys917Thr, ex. 7 Gly1119Ala, ex.7 RArg34Cys, ex. 9 Arg402Cys, ex.11 Heterozygous exonic rearrange ment, ex. 2, 4, 6, 7, 8	Novel Novel Novel Novel Novel Novel Novel Known Known	259 173 YOPD; 51 LOPD; 35 F PD	55	8.5 14.3% FPD; 6.9% YOPD; 5.9% LO PD	NI & SI	Chaudhary et al., 2006
	Gln34Arg, ex. 2 Arg42Cys, ex. 2 Arg42His, ex.2 Tyr143Cys, ex. 4 Arg334Cys ex. 9 IVS8–20–16del CTGCT IVS2+25 T>C IVS2+35 G>A IVS3-20 T>C IVS7-35 G>A IVS7-68 C>G IVS8–21–17del TCTGC IVS8-22A>G Ala339Ala, ex. 9 c.1514C4A, 3'UTR exon del. 3 & 4	Novel Novel Known Novel Known Novel Known Known Known Known Known Known Known Known Known Known	138	100	7.24	EI	Biswas et al., 2006
	Homozygous deletion in ex. 1, 3, 12 Gly1083Trp	Novel Novel	69	43	39.1	NI &WI	Vanish et al., 2010
	c.1239G>C c.171+25T>C c.202A>G c.601G>A 7 c.798_799insA, ex.7	Known Known Known Known Novel	16 (F)	-	68	SI	Padmaja et al.,2012

Gene, Position First discovered	Mutations, Location	Status	Sample size		% Mutation Frequency[$]	Population[*] NI/SI/EI/WI	References
			Cases (F/YO/LO)[#]	Controls			
	Only ex. 3 screened	-	16 (YO)	-	-	SI	Thirugnanasambandam et at., 2016
	S167N, ex. 4	Known	23	23	8.69	SI	Dhivya et al., 2016
PINK1, 1p35–p36 (PTEN Induced Putative Kinase 1) Valente et al., 2004	Leu63Leu, ex. 1 IVS1-14 G > A IVS1-93 C > T IVS1-65 C > G IVS1-7 A > G Pro179Pro, ex. 2 Gly197Gly, ex. 2 IVS2-77 G > A Arg246Gln, ex. 3 Arg276Gln, ex. 4 Thr282Thr, ex. 4 IVS4þ72 G > C Ala340Thr, ex.5 IVS6-25 T > C c.1786G >A, 3'UTR	Known Novel Known Novel Known Novel Novel Known Novel Known Novel Novel Known Novel Novel	250	205	2.0	EI	Biswas et al., 2010
DJ-1, 1p36 (Parkinsonism Associated Deglycase) Bonifati et al., 2003	IVS4-30T>G IVS4z45G>A IVS4-46G>A IVS4-98G>A IVS5z31G>A IVS5z69G>C	Known Known Known Known Known Novel	150	150	-	EI	Sanyal et al., 2011
	g. −21 G > A, promoter c.-99T > C, ex. 1a g. 168 185 del g. 168 185 dup g. 213 G > T, ex. 1b IVS1-22 T del IVS2-109 C > T Val35Ile, exon 3 IVS4 + 30 T > G 4IVS4 + 45 G > A IVS4 + 46 G > A IVS4 + 47 G ins IVS4 + 47 A ins IVS4 + 81 C ins IVS4-98 G > A IVS4-4 C > T Arg98Gln, ex. 5 IVS5 + 31 G > A IVS5-216 G > A IVS6-593 T > C Stop + 38 G > A, 3' UTR	Known Known Novel Novel Known Novel Known Novel Known Known Known Novel Novel Novel Known Novel Known Known Known Known Novel	308	248	3.9	EI	Sadhukhan et al., 2012
	I105P, exon 5 Arg98Gln, exon 5	Novel Known	100	114	~5%	NI & SI	Abbas et al., 2016
PODXL, 7q32.3 (Podocalyxin Like) Sudhaman et al., 2016	c.89_90insGTCGCCCC, ex.1 Pro429Thr, ex.7 Ser373Asn, ex. 5 Arg294Gln, ex. 3	Novel Novel Novel Novel	288	186	-	NI & SI	Sudhaman et al., 2016

Gene, Position First discovered	Mutations, Location	Status	Sample size		% Mutation Frequency$	Population* NI/SI/EI/WI	References
			Cases (F/YO/LO)#	Controls			
SYNJ1, 21q22.11 (Synaptojanin 1) Krebs et al., 2013; Quadri et al., 2013	Arg459Pro, ex.11	Novel	285	-	-	NI & SI	Kirola et al., 2016

F- Familial; YO- Young onset; LO- Late onset; $ frequency of mutation calculated on exonic basis only* NI/SI/EI/WI- north, south, east and west Indian; ex.@- exon; IVS**- Intervening sequence; @@- since the mutation profile is not available to the author for conformation the frequencies can be deduced with confidence

Discovery genomics in PD in India

From the preceding paragraphs and also Table 2, it is clear that the commonly implicated genes such as *SNCA, LRRK2, PARKIN, PINK1,* and *DJ1* have been screened to some extent in the Indian PD families and sporadic PD cohorts. Based on the available data, the most notable finding is that mutations in these known genes account for <30% of all familial PD cases tested to date in the country, similar to what is reported globally. This implies that more genes remain to be identified. India has a vast and informative patient resource including small and large PD families, young onset and late onset sporadic PD cases. This provides a good opportunity for discovery genomics and to make notable contributions to global knowledge in this debilitating neurological condition. With this in view, the author's laboratory focussed on novel disease causal gene identification in the last 4-5 years. Revisiting informative PD families available in the laboratory (previously shown to be mutation negative for known PD genes) using the contemporary WES approaches is hypothesized to be a powerful strategy for causal gene discoveries. Several families with either ADPD or ARJP have been used in these efforts. Two novel causal genes namely *RIC3* and *PODXL,* thus identified in an ADPD and ARJP family each and a novel mutation in *SYNJ1,* a recently discovered gene in an ARJP family are briefly described below.

RIC3: WES performed in a 14-member well characterized south Indian family with ADPD followed by standard bio-informatic analysis, revealed a heterozygous missense variant c169 C>A, pP57T in exon 2 of *RIC3* (11p15), in nine affected members (with <30 years of age at onset in a third generation proband) but not in the five unaffected. On screening for additional mutations in sporadic PD cases available with the investigators another heterozygous mutation c.502G>C, p. V168L in exon 4 was identified in a young onset case of south Indian origin. These mutations were notably absent in ~800 chromosomes from the population matched controls screened. Functional characterization of the two variants in this chaperone of nicotinic acetylcholine receptors (NAchRs) in PC12 cells differentiated into neurons showed that there was decreased co-localization of this chaperone with endogenously expressed CHRNA7. Furthermore, a decrease in the number of these receptor molecules transported to the plasma membrane was also observed. Additional analysis showed that these two mutations have a dominant negative effect. Interestingly, based on GWASs, *RIC3* has been reported to confer risk in Schizophrenia, bipolar disorder with psychosis, multiple sclerosis, and cognition.[32–34] However, the first study providing genetic evidence for the involvement of *RIC3* in PD pathogenesis has been from the laboratory.[35] Though, the mechanism underlying its pathogenic role is not known, it is hypothesized to be via the cholinergic anti-inflammatory pathway and this is being currently investigated.

PODXL: Using the hypothesis free WES approach in a north Indian family with eight children, three of whom were diagnosed with very early onset PD (<20yrs), we discovered a novel homozygous frameshift mutation c.89_90insGTCGCCCC in exon 1 of podocalyxin-like gene (*PODXL*, 7q32-33) resulting in a loss of protein. Screening of additional PD samples in the laboratory identified three more novel heterozygous missense mutations namely c.1285C>A, c.1118G>A, and c.881G>A in three unrelated cases (one from a PD family, one young onset, and one late onset PD), all of north Indian origin. These mutations were absent in ~ 800 control chromosomes screened. Functional analysis of this gene, which has been primarily shown to be involved in neuronal development in mice models,[36] showed significant differences in neurite branching pattern and neurite length in PC12 cells differentiated into neurons. This is the first evidence for the involvement of a gene from the neurodevelopmental/neuroinflammatory pathway in PD etiology.[37]

SYNJ1: In yet another study, WES was performed on a consanguineous ARJP family from eastern part of India, with five children, of whom two were affected with PD at 12 and 18 years of age. A novel homozygous missense mutation(c.1376C > G, p.Arg459Pro) in exon 11 of a recently discovered ARJP gene namely *SYNJ1*[38,39] was identified. The index mutation was present in a homozygous state in both the affected siblings while both the parents and unaffected siblings were heterozygous carriers. Complete screening of exon 11 of *SYNJ1* in the laboratory PD cohort did not reveal the index or any additional mutations. Of note, this is only the second mutation in *SYNJ1* to be reported to date.[40] Interestingly, *SYNJ1* is very well investigated in other neurological disorders including bipolar disorder, schizophrenia, Down's syndrome etc. This gene has a range of functions at nerve terminals depending on the versatile nature of its protein. Mostly, SYNJ1 is involved in synaptic vesicle trafficking at the nerve terminals and maintains the coating and decoating of the proteins.

Pathobiology of PD: Novel insights and the way forward

Almost all the previously reported genes for PD are involved in protein aggregation and clearance via misfolded proteins and ubiquitin proteasomal activity (*SNCA+PARKIN*), or reactive oxygen species (ROS) dependent (*PINK1, DJ1* etc.) and ROS independent (*LRRK2+SYNJ1* etc.) mitochondrial function and oxidative stress pathways. In other words, it is evident that the mechanism leading to dopaminergic neuronal cell death could be varied in different PD patients. However, almost all PD cases with mutations in these genes have been reported to be Levodopa responsive[41] for managing the motor symptoms. This suggests that depletion of dopaminergic neurons could be the ultimate pathology. But what remains poorly understood to date is the pathogenic mechanism(s) underlying disease initiation and subsequent progression. What is also notably lacking is any/new knowledge on early biomarkers of disease onset and its deterioration. Consequently, global progress in attempts for early diagnosis, retarding the rate of disease progression and all other aspects of personalized medicine is very negligible. In this context, the two novel genes that we have identified from our studies and described above are from hitherto unreported pathways. *RIC3* is a known chaperone of neuronal nicotinic acetylcholine receptor subunit α-7 and is expressed abundantly in the brain. Mutations in this gene were observed to lead to a receptor density alteration in the PC12 cells differentiated into neurons. This provides the first ever evidence for the role of cholinergic pathway in PD etiology. Though the actual mechanism by which *RIC3* may contribute to PD pathology remains to be investigated while non-motor symptoms such as hallucinations and depression observed in this family can be well explained.

CHRNA7, an acetylcholine receptor regulates Ca++ influx inside the neurons, which actually helps in downstream signalling of many molecules and regulates cognition, memory, and attention. The importance of RIC3, together as a modulator with CHRNA7 enhances the activation of Ca++ influx, which has been demonstrated experimentally.[42] The haplo-insufficiency of CHRNA7 in dopaminergic neurons in PD has also been demonstrated.[43,44] On the other hand, *PODXL* is a neuronal adhesion molecule which expresses highly in the brain and it is known to be involved in neuronal plasticity and development, as demonstrated already in many in vivo studies in mice.[45] Differences between NCAM and PSA-NCAM are shown to lead to microglial cell activation that results in neuroinflammation in the brain.[46] Thus, the role of *PODXL* seems to be via neurodevelopmental and neuroinflammatory pathways in PD etiology and this shows the importance of these pathways in PD pathogenesis with specific reference to neurodevelopmental pathway. Furthermore, mutation observed in another ARJP gene namely *SYNJ1* in a consanguineous East Indian family with PD seems to contribute via synaptic vesicle trafficking. *SYNJ1* is a phosphoinositide phosphatase coding gene, which is highly expressed at brain nerve terminals and involved in clathrin coat-mediated endocytosis in adult brain.[38,47] This phosphatase interacts with different proteins including LRRK2 and PARKIN the well-known PD associated proteins and is responsible for signalling, endocytosis, and actin nucleations.[48] Of note, the backward falling phenotype witnessed in the two affected siblings in our study has also been reported in the unc-26 mutant in *Caenorhabditis elegans*[49], which is an ortholog of human *SYNJ1*.

These novel insights from our findings in Indian PD families together with the well investigated pathways of the previously known PD genes are presented in an updated model for PD pathogenesis. Efforts to establish the role of newly identified genes *PODXL* and *RIC3* in PD etiology using iPSCs and CRISPR-Cas9 mediated genome editing technology with long term implications for personalized medicine/treatment are underway. Efforts for additional gene discoveries using more PD families which may provide novel insights into PD pathogenesis are also underway.

Thus, early genetic studies in PD had demonstrated association of genetic variants in *DRD4, DBH,* and *NAT2* with sporadic PD among Indians. Subsequent work on familial forms of PD identified novel/known mutations in autosomal recessive PD genes such as *PARKIN, PINK1,* and *DJ1* but not in ADPD genes namely *SNCA, LRRK2,* and *VPS35*. Most notably, two putative causal genes namely *RIC3* for ADPD and *PODXL* for ARJP providing leads for involvement of novel disease related pathways have been discovered in Indian PD families. These findings warrant continuation of our efforts in additional PD families to contribute to global discovery genomics and to provide further insights into disease biology.

Majority of PD cases globally are sporadic in occurrence and possibly caused by gene-environment interactions but etiological components are not well understood. However, in ~10-15% of cases which show familial clustering and a monogenic inheritance pattern, ~20 genes have been discovered and a population specific pattern of mutations therein is noted. Novel/known mutations in autosomal recessive PD genes such as *PARKIN, PINK1, DJ1,* and *SYNJ1* but not in autosomal dominant genes namely *SNCA, LRRK2,* and *VPS35* have been observed in a few Indian PD families. However, using contemporary next generation sequencing approaches on informative families, two novel putative causal genes namely *RIC3* for autosomal dominant and *PODXL* for autosomal recessive juvenile Parkinsonism have been identified. These findings have implicated cholinergic and neurodevelopmental/

neuroinflammatory pathways in PD etiology for the first time. Continued efforts to analyse additional Indian PD families are expected to make significant contributions to discovery genomics, improved understanding of disease biology, and possibly new therapeutics.

References

1. GC Davis, AC Williams, SP Markey, MH Ebert, ED Caine, CM. Reichert, IJ Kopin. Chronic parkinsonism secondary to intravenous injection of meperidine analogues. Psychiatry Res. 1 (1979) 249–254. doi:10.1016/0165-1781(79)90006-4.
2. Y Mizuno, S Ikebe, N Hattori, et al. Role of mitochondria in the etiology and pathogenesis of Parkinson's disease. Biochim. Biophys. Acta. 1271 (1995) 265–274. doi:10.1016/0925-4439(95)00038-6.
3. JR Cannon, JT Greenamyre. Gene-environment interactions in Parkinson's disease: Specific evidence in humans and mammalian models. Neurobiol. Dis. 57 (2013) 38–46. doi:10.1016/j.nbd.2012.06.025.
4. L Polito, A Greco, D Seripa. Genetic Profile, Environmental Exposure, and Their Interaction in Parkinson's Disease. Parkinsons. Dis. 2016 (2016). doi:10.1155/2016/6465793.
5. M Gourie-Devi. Epidemiology of neurological disorders in India: Review of background,\nprevalence and incidence of epilepsy, stroke, Parkinson's disease and\ntremors. Neurol. INDIA. 62 (2014) 588–598. doi:10.4103/0028-3886.149365.
6. S Razdan, RL Kaul, A Motta, S Kaul, RK Bhatt. Prevalence and pattern of major neurological disorders in rural kashmir (India) in 1986. Neuroepidemiology. 13 (1994) 113–119. doi:10.1159/000110368.
7. NE Bharucha, AE Bharucha, EP Bharucha. Prevalence of peripheral neuropathy in the Parsi community of Bombay. Neurology. 41 (1991) 1315–7. doi:10.1212/WNL.41.8.1315.
8. O Corti, S Lesage, A Brice. What genetics tells us about the causes and mechanisms of Parkinson's disease. Physiol Rev. 91 (2011) 1161–1218. doi:10.1152/physrev.00022.2010.
9. A Karic, R Terzic, A Karic, B Peterlin. Identifying candidate genes for Parkinson's disease by integrative genomics method. Biochem. Medica. 21 (2011) 174–181. http://ovidsp.ovid.com/ovidweb.cgi?T=JS&PAGE=reference&D=medl&NEWS=N&AN=22135858.
10. RC Juyal, M Das, S Punia, et al. Genetic susceptibility to Parkinson's disease among South and North Indians: I. Role of polymorphisms in dopamine receptor and transporter genes and association of DRD4 120-bp duplication marker. Neurogenetics. 7 (2006) 223–229. doi:10.1007/s10048-006-0048-y
11. S Punia, M Das, M Behari, et al. Role of polymorphisms in dopamine synthesis and metabolism genes and association of DBH haplotypes with Parkinson's disease among North Indians. Pharmacogenet. Genomics. 20 (2010) 435–41. doi:10.1097/FPC.0b013e32833ad3bb
12. S Chaudhary, M Behari, M Dihana, et al. Association of N-acetyl transferase 2 gene polymorphism and slow acetylator phenotype with young onset and late onset Parkinson's disease among Indians. Pharmacogenet Genomics. 15 (2005) 731–735. doi:01213011-200510000-00007 [pii]
13. J Simón-Sánchez, C Schulte, JM Bras, et al. Genome-wide association study reveals genetic risk underlying Parkinson's disease. Nat. Genet. 41 (2009) 1308–12. doi:10.1038/ng.487

14. C Labbé, O a Ross. Association studies of sporadic Parkinson's disease in the genomic era. Curr. Genomics. 15 (2014) 2–10. doi:10.2174/1389202914666131210212745

15. G Gyapay, J Morissette, A Vignal, et al. The 1993-94 Genethon human genetic linkage map. Nat Genet. 7 (1994) 246–339. doi:10.1038/ng0694supp-246

16. MH Polymeropoulos, C Lavedan, E Leroy, et al. Mutation in the alpha-synuclein gene identified in families with Parkinson's disease. Science. 276 (1997) 2045–7. doi:10.1126/science.276.5321.2045

17. T Kitada, S Asakawa, N Hattori, et al. Mutations in the parkin gene cause autosomal recessive juvenile parkinsonism. Nature. 392 (1998) 605–608. doi:10.1038/33416

18. C Paisán-Ruíz, S Jain, EW Evans, et al. Cloning of the gene containing mutations that cause PARK8-linked Parkinson's disease. Neuron. 44 (2004) 595–600. doi:10.1016/j.neuron.2004.10.023

19. EM Valente, S Salvi, T Ialongo, et al. PINK1 mutations are associated with sporadic early-onset Parkinsonism. Ann. Neurol. 56 (2004) 336–341. doi:10.1002/ana.20256

20. V Bonifati, P Rizzu, MJ van Baren, et al. Mutations in the DJ-1 gene associated with autosomal recessive early-onset parkinsonism. Science. 299 (2003) 256–259. doi:10.1126/science.1077209

21. L Xu, J Pu. Alpha-Synuclein in Parkinson's Disease: From Pathogenetic Dysfunction to Potential Clinical Application. Parkinsons. Dis. 2016 (2016). doi:10.1155/2016/1720621

22. S Chaudhary, M Behari, M Dihana, et al. Parkin mutations in familial and sporadic Parkinson's disease among Indians. Parkinsonism Relat. Disord. 12 (2006) 239–45. doi:10.1016/j.parkreldis.2005.12.004

23. Y Mizuno, N Hattori, H Mori, T Suzuki, K Tanaka. Parkin and Parkinson's disease. Curr. Opin. Neurol. 14 (2001) 477–482. http://www.ncbi.nlm.nih.gov/entrez/query.fcgi?cmd=Retrieve&db=PubMed&dopt=Citation&list_uids=11470964.

24. K Kalinderi, S Bostantjopoulou, L Fidani. The genetic background of Parkinson's disease: Current progress and future prospects. Acta Neurol. Scand. (2016). doi:10.1111/ane.12563

25. S Lubbe, HR Morris. Recent advances in Parkinson's disease genetics. J. Neurol. 261 (2014) 259–266. doi:10.1007/s00415-013-7003-2

26. CM Lill, JT Roehr, MB McQueen, et al. Comprehensive research synopsis and systematic meta-analyses in Parkinson's disease genetics: The PDgene database. PLoS Genet. 8 (2012). doi:10.1371/journal.pgen.1002548

27. A Verstraeten, J Theuns, C Van Broeckhoven. Progress in unraveling the genetic etiology of Parkinson disease in a genomic era. Trends Genet. 31 (2015) 140–149. doi:10.1016/j.tig.2015.01.004

28. S Nagar, RC Juyal, S Chaudhary, et al. Mutations in the alpha-synuclein gene in Parkinson's disease among Indians. Acta Neurol Scand. 103 (2001) 120–122. http://www.ncbi.nlm.nih.gov/entrez/query.fcgi?cmd=Retrieve&db=PubMed&dopt=Citation&list_uids=11227130.

29. S Punia, M Behari, ST Govindappa, et al. Absence/rarity of commonly reported LRRK2 mutations in Indian Parkinson's disease patients. Neurosci. Lett. 409 (2006) 83–88. doi:10.1016/j.neulet.2006.04.052

30. S Sudhaman, M Behari, ST Govindappa, UB Muthane, RC Juyal, BK Thelma. VPS35 and EIF4G1 mutations are rare in Parkinson's disease among Indians. Neurobiol. Aging. 34 (2013). doi:10.1016/j.neurobiolaging.2013.04.025

31. Bittles. The impact of consanguinity on the Indian population. Indian J. Hum. Genet. 8 (2002) 45–51.

32. EG Severance, RH Yolken. Lack of RIC-3 congruence with beta2 subunit-containing nicotinic acetylcholine receptors in bipolar disorder. Neuroscience. 148 (2007) 454–60. doi:10.1016/j.neuroscience.2007.06.008

33. PA Gourraud, M Sdika, P Khankhanian, et al. A genome-wide association study of brain lesion distribution in multiple sclerosis. Brain. 136 (2013) 1012–24. doi:10.1093/brain/aws363.

34. JS Yokoyama, DS Evans, G Coppola, JH Kramer, GJ Tranah, K Yaffe. Genetic modifiers of cognitive maintenance among older adults. Hum. Brain Mapp. 35 (2014) 4556–4565. doi:10.1002/hbm.22494

35. S Sudhaman, UB Muthane, M Behari, ST Govindappa, RC Juyal, BK Thelma. Evidence of mutations in RIC3 acetylcholine receptor chaperone as a novel cause of autosomal-dominant Parkinson's disease with non-motor phenotypes. J. Med. Genet. (2016). doi:10.1136/jmedgenet-2015-103616

36. N Vitureira, R Andrés, E Pérez-Martínez, et al. Podocalyxin is a novel polysialylated neural adhesion protein with multiple roles in neural development and synapse formation. PLoS One. 5 (2010). doi:10.1371/journal.pone.0012003

37. S Sudhaman, K Prasad, M Behari, UB Muthane, RC Juyal, BK Thelma. Discovery of a frameshift mutation in podocalyxin-like (PODXL) gene, coding for a neural adhesion molecule, as causal for autosomal-recessive juvenile Parkinsonism. J. Med. Genet. 53 (2016) 450–6. doi:10.1136/jmedgenet-2015-103459

38. CE Krebs, S Karkheiran, JC Powell, et al. The Sac1 domain of SYNJ1 identified mutated in a family with early-onset progressive Parkinsonism with generalized seizures. Hum. Mutat. 34 (2013) 1200–1207. doi:10.1002/humu.22372

39. M Quadri, M Fang, M Picillo, et al. Mutation in the SYNJ1 gene associated with autosomal recessive, early-onset parkinsonism. Hum. Mutat. 34 (2013) 1208–1215. doi:10.1002/humu.22373

40. L Kirola, M Behari, C Shishir, BK Thelma. Identification of a novel homozygous mutation Arg459Pro in SYNJ1 gene of an Indian family with autosomal recessive juvenile Parkinsonism. Parkinsonism Relat. Disord. (2016) 5–9. doi:10.1016/j.parkreldis.2016.07.014

41. D Salat, E Tolosa. Levodopa in the treatment of Parkinson's disease: Current status and new developments. J. Parkinsons. Dis. 3 (2013) 255–269. doi:10.3233/JPD-130186

42. F Castelán, M Castillo, J Mulet, et al.Molecular characterization and localization of the RIC-3 protein, an effector of nicotinic acetylcholine receptor expression. J. Neurochem. 105 (2008) 617–627. doi:10.1111/j.1471-4159.2007.05169.x

43. D Kalamida, K Poulas, V Avramopoulou, et al. Muscle and neuronal nicotinic acetylcholine receptors: Structure, function and pathogenicity. FEBS J. 274 (2007) 3799–3845. doi:10.1111/j.1742-4658.2007.05935.x

44. M Treinin. RIC-3 and nicotinic acetylcholine receptors: Biogenesis, properties, and diversity. Biotechnol. J. 3 (2008) 1539–1547. doi:10.1002/biot.200800179

45. D Muller, C Wang, G Skibo, et al.PSA-NCAM is required for activity-induced synaptic plasticity. Neuron. 17 (1996) 413–422. doi:10.1016/S0896-6273(00)80174-9

46. H Cremer, G Chazal, C Goridis, A Represa. NCAM is essential for axonal growth and fasciculation in the hippocampus. Mol. Cell. Neurosci. 8 (1997) 323–35. doi:10.1006/mcne.1996.0588

47. PS McPherson, EP Garcia, VI Slepnev, et al. A presynaptic inositol-5-phosphatase., Nature. 379 (1996) 353–357. doi:10.1038/379353a0

48. V Drouet, S Lesage. Synaptojanin 1 mutation in Parkinson's disease brings further insight into the neuropathological mechanisms. Biomed Res. Int. 2014 (2014) 289728. doi:10.1155/2014/289728

49. TW Harris, E Hartwieg, HR Horvitz, EM Jorgensen. Mutations in synaptojanin disrupt synaptic vesicle recycling. J. Cell Biol. 150 (2000) 589–599. doi:10.1083/jcb.150.3.589

Chapter 5

Biomarkers in Parkinson's Disease

Soaham Desai, Devangi Desai, Siddharth Singh

Parkinson's Disease (PD) affects nearly 1% population worldwide. The prevalence of PD is expected to rise in the future with increasing aging population of world.[1] PD is characterized by progressive reduction in dopaminergic neurons in the nigrostriatal network in brain, leading to its classical manifestations i.e. bradykinesia, rigidity, and tremors at rest.[2,3] PD is a progressive multisystem/ multi-organ disease with central nervous system (CNS) and extra-CNS manifestations, involving not only the brain but also the gastrointestinal system, pancreas, cardiac function, salivary glands, and the skin.[4,5] Motor manifestations of PD show up only after there has been more than 70% dopaminergic loss in nigrostriatal neurons.[2] Thus, the neuropathological degeneration has already occurred significantly by the time clinical diagnosis is made, making disease modification interventions rather impossible. Non-motor symptoms associated with PD actually begin much before the motor symptoms. These symptoms occur due to loss of noradrenergic, non-dopaminergic, and serotonergic pathways in brainstem and frontal lobe.[3] Therefore, there is a need to diagnose PD at an early stage before clinical symptoms begin and a good number of dopaminergic neurons are still available which can be protected from damage by the process of neuro-degeneration. Identifying biomarkers of the disease can help in early identification of people having a higher probability of developing PD, make early diagnosis, prognosticate, and monitor therapeutic response in a better manner.

A biomarker is a characteristic which can be measured in an objective way and evaluated to indicate normal biological or pathogenic processes as well as responses to any treatment.[6]

What can be considered as an "ideal" biomarker?

a. It should have high sensitivity and specificity to get validated by neuropathologic evaluation,
b. It should have satisfactory test–retest reproducibility,
c. It should be easily accessible for evaluation,
d. It should not be very expensive,
e. It should provide a reliable correlation with the severity of disease and its progression, and

DOI: 10.1201/9781003505488-5

f. It should not be affected by a change in the age, compensatory mechanisms, or treatments which the patient receives.[7]

Unfortunately, not a single biomarker studied till now can be said to be an ideal biomarker for PD. Also, it is not possible that we will have a unique solitary biomarker which will satisfy the credentials described above. Studies involving biomarker evaluation are also difficult due to diagnostic difficulty in clinical practice and impracticality of neuropathological studies in living individuals. The rate of misdiagnosis of PD is often high even in the hands of expert physicians with formal training in management of patients with movement disorders.[8] This is due to the fact that atypical parkinsonian syndromes have similar initial clinical features in the first 6-10 months. Since, it is hard to define Parkinson disease at an early stage; it is tough to validate biomarkers and tougher still to identify single candidate biomarkers. Therefore, it is necessary to search for and perform necessary validation to find a combination of biomarkers which can be used to ensure an early diagnosis, which could then allow therapies which may help in modifying the disease course and progression.

Need for biomarkers in Parkinson's Disease

- To confirm and support its diagnosis,
- To identify patients before motor symptoms appear, so that neuro-protective therapy can be instituted before major neuronal loss occurs,
- To track the progress of disease,
- To predict the speed of progression,
- To identify possible mechanisms of action of drugs for treatment and research focused on those mechanisms; and
- To address efficacy of treatment.

Types of biomarkers of PD

A wide spectrum of different types of biomarkers have been proposed and studied in PD. These utilize following approaches:

- Clinical manifestations assessment (motor/non-motor),
- In vitro testing (e.g. of blood serum, cerebrospinal fluid (CSF), etc.),
- In vivo tests approach of suspected patients (e.g. Imaging and electrophysiology),
- Biomarkers utilizing network-based approach, and
- Omics approach (utilizing genomics, proteomics and metabolomics)

There are various initiatives with focused research to identify and validate various biomarkers for PD. Parkinson's Disease Biomarkers Program (PDBP) by National Institute of Neurological Disorders and Stroke (NINDS) and Parkinson's progression Markers Initiative (PPMI) are two such initiatives working on evaluating role of different biomarkers using strict parameters for bio-sampling, archiving, storage, and analysis therefore leading to more validated results.

An approach to classify the spectrum of different biomarkers is depicted in Table 1 below:

Table 1: *Spectrum of different types of biomarkers in PD*

1. Clinical
2. Biochemical
3. Immunological
4. Genetic
5. Pathological: Tissue based: e.g. Blood, nasal mucosa, salivary gland, colon, and skin
6. Pharmacological
7. Imaging
8. Systems and networking based
9. Omics approach based

A. Clinical biomarkers of PD

These are clinical features which may precede development of classical motor features (tremor, bradykinesia, and rigidity). Several non-motor clinical features precede motor manifestations by several years in patients with PD. They have a role in predicting individuals who are susceptible to develop disease and can be used as biomarkers for initiating neuroprotective and disease modifying agents. Reduced smell sensation, sleep disorders, constipation, depression, and problems in frontal subcortical executive function are the predominant non-motor symptoms which have been evaluated as biomarkers of PD. There are other less studied non-motor symptoms too. Table 2 below enlists all such possible biomarkers of PD. Not all of these proposed clinical features have been studied adequately.

Table 2: *List of clinical biomarkers in PD*

Non-motor Markers	Motor clinical markers
Loss of smell	Drooling
Depression	Impaired handwriting, micrographia
Circadian rhythm of sleep abnormalities	Walking difficulty
REM sleep behaviour disorder (RBD)	Frozen shoulder
Visuo-spatial and colour discrimination abnormality	Postural changes
Bladder control issues and nocturia	Falls
Constipation	Muscle rigidity
Reduction in sympathetic skin response	Slowness
Altered cardiac sympathetic innervation	Body tremors
Postprandial fullness	
Deafness	
Dyscalculia	
Dysgeusia	
Impulse control disorders	

1. Rapid eye movement (REM)-sleep behaviour disorder (RBD)

RBD is linked with the highest chances of occurrence of PD across various studies. In an initial study on RBD, 38% of patients were reported to develop PD over next five years.[9] On further follow-up, 81% of these patients were found to develop some neurodegenerative disease suggesting that RBD can be a major predictor of neurodegenerative disorders.[9] Other studies have also described the relation of RBD preceding PD by decades and have also suggested its strong association with PD associated cognitive impairment, dysautonomia, hyposmia, and falls as compared to tremors.[10,11,12] RBD is reported to be predictive of alpha-synucleinopathy (SNCA) and should never be considered benign. The association is high enough to suggest prophylactic therapy for prevention/delay development of PD/Dementia with Lewy Bodies (DLB) in patients with idiopathic REM-sleep behaviour disturbances. RBD is best diagnosed by a video-polysomnography. Questionnaires have also been suggested to diagnose RBD, but these are not sensitive or specific for the diagnosis compared to video-polysomnography. The underlying pathogenesis is postulated to be due to SNCA pathology in sublaterodorsal nucleus, pre-coeruleus complex, and magnocellular reticular formation and its projection areas.

2. *Olfactory dysfunction*

Hyposmia/anosmia is found in nearly 90% of individuals with PD.[13] Olfactory dysfunction can be a predecessor of a clinical PD symptoms by 2-7 years. Olfactory loss in PD occurs due to central pathology. There is no loss of olfactory epithelium in patients with PD. Testing modules like University of Pennsylvania Smell Identification Test (UPSIT) can determine reduction of sense of smell clinically. Clinical studies have demonstrated impaired olfactory function in up to 85% of PD patients. In the Honolulu Asia Aging population study, assessing olfaction in healthy Japanese-American people, the population in the lowest tertile of olfactory dysfunction had 5.2 times increased chances of having clinical PD on follow up and 11 times risk of having incidental Lewy body pathology in brain on autopsy in the absence of clinical features of PD.[14,15] In the Parkinson Associated Risk Study (PARS), abnormal olfaction was linked to other markers of preclinical PD, like constipation, RBD, anxiety, depression, and mild motor symptoms.[16,17] Hyposmia is very rare in Progressive Supranuclear Palsy hence is a very good clinical tool for differentiating it from PD, though in multiple system atrophy (MSA) mild degree of olfactory dysfuncion may be seen. Olfaction is also preserved in PARK 2 (early onset genetic PD, which does not have SNCA pathology).

3. *Depression, dysthymia, anxiety, apathy and personality changes*

Psychiatric disorders are frequently associated with PD. Approximately fifty percent of patients who have PD also have depression or anxiety. Depression, sleep disturbances, anxiety, and apathy have been linked to preclinical phase of PD.[18] A premorbid "Parkinsonian personality" (i.e. tense, serious, honest, being introvert since childhood, nervous, cautious, inflexible, obsessive-compulsive, industrious, and are the ones who generally avoid impulsive, novel seeking behaviours or risk taking behaviours like smoking) has been suggested to be related to PD, but these personality traits are very nonspecific and lack evidence to link them to PD. Involvement of frontal dopaminergic projections, and abnormalities in mesocortical-prefrontal areas which are involved in motivation, reward, and stress response are believed to be the cause of depression, anxiety, dysthymia, and apathy in PD. Family history of depression and onset of symptoms from the right side of the body have also been found to be associated with depression in PD.

4. *Restless limb syndrome (RLS)*

Sleep disorders like excessive daytime sleepiness, sleep attacks, insomnia, restless limb syndrome (RLS), and RBD are quite common in patients with PD, negatively impacting their quality of life. Neurodegenerative processes within sleep regulatory brain circuitries, dopaminergic agents, antidepressant use, and depression are the possible reasons for these disorders. As RBD is a possible biomarker for PD, it was also thought that RLS could also be a pre-symptomatic biomarker for PD. However, a recent study assessing the association between known RLS genes and development of PD did not find any significant association between the two, suggesting that RLS and PD may be distinct entities despite the fact that RLS is common in PD, it actually does not predetermine occurrence of PD.[19]

5. Dysautonomia

Constipation is an early manifestation related to dysautonomia in PD and can precede motor manifestations by many years. Other manifestations like orthostatic hypotension, sexual dysfunction, and urinary disturbances are signs of significant dysautonomia and are usually seen in advanced PD. Presence of latter autonomic features early in disease is indicative of atypical Parkinsonism.[20] Other tests utilizing autonomic dysfunction in PD have been described later along with in vivo testing methodologies.

6. Other clinical biomarkers

Impairment of saccadic eye movements have been suggested but have not been sensitive or specific for PD.[21]

B. In vitro tests as biomarkers of PD

These tests utilize a body product evaluated in laboratory to look for various markers of PD. These have been tested in blood serum, cerebrospinal fluid, saliva, urine, skin, nasal endothelium, submandibular glands, and rectal/colonic tissue utilizing system biology techniques. The studied markers are genetic markers, inflammatory markers, degenerative protein markers, neurochemical markers, cytokines, etc.

Table 3 summarizes important in vitro biomarkers studied in PD utilizing blood and CSF. No significant association has been found using tests on saliva, urine, skin, or nasal endothelium.

Table 3: *List of in vitro biomarkers*

a. Blood based biomarkers

• Neurotrophic factors	High IGF, low vitamin D, low BDNF GH response to arginine (to D/DMSA) EGF levels (predict cognitive decline)
• Oxidative stress	High Uric acid level protective Bilirubin, ceruloplasmin Coenzyme Q10 level (disability)
• Metabolic factors	Homocysteine and its metabolites Apolipoprotein (a)
• Iron and heavy metals	Iron and manganese
• Inflammatory markers	PDGF subunit B, prolactin
• Genetic biomarkers	Alpha synuclein level SNCA (encoding alpha synuclein) LRRK-2, PARKIN, DJ1

b. CSF biomarkers

Degeneration markers (Protein aggregate)	Altered Aβ peptides Decreased alpha synuclein High tau protein

• Inflammatory markers	IL-1beta, IL-6, and IL-8 levels
• Genetic biomarkers	Higher DJ1
• Mitochondrial dysfunction	SOD-1, DJ1, Osteopontin

c. Gut and other systems-based biomarkers

• Skin biopsy	For alpha synuclein deposits and Lewy body pathology
• Nasal endothelium	For alpha synuclein deposits
• Gastrointestinal tract	For alpha synuclein deposits, reduced Prevotellaceae bacteria abundance of Enterobacteriaceae SIBO
• Salivary gland	Submandibular gland biopsy for alpha synuclein staining
• Gut mucosal biopsy: Colon, rectum, stomach duodenum	Alpha synnuclein staining

[IGF: insulin like growth factor, BDNF: brain derived neurotrophic factor GH: growth hormone, EGF: epidermal growth factor, PDGF: platelet derived growth factor, SNCA: synuclein alpha gene, LRRK: leucine rich repeat kinase, SIBO: small intestinal bacterial overgrowth]

a) *Blood based biomarkers*

Genetic biomarkers

Genome-wide association studies (GWAS) have isolated sixteen different genes for PD. Multiple genes like SNCA (encoding for alpha synuclein), LRRK-2, Parkin, DJ-1, etc. are crucial in that they provide more information regarding pathophysiology of PD. Such genetic markers may have a significant role in predicting PD pre-clinically in high risk individuals.

- *SNCA:* It is an important component of Lewy bodies. Lewy bodies are the neuropathological cytoplasmic inclusions which have an important role in PD pathology.[22] Some studies have reported increase[23] while others have reported decrease[24] in its levels in patients with PD versus controls. This discrepancy is possibly due to high abundance of SNCA within RBCs which can act as a confounding factor. Therefore, currently there are attempts to determine oligomeric/total/phosphorylated forms of SNCA in plasma which could be more accurate markers.[25, 26, 27]
- *DJ-1:* The gene which encodes for DJ-1 (formerly known as PARK7) has been identified in familial PD therefore leading to study of DJ-1 as a candidate biomarker.[28] Various studies with DJ-1 protein and their variants using techniques like 2d-Gel electrophoresis and immunoblotting have been done. Total Plasma DJ-1 was not much different in PD or control subjects but 4-HNE modification (which is indicative of oxidative stress) was significantly different in PD patients (HNE-modified DJ-1 isoform 4 is decreased, but isoform 6 is raised in patients with PD in late stage). The levels of these biomarkers are being studied to differentiate individuals with PD from those without PD as well as the patients with PD in different stages of disease.[29]

Other blood-based biomarkers

Various other blood-based biomarkers have been studied to look for how they are associated with PD. Table 3 enlists these biomarkers.

- *Uric acid:* Oxidative stress plays an important role in loss of dopaminergic neurons in PD. Uric acid due to its antioxidant properties and identification of low levels in substangia nigra in patients with PD undergoing post-mortem studies has led to studies of uric acid serum levels as a biomarker in PD.

- *Epidermal growth factor (EGF):* Using proteomics-based approach and utilizing a multiplex immunoassay technique, 70 PD patients and 102 proteins were tested and analysed simultaneously. EGF emerged as the top analyte and higher EGF levels were indicator of better cognitive levels.[30]
- *Apolipoprotein A1:* Significant association between Serum ApoA1 levels and DAT uptake has been seen and decreased ApoA1 has shown more severe DAT deficit.[30] This is an important observation as ApoA1 is modifiable by statins treatment.[31]
- *Haemoglobin:* Haemoglobin forms a complex with SNCA thus reducing levels of neuronal Hb. Neuronal Haemoglobin may contribute to elevated iron levels in substantia nigra.

b) *CSF based biomarkers of PD*

CSF surrounds brain and reflects the milieu surrounding the pathogenic process. It is easily obtainable and minimally invasive. It is one of the most cost-effective screening markers available for PD patients.

Following markers have been studied in CSF:

- **Markers of protein aggregation:** SNCA is an intracellular protein found in the Lewy bodies. Patients with synucleinopathies have lower levels of SNCA in CSF as compared to individuals without synucleinopathies. However, SNCA is reduced in all synucleinopathies and so it cannot be used to distinguish PD from other synucleinopathies.[30] A study with SNCA was done by Hong et al. in 2010 involving 299 subjects (117 PD patients, 50 patients with AD and 132 healthy controls).[32] No relation of severity was detected in the study. Further modifications in its estimation (with calculations of oligomeric and total syn-nuclein) are hypothezised to improve the reliability of CSF SNCA testing in the future.
- **Alpha synuclein RT-QuIC** is a new real time quaking induced co-inversion (RT-QuIC) based quantitative test to determine the SNCA accumulation in CSF which is being evaluated for its role as a biomarker.
- **Markers of mitochondrial dysfunction and oxidative stress**

  a) DJ-1 is a mitochondrial related anti-oxidant and reflects the oxidative stress which is associated in pathology of PD. DJ-1 levels were increased in CSF of individuals with PD compared to individuals without PD.[33] But it was also found to be increased in MSA patients in another study.[34, 35]

  b) Osteopontin is a protein related to mitochondrial dysfunction as well as oxidative stress. Osteopontin is raised in CSF of patients with PD but not in controls.[36]

  c) Superoxide dismutase-1 (SOD-1) is decreased in PD, Huntington's disease, Alzhiemier Disease[AD], and Amyotrophic Lateral Sclerosis [ALS] but the oxidatively modified version of SOD-1 is increased in PD.[37]
- **Inflammatory markers in CSF:** Inflammatory markers are detected in CSF of PD pointing to an inflammatory pathogenesis of PD. IL-1beta, IL-6, and IL-8 were found along with decreased complement components due to activation of the complement pathway.[38]

c) *Gut and other system-based biomarkers*

Association of gut in Parkinson's disease

Alteration in the composition of gut microbiological system causes abnormalities in intestinal function and permeability. This results in involvement of gastrointestinal epithelial cells which also have a role in functioning of the intestinal nerve function as well as immune system.[39,40] Intestinal inflammation and changes in the gut microorganism may lead to misfolding of SNCA leading to development of PD.[41] Although a causal relationship between alteration in gut flora and PD genesis has not been clearly explained, studies have demonstrated that alterations in the gastrointestinal microbiota occur before or during the evolution of PD.[42]

Individuals with PD have decreased population of Prevotellaceae bacteria as compared to healthy controls. Also, Enterobacteriaceae levels are greater in the gut amongst patients with a subtype of PD with Postural Instability-Gait Disturbance (PIGD) as compared to those with tremor-dominant PD.[43] Prevotellaceae bacteria have a role in mucin synthesis and formation of certain short-chain fatty acids (SCFA) with important role in neural function.[44] Decrease in the number of Prevotellaceae leads to reduced mucin synthesis, which in turn causes reduction in mucin synthesis worsening intestinal permeability, exposing the intestinal system to bacterial endotoxins, triggering, and maintaining SNCA hyperfunction in the colon, finally leading to its misfolding.[45,46] Studies have shown high rates of small intestinal bacterial overgrowth (SIBO) in PD, (amounting to 54% to 67%)[47] which could be due to impaired GI motility seen in PD patients. Also, unexpectedly SIBO was seen to be associated with motor symptoms which could also be due to disruption of small intestinal integrity by SIBO and affecting immune stimulation and alteration of L-Dopa absorption.[48]

Colonic, rectal or gastroduodenal biopsy

A number of studies have been conducted utilizing biopsy of gastrointestinal tract to look for alpha syn-nuclein deposition in patients with PD. Initial findings of these deposits in autopsy samples of patients with PD led to further studies comparing depositions of alpha synuclein in healthy controls compared to patients with PD. However, these tests found that alpha synuclein deposition in the mucosa of the colon is not a very sensitive biomarker for PD.[49]

Skin biopsy and nasal endothelium studies

Recent studies have evaluated Lewy pathology in the skin of patients with PD to look for abnormal accumulation of SNCA, however no significant correlation was observed to suggest that it can be used for routine clinical use.[50] Similarly studies on nasal endothelium also did not reveal significant positive associations to recommend its use.[50, 51, 52, 53]

Salivary gland biopsy studies

Submandibular salivary gland biopsy studies on autopsy and in patients with advanced PD have suggested that SNCA deposits can be found in salivary gland of patients with PD. A study of patients in initial stages of PD also found SNCA deposits in nearly three fourth of patients with PD.[51]

Large multicentre studies assessing well-defined biopsy histology by blinded panel of expert pathologists would be required to identify potential of peripheral tissue biopsy as biomarker of PD in clinical practice. A review assessing the role of peripheral tissue biopsies as biomarker of PD concluded that peripheral phosphorylated as well as alpha synuclein is not yet sensitive or specific enough to be accepted it as a valuable biomarker of PD in routine clinical practice.[52]

C. Neuroimaging as a biomarker in PD

With the advent of new techniques in neuroimaging like functional MRI and PET, it is now possible to detect neural structural and functional pathology. These tests are being studied to assess their role as biomarker in pre-clinical PD. These neuroimaging markers may have different indications, like early identification of PD, differentiating PD from secondary Parkinsonism, differentiating PD from atypical Parkinsonism, subtyping different non-motor types of PD based on predominant metabolic function derangement, assessing role of possible neuroprotective agents in terms of metabolic neurotransmitter functions, and tracking disease progression serially. Neuroimaging tests may assess a particular metabolic pathway and its function or assess brain structure and predict PD and its characteristics.

Table 4 enlists the different neuroimaging tests in terms of their role and mechanism.

Table 4: *Imaging tests which can be used as biomarkers of PD*

A. Tests assessing metabolic functions of the brain	
1. Assessing presynaptic dopamine terminal	1. 18 F-DOPA / 11 C-DOPA PET 2. Tropane based PET and SPECT (DAT SCAN) 3. VMAT2 density measurement 4. 11 Raclopride PET
2. Assessing serotonin syndrome	1. 11C-WAY100633 PET
3. Assessing cholinergic system	1. 123 I-Benzovesamicol SPECT 2. 11C-MPHA PET
Tests assessing brain structure	
1. 18 F DOPA PET	Reduced putamen18 F DOPA associated with PD
2. 11-C-PK 11195 PET	It estimates microglial function. Its role is being evaluated in PD and cortico basal ganglionic degeneration[CBGD]
3. T1 MRI based automated ROI based study	Computer based PD diagnosis using percentage reduction in grey matter, white matter, and gain in CSF volume
B. Conventional imaging tests	
1. MRI Brain 2. Voxel based morphometry 3. Diffusion tensor imaging	
C. Transcranial ultrasound Hyper-echogenicity of SNr	Asymmetric SNr [substantia nigra] hyper-echogenicity is associated with PD.
D. Cardiac MIBG scan	
Decreased pre-synaptic myocardial uptake of I-123 meta-iodo-benzyl-guanidine (MIBG)	Differentiates PD from other atypical parkinsonism

A. *Test assessing metabolic functions of the brain*

These can be sub-classified on the basis of the metabolic pathway they utilize for imaging:

- **Tests assessing dopamine terminals:**
  - 18 F-DOPA/11C-DOPA PET measure presynaptic dopamine decarboxylase activity. 40-50% loss of posterior putamen Dopa-decarboxylase activity coincides with onset of symptoms. Metabolic imaging including use of glucose metabolic PET and network analysis is being assessed to definitively diagnose PD compared to PSP, MSA, and CBGD.[53]
  - Tropane based PET and SPECT tracers assess availability of pre-synaptic dopamine transporters (DATs) using the ^{123}I-ioflupane also known as DaT SCAN which reveals normal levels in drug induced parkinsonism and essential tremor but reduced levels in PD and other neurodegenerative disorders with a good sensitivity (87-98%) and specificity (80-100%) in differentiation. DaT-SPECT however does not differentiate PD from PSP and MSA.[54]
  - VMAT2 density in dopamine terminals 11 C-Tetra hydrobenazine is used for the same.[55]
  - Ability of dopamine terminals to release dopamine 11 Raclopride-PET with amphetamine/ L-DOPA challenge measures changes in D2-receptor availability.[56]

- **Tests assessing serotonin system**

  11C-WAY100633 PET is used to study median raphe serotonin HT1a levels in midbrain which reflects function of serotonergic cell bodies. This is closely related with the degree of tremor, rigidity, and slowness of patients with PD. However, it does not correlate well with depression in subjects with PD and hence, suggests no direct roles of serotonergic pathway in PD.[53] Studies have proven that it is more likely that noradrenergic and limbic pathways have association with development of depression in individuals with PD.

- **Tests assessing cholinergic system**
  - Assessing presynaptic cholinergic terminal with 123 I-Benzo-vesamicol SPECT reflecting the vesicular transport
  - Assessing post synaptic cholinergic terminal with 11C-MPHA PET which reflects the muscarinic receptor availability.[57]

B. *Imaging modalities assessing brain structure*

- 18 F-DOPA PET is used to detect decreased Putaminal 18-FDP uptake which denotes high likelihood of PD.
- 11-C-PK 11195 PET is a technique used as marker of activation of microglia in PD. It correlates with increased activation in basal ganglia, midbrain, and frontal cortex. Microglia form 20-30% of white matter of brain. They are a part of its natural defence mechanism. Local injury and inflammation swell up the microglia and activate them with expression of BDZ sites on surface which bind 11C-PK 11195.[58,59]
- An automated regions of interest (ROIs) based study proposed computer aided diagnosis of PD by studying five ROIs (substantia nigra (SN), hippocampus, thalamus, midbrain, and frontal lobe (FL)) using multivariate feature selection approach (MIFST) to correlate selected features in the ROIs. In a study from India, loss in grey matter (GM) volume and white matter (WM) volume and increase in CSF volume was determined in individuals with PD as

compared to people without PD. By using SN volume for grey matter (GM) and Substantia Nigra + Frontal Lobe (SN+FL) volume for white matter (WM), a measurement by computer-based decision system; individuals with PD could be differentiated with good sensitivity and specificity. In future, this decision support system may be used by clinicians to diagnose PD with high accuracy.[60]

C. *Use of conventional MRI and other technique*

Conventional MRI is used to identify Parkinsonism secondary to structural lesions like vascular Parkinsonism, tumours, hydrocephalus, and infection. Abnormal T2 hypointensities in putamen and pons can be used to differentiate between PD and MSA.[61] Voxel based morphometry and diffusion tensor imaging have been used to diagnose PD, atypical Parkinsonism, and healthy controls with good accuracy.[62]

D. *Other miscellaneous biomarkers utilizing imaging modalities*

- **Transcranial USG:** Temporal bone acoustic window is used to assess the structure of substantia nigra. Patients with PD demonstrate asymmetrical bilateral increased echogenicity of SN.[63] This is suspected to be related to levels of ferritin and iron and consequential damage caused by iron derived free radicals. Though a good test, high inter-observer variability, and difficulty in obtaining sufficient acoustic window are drawbacks in using this technique routinely.
- **Cardiac MIBG scintigraphy:** PD patients have many signs of autonomic dysfunction like bladder involvement, sweating, and constipation which are found to precede extrapyramidal symptoms. Also, many post mortem studies of PD patients have revealed Lewy body deposition in the enteric and cardiac plexus. Hence, non-invasive techniques to evaluate cardiac plexus were envisioned. ^{123}I-meta-iodobenzylguanidine (MIBG) is a radio-labelled compound with adrenergic action. The level of MIBG uptake depends on the postganglionic sympathetic function and its uptake is reduced when the sympathetic network is dysfunctional. Thus, MIBG scintigraphy allows us to quantify postganglionic sympathetic cardiac innervation. Decreased myocardial uptake of MIBG is an important feature of PD.[58] Low myocardial MIBG, if present in patients with mild cognitive impairment, constipation, postural hypotension, depression/ anxiety or RBD can be considered as an early preclinical biomarker of PD.[64]

Neuroimaging to track disease progression

1) *Dopaminergic system*: DDC PET using functional ligands shows mean annual decline in dopamine function in putamen and caudate. This decline in dopamine function is faster in individuals with PD compared to typical decline observed in normal ageing.[65,66]
2) *Serotonergic system*: PET ligand 11C-DASB, which has selective binding to serotonin transporter, provides the value of the serotonin function. Individuals with PD have loss in serotonin function in selected regions of the brain (basal ganglia, hypothalamus, and anterior cingulate cortex) as the disease progresses.
3) *Newer techniques underway*: Several MRI techniques are currently being explored for their role in tracking PD progression such as measuring cortical thickness, functional connectivity, diffusivity, regional cerebral blood flow, and fractional anisotropy. There is significant hope of clinical benefits when these techniques have been validated and ready to be put in clinical usage.

D. Biomarkers of PD utilizing network based approach

These consist of use of spatial covariance techniques to study network level abnormalities and detect metabolic abnormalities at voxel level.

1) **Parkinson disease related pattern (PDRP):** Spatial covariance of the cortico-striato-pallido-thalamo-cortical loops was studied and PDRP had increased activity in pallido thalamic and pontine levels versus reduced activity in pre-motor cortical area, supplementary motor cortex, and parietal association regions.[67]
   - PDRP can differentiate between PD and parkinsonian plus syndromes.
   - PDRP also correlates with motor scores on questionnaires of patients with PD[68] and also with symptom duration.[69]
   - MSA-related pattern consists of bilateral decreased metabolism in the cerebellum and putamen.
   - PSP patients' studies reveal reduced metabolism in upper brainstem (midbrain) and prefrontal cortex.

2) **Cognitive pattern related to Parkinson disease (PDCP)**: It reflects reduced metabolism in medial frontal and parietal association areas and increased metabolism in cerebellar vermis and dentate nuclei. This correlates with the dysfunction in cognition in PD which often predates motor symptoms.

   PDRP is also modulated by treatment, whereas PDCP is not. Hence, PDRP may also be found to be useful in future to assess the role of various therapeutic modalities while PDCP could be used to monitor the effects of neuro-protective modalities.[70]

E. Utilizing "Omics" approach to define biomarkers of PD

1. **Genomics:** It is the study using the identification of genes and gene expression using microarrays to simultaneously check thousands of genes to look for significant associations with PD. Genome wide association studies (GWAS) have discovered links to diseases with single nucleotide polymorphisms (SNPs). Despite GWAS, the puzzle of genetics of PD is still unsolved. The GWAS are modelled linearly and so they analyze usually only one SNP at an instance and so the genomic and epigenomic factors of each SNP are often ignored. GWAS have been used to identify missense mutations in 28 genes of SNCA gene including (that encoding for syn-nuclein),[71]PARKIN gene mutation linked with recessive PD, LRKK2, and PINK1 mutations.[72] However, these advances are helpful in isolating factors causing rare familial PD, the genetic factors which influence sporadic PD have still to be fully revealed.
2. **Proteomics:** It is the study of the structure and functions of proteins, recognising and measuring the changes in proteins. These consist of changes (quantitative and qualitative) changes in proteins and the post-translational modifications in the proteins. Methods used are liquid chromatography based tandem mass spectrometry (LCG-TMS), 2-Dimensional gel electrophoresis and immuno-assays. Mutations studied in GWAS and others as part of genomics encode for proteins and molecular processes that lead to development of PD, which could be linked to cytoskeleton structure–dynamics, oxidative stress (α−synuclein protein , DJ-1 protein), mitochondrial activity (DJ-1 protein and PINK1 protein), intracellular signalling

(DJ-1, LRRK2) protein aggregation and calcium homeostasis (parkin, SNCA, UCHL1).[73, 74] These proteins are increasingly being used as biomarkers and detected in plasma and CSF of patients of PD. CSF SNCA levels are inversely related to the stage of patients with PD. DJ-1 levels are high in CSF and plasma of individuals with PD and their levels are associated with the severity of the disease. Other proteins like Glial related proteins glial fibrillary acidic protein (GFAP), glial maturation factor beta (GMFB), sorcin (SRI) and galectin 1 (LGALS1) are increased in individuals with PD. These proteins are also being studied to assess their role as biomarker of neurotoxicity.[75]

3. **Metabolomics:** It is the study of chemical metabolic processes identifying and quantifying small molecule metabolites to give an assessment of the existing metabolic state. The methods used are liquid chromatography electrochemical array (LCECA) or mass spectrometry. Metabolomics have found that increased serum levels of uric acid may suggest low chance of development of PD. Increased glutathione was found in plasma of individuals with PD probably suggesting oxidative damage. A recent study of metabolic profiles has shown differences between idiopathic PD patients, LRKK-2 related PD and controls with different metabolome found in all three.[76] Thus, metabolomics may provide critical information on cellular processes, molecular interactions, and metabolic pathways associated with PD. Metabolomics may provide clues to the etio-pathogenesis of PD, lead to discovery of newer targets for its treatment, and identify further biomarkers in the future.

A recent study using metabolomics has suggested that metabolomic biomarkers can act as strong correlates of Parkinson disease progression. Forty-nine randomly selected placebo-treated "Dopa naïve" participants with early PD were included in that study. Baseline unified Parkinson's disease rating scale (UPDRS) (part 2 and 3) scores were collected along with CSF and plasma specimens serially. This was a unique study relating progression of disease to metabolomics in treatment-naive patients. Three separate mass spectrometers were used for metabolomic profiling. The strongest correlation of change in UPDRS scores (marker of disease progression) was seen with CSF concentration of benzoate. Homovanillate (terminal dopamine metabolite) was found to poorly correlate with disease progression. Purine metabolites like inosine and theobromine were found to have inverse correlation to disease progression.[77]

However, metabolome is a heterogeneous mixture of compounds with a range of turn-over rates, stabilities, solubilities, and volatilities. Issues with handling of samples and procedures involved in preparation may disturb the biological sample composition leading to inaccurate results. This is one of the reasons for different discrepancies across different studies evaluating the role of metabolomics in PD in the current medical literature.

Biomarkers and their various possible uses across different spectrum/role in PD

Since an ideal biomarker of PD would never be available, a better way is to define certain definite goal or purpose for which a biomarker is needed and see how different biomarkers in combination could fare in solving a particular purpose, e.g.: early identification, establish diagnosis, help in differentiate from other atypical parkinsonism, assess disease progression serially, identify non-motor subtypes, favour drug safety profile, etc. (Table 5)

Table 5: *PD biomarkers and their roles*

Early identification of prodromal PD	Diagnosis of PD	Assessing progression of PD
Clinical biomarkers		
Family history PD or tremor	RBD, insomnia: Association with cognitive decline	**Markers suggesting early cognitive decline:** MCI Apathy Hallucinations Gait freezing
Preceding two years Lack of smoking, anhedonia, apathy, memory problems, inattention	Psychiatric comorbidity Memory problems Inattention	
Preceding 2-10 years Anosmia Sweating Fatigue Pain Mood fluctuation Constipation	**Important role of excluding "Red flags" which suggest parkinsonian plus syndrome:** Early falls Abnormal extraocular movement Absent optokinetic nystagmus Early cognitive decline Hallucinations Severe autonomic dysfunction Apraxia Alien limb phenomenon Focal dystonia	**Marker suggestive of less cognitive decline** Tremor predominant form
Preceding >10years Constipation RBD Postprandial fullness Excess daytime sleepiness		Tremor predominant form associated with less depression and less hallucinations
Biochemical biomarkers		
Associated with PD: High IGF1 Low vit D	**Associated with PD:** Raised homocysteine Low BDNF High IGF1	Serum BDNF with severe motor impairment and cognitive decline Serum EGF with cognitive decline
Protection: High uric acid	CSF biomarkers of PD: High tau, DJ1 Altered Aβ peptides Low α synuclein	α synuclein: Role contrasting results across studies CSF tau and Aβ peptides associated with cognitive decline
Imaging biomarkers		
DaT scan can detect early striato-nigral degeneration	DaT, 18 flourodopa or VMAT scans can differentiate between degenerative parkinsonian syndromes and essential tremor/ 18 F-DOPA secondary parkinsonism. These cannot differentiate PD from other parkinsonism subtypes.	Grey matter volume, caudate volume and cortical thickness measurements can correlate with cognitive decline in PD.
Transcranial ultrasound showing substantia nigra hyperechogenicity associated with PD development at 3-5-year follow-up	Functional MRI [FMRI] can differentiate between tremor predominant and rigidity predominant subtypes of PD. Few imaging signs like "humming bird sign" and "hot cross bun sign" can help identifying specific syndromes.	Basal ganglia iron assessment by MRI relaxation rate *R2 can correlate with disease progression.

Role of biomarkers in PD

1. *Role of biomarkers in identification of early PD in its prodromal/premotor phase (before motor symptoms appear)*

   Study of use of biomarkers in prodromal phase of PD to predict the risk of development of PD in individuals at risk, a few years before development of clinical symptoms can be helpful. By this, at risk individuals can be identified and treated by various neuroprotective modalities [eg IGF1 levels]. Currently, transcranial USG, DAT imaging for striato-nigral degeneration, and genetic testing for known gene mutations are the most promising biomarkers for the use in prodromal phase of PD.

2. *Role in improving diagnostic certainty of PD and differentiating from secondary and atypical Parkinsonism*

   Biomarkers can also be useful in diagnosing PD with more certainty, especially in differentiating PD from other atypical Parkinsonism. Cardiac MIBG scintigraphy[78] and growth hormone response to arginine may help in differentiating PD from MSA.[79] DAT scan may help in differentiating PD from other tremor syndromes and secondary Parkinsonism, but it does not differentiate between PD and other atypical Parkinsonism. Diffusion weighted imaging of Putamen is found to be abnormal in MSA with increased nigral echogenicity in PD. The role of other biomarkers in differentiating various types of Parkinsonism is still uncertain.

3. *Role in assessing progression of PD, to assess for severity of disease, its staging as well as for use in studies evaluating role of agents in neuroprotection*

   Biomarkers may also be used in assessing PD progression over the years. Higher serum BDNF and EGF levels are found to correlate with higher cognitive deficits and motor disability in individuals with PD.[80,81] Uric acid and Coenzyme Q10 have been found to correlate inversely with the motor burden of PD.[82,83] Higher phosphorylated tau protein level in CSF correlates with cognitive decline in PD.[84] Cortical thickness measurement and caudate nucleus volume can predict of the cognitive dysfunction in patients with PD. Iron content in basal ganglia on T2 relaxometry on MRI brain has also been found to be associated with clinical progression of PD.

To conclude, in this chapter we have reviewed different biomarkers of PD and certain clinical circumstances where these biomarkers will have important role in the future. At this time, no single "ideal" biomarker of PD is available, and it is highly unlikely that one will be available in near future. However, new studies involving multiple centres, utilizing a combination strategy using different types of biomarkers is proving good pathway for use in PD in its prodromal phase, diagnosis as well as study its progression in the future.

Key points

1. There is no "ideal" biomarker for PD.
2. Anosmia, RLS, constipation, and RBD are important clinical biomarkers.
3. A variety of in vitro and imaging biomarkers are currently under evaluation.
4. A combination of different types of biomarkers can be used to identify prodromal PD, to diagnose PD with certainty, and to assess its progression.

References

1. Kowal SL, Dall TM, Chakrabarti R, Storm MV, Jain A. The current and projected economic burden of Parkinson's disease in the United States. Mov Disord. 2013; 28(3);311-8. doi: 10.1002/mds.25292

2. Cheng HC, Ulane CM, Burke RE. Clinical progression in Parkinson disease and the neurobiology of axons. Ann Neurol. 2010;67(6);715-25.
3. Jellinger KA. Neuropathobiology of non-motor symptoms in Parkinson disease. J Neural Transm (Vienna). 2015;122(10);1429-40.
4. Jellinger KA. Neuropathology of sporadic Parkinson's disease: evaluation and changes of concepts. Mov Disord. 2012;27(1);8-30.
5. Titova N, Padmakumar C, Lewis SJ, Chaudhuri KR. Parkinson's: a syndrome rather than a disease? J Neural Transm (Vienna). 2016. doi:10.1007/s00702-016-1667-6
6. Biomarkers Definition Working Group Biomarkers and surrogate endpoints: preferred definitions and conceptual framework. Clin Pharmacol Therapeutics. 2001;69;89–95.
7. Wang J, Hoekstra JG, Zuo C, et al. Changes in network activity with the progression of Parkinson's disease. Brain. 2007;130(7);1834-1846.
8. Biomarkers of Parkinson's disease: current status and future perspectives. Drug Discov Today. 2013;18(3-4);155-62. doi: 10.1016/j.drudis.2012.09.00
9. Meara J, Bhowmick BK, Hobson P. Accuracy of diagnosis in patients with presumed Parkinson's disease. Age Ageing. 1999;28(2);99-102.
10. Schenck CH, Boeve BF, Mahowald MW. Delayed emergence of a parkinsonian disorder or dementia in 81% of older men initially diagnosed with idiopathic rapid eye movement sleep behavior disorder: a 16-year update on a previously reported series. Sleep Med. 2013;14(8);744-8.
11. Monderer R, Thorpy M. Sleep disorders and daytime sleepiness in Parkinson's disease. Curr Neurol Neurosci Rep. 2009;9(2);173-80.
12. Gagnon JF, Vendette M, Postuma RB, et al. Mild cognitive impairment in rapid eye movement sleep behavior disorder and Parkinson's disease. Ann Neurol. 2009;66(1);39-47.
13. Haehner A, Hummel T, Hummel C, Sommer U, Junghanns S, Reichmann H. Olfactory loss may be a first sign of idiopathic Parkinson's disease. Mov Disord. 2007;22(6);839-42.
14. Ross GW, Petrovitch H, Abbott RD, et al. Association of olfactory dysfunction with risk for future Parkinson's disease. Ann Neurol. 2008;63(2);167-73.
15. Ross GW, Abbott RD, Petrovitch H, et al. Association of olfactory dysfunction with incidental Lewy bodies. Mov Disord. 2006;21(12);2062-7.
16. Ponsen MM, Stoffers D, Twisk JW, Wolters ECh, Berendse HW. Hyposmia and executive dysfunction as predictors of future Parkinson's disease: a prospective study. Mov Disord. 2009;24(7);1060-5. doi: 10.1002/mds.22534
17. Evatt ML, Chaudhuri KR, Chou KL, et al. Dysautonomia rating scales in Parkinson's disease: sialorrhea, dysphagia, and constipation--critique and recommendations by movement disorders task force on rating scales for Parkinson's disease. Mov Disord. 2009;24(5);635-46.
18. Aarsland D, Brønnick K, Alves G, et al. The spectrum of neuropsychiatric symptoms in patients with early untreated Parkinson's disease. J Neurol Neurosurg Psychiatry. 2009;80(8);928-30.
19. Gan-Or Z, Alcalay RN, et al. Genetic markers of Restless Legs Syndrome in Parkinson disease. Parkinsonism Relat Disord. 2015;21(6);582-5.

20. Evatt ML, Chaudhuri KR, Chou KL, et al. Dysautonomia rating scales in Parkinson's disease: sialorrhea, dysphagia, and constipation--critique and recommendations by movement disorders task force on rating scales for Parkinson's disease. Mov Disord. 2009;24(5);635-46.

21. Barker RA, Michell AW, "The eyes have it". Saccadometry and Parkinson's disease. Exp Neurol 219. 2009;382–384.

22. Spillantini MG, Schmidt ML, Lee VM, Trojanowski JQ, Jakes R, Goedert M. Alpha-synuclein in Lewy bodies. Nature. 1997;388(6645);839-40.

23. Lee PH, Lee G, Park HJ, Bang OY, Joo IS, Huh K. The plasma alpha-synuclein levels in patients with Parkinson's disease and multiple system atrophy. J Neural Transm (Vienna). 2006;113(10);1435-9. PubMed PMID: 16465458.

24. Li QX, Mok SS, Laughton KM, et al. Plasma alpha-synuclein is decreased in subjects with Parkinson's disease. Exp Neurol. 2007;204(2);583-8.

25. El-Agnaf OM, Salem SA, Paleologou KE, et al. Detection of oligomeric forms of alpha-synuclein protein in human plasma as a potential biomarker for Parkinson's disease. FASEB J. 2006;20(3);419-25.

26. Tinsley RB, Kotschet K, Modesto D, et al. Sensitive and specific detection of α-synuclein in human plasma. J Neurosci Res. 2010;88(12);2693-700.

27. Foulds PG, Mitchell JD, Parker A, et al. Phosphorylated α-synuclein can be detected in blood plasma and is potentially a useful biomarker for Parkinson's disease. FASEB J. 2011;25(12);4127-37.

28. Bonifati V, Rizzu P, van Baren MJ, et al. Mutations in the DJ-1 gene associated with autosomal recessive early-onset parkinsonism. Science. 2003;299(5604);256-9.

29. Lin X, Cook TJ, Zabetian CP, et al. DJ-1 isoforms in whole blood as potential biomarkers of Parkinson disease. Sci Rep. 2012;2;954.

30. Qiang JK, Wong YC, Siderowf A, et al. Plasma Apolipoprotein A1 as a Biomarker for Parkinson's Disease. *Annals of neurology*. 2013;74(1);119-127. doi:10.1002/ana.23872

31. Grover A, et al. Correlation of compliance to statin therapy with lipid profile and serum HMGCoA reductase levels in dyslipidemic patients, Indian Heart J. 2016. http://dx.doi.org/10.1016/j.ihj.2016.07.007

32. Hong Z, Shi M, Chung KA, et al. DJ-1 and alpha-synuclein in human cerebrospinal fluid as biomarkers of Parkinson's disease. Brain. 2010;133(3);713-26. doi:10.1093/brain/awq008

33. van Dijk KD, Teunissen CE, Drukarch B, et al. Diagnostic cerebrospinal fluid biomarkers for Parkinson's disease: a pathogenetically based approach. Neurobiol Dis. 2010;39(3);229-41

34. Waragai M, Wei J, Fujita M, et al. Increased level of DJ-1 in the cerebrospinal fluids of sporadic Parkinson's disease. Biochem Biophys Res Commun. 2006;345(3);967-72.

35. Herbert M K, Eeftens J M, Aerts M B, et al. (2014). CSF levels of DJ-1 and tau distinguish MSA patients from PD patients and controls. Parkinsonism Relat. Disord. 20, 112–115.

36. Maetzler W, Berg D, Schalamberidze N, et al. Osteopontin is elevated in Parkinson's disease and its absence leads to reduced neurodegeneration in the MPTP model. Neurobiol Dis. 2007;25;473–482.

37. Butterfield DA, Perluigi M, Reed T, et al. Redox Proteomics in Selected Neurodegenerative Disorders: From Its Infancy to Future Applications. Antioxidants & Redox Signaling. 2012;17(11);1610-1655. doi:10.1089/ars.2011.4109

38. Lindqvist D, Hall S, Surova Y, et al. Cerebrospinal fluid inflammatory markers in Parkinson's disease—associations with depression, fatigue, and cognitive impairment. Brain Behav Immun. 2013;33:183-9. doi: 10.1016/j.bbi.2013.07.007. PubMed PMID: 23911592.

39. Mulak A, Bonaz B. Brain-gut-microbiota axis in Parkinson's disease. World J Gastroenterol. 2015;21(37);10609-20.

40. Hollister EB, Gao C, Versalovic J. Compositional and functional features of the gastrointestinal microbiome and their effects on human health. Gastroenterology. 2014;146;1449–1458.

41. Devos D, Lebouvier T, Lardeux B, et al. Colonic inflammation in Parkinson's disease. Neurobiol Dis. 2013; 50;42–48.

42. Olanow CW, Wakeman DR, Kordower JH. Peripheral alpha-synuclein and Parkinson's disease. Mov Disord. 2014;29;963–966.

43. Scheperjans F, Aho V, Pereira PA, et al. Gut microbiota are related to Parkinson's disease and clinical phenotype. Mov Disord. 2015;30;350–358.

44. Arumugam M, Raes J, Pelletier E, et al. Enterotypes of the human gut microbiome. Nature. 2011;473;174–180.

45. Pfeiffer R. Beyond here be dragons: SIBO in Parkinson's disease. Mov Disord. 2013;28;1764–1765.

46. Niehaus I, Lange JH. Endotoxin: is it an environmental factor in the cause of Parkinson's disease? Occup Environ Med. 2003;60;378.

47. Fasano A, Bove F, Gabrielli M, et al. The role of small intestinal bacterial overgrowth in Parkinson's disease. Mov Disord. 2013;28;1241–1249.

48. Tan AH, Mahadeva S, Thalha AM, et al. Small intestinal bacterial overgrowth in Parkinson's disease. Parkinsonism Relat Disord. 2014;20;535–540.

49. Visanji NP, Marras C, et al. Colonic mucosal a-synuclein lacks specificity as a biomarker for Parkinson's disease. Neurology. 2015;84(6);609-16.

50. Miki Y, Tomiyama M, et al. Clinical availability of skin biopsy in the diagnosis of Parkinson's disease. Neurosci Lett. 2010;469(3);357-9.

51. Adler CH, Dugger BN, et al. Peripheral Synucleinopathy in Early Parkinson's Disease: Submandibular Gland Needle Biopsy Findings. Mov Disord. 2016;31(2);250-6.

52. Schneider SA, Boettner M, Alexoudi A, Zorenkov D, Deuschl G, Wedel T. Can we use peripheral tissue biopsies to diagnose Parkinson's disease? A review of the literature. Eur J Neurol. 2016;23(2);247-61.

53. Holtbernd F, Eidelberg D. The Utility of Neuroimaging in the Differential Diagnosis of Parkinsonian Syndromes. Seminars in neurology. 2014;34(2);202-209. doi:10.1055/s-0034-1381733

54. Cilia R, Marotta G, Benti R, Pezzoli G, Antonini A. Brain SPECT imaging in multiple system atrophy. J Neural Transm (Vienna). 2005;112(12);1635-45.

55. Lin KJ, Weng YH, Hsieh CJ, Lin WY, Wey SP, et al. (2013) Brain Imaging of Vesicular Monoamine Transporter Type 2 in Healthy Aging Subjects by ^{18}F-FP-(+)-DTBZ PET. PLOS ONE 8(9): e75952. doi: 10.1371/journal.pone.0075952

56. Loane C, Politis M. Positron emission tomography neuroimaging in Parkinson's disease. American Journal of Translational Research. 2011;3(4);323-341.

57. Lozza C, Baron JC, Eidelberg D, Mentis MJ, Carbon M, Marie RM. Executive processes in Parkinson's disease: FDG-PET and network analysis. Hum Brain Mapp. 2004;22;236–245.

58. Henkel K, Karitzky J, Schmid M, et al. (2004), Imaging of activated microglia with PET and [^{11}C] PK 11195 in corticobasal degeneration. Mov. Disord., 19: 817–821. doi:10.1002/mds.20040

59. Politis M, Su P, Piccini P. Imaging of Microglia in Patients with Neurodegenerative Disorders. Front. Pharmacol. 2012;3;96. doi:10.3389/fphar.2012.00096

60. Rana Bharti, Juneja Akanksha, Saxena Mohit, et al. Regions-of-Interest based Automated Diagnosis of Parkinson's Disease using T1-weighted MRI. Expert Systems with Applications. 2015;42(9);4506–4516.

61. Schocke MF, Seppi K, Esterhammer R, et al. Diffusion-weighted MRI differentiates the Parkinson variant of multiple systematrophy from PD. Neurology. 2002;58(4);575-80.

62. Focke NK, Helms G, Scheewe S, et al. Individual voxel-based subtype prediction can differentiate progressive supranuclear palsy from idiopathic Parkinson syndrome and healthy controls. Hum Brain Mapp. 2011;32(11);1905-15. doi:10.1002/hbm.21161

63. Berg D, Behnke S, et al. Enlarged hyperechogenic substantia nigra as a risk marker for Parkinson's disease. Mov Disord. 2013;28(2);216-9.

64. Sakakibara R, Tateno F, Kishi M, Tsuyusaki Y, Terada H, Inaoka T. MIBG myocardial scintigraphy in pre-motor Parkinson's disease: a review. Parkinsonism Relat Disord. 2014;20(3):267-73.

65. Badoud S, Nicastro N, Garibotto V, Burkhard PR, Haller S. Distinct spatiotemporal patterns for disease duration and stage in Parkinson's disease. Eur J Nucl Med Mol Imaging. 2016;43(3);509-16. doi:10.1007/s00259-015-3176-5

66. Poston KL, Eidelberg D. FDG PET in the Evaluation of Parkinson's disease. PET clinics. 2010;5(1);55-64. doi:10.1016/j.cpet.2009.12.004

67. Polymeropoulos MH, Lavedan C, Leroy E, et al. 1997. Mutation in the α-synuclein gene identified in families with Parkinson's disease. Science 276: 2045–2047.

68. Lin MK, Farrer MJ. Genetics and genomics of Parkinson's disease. Genome Med. 2014;6(6);48. doi: 10.1186/gm566. Review

69. Lotharius J, Brundin P. Pathogenesis of Parkinson's disease: dopamine, vesicles and alpha-synuclein. Nat Rev Neurosci. 2002;3(12);932-42. Review

70. Bonifati V, Oostra BA, Heutink P. Unravelling the pathogenesis of Parkinson's disease--the contribution of monogenic forms. Cell Mol Life Sci. 2004;61(14);1729-50.

71. Saracchi E, Fermi S, Brighina L. Emerging candidate biomarkers for Parkinson'sdisease: a review. Aging Dis. 2013;5(1);27-34. doi:10.14366/AD.2014.050027. Review

72. Johansen KK, Wang L, Aasly JO, White LR, Matson WR, Henchcliffe C, Beal MF, Bogdanov M. Metabolomic profiling in LRRK2-related Parkinson's disease. PLoS One.2009;4(10);e7551. doi: 10.1371/journal.pone.0007551

73. Chai C, Lim K L. Genetic Insights into Sporadic Parkinson's Disease Pathogenesis. Current Genomics. 2013;14(8);486-501. doi:10.2174/1389202914666131210195808

74. Martin I, Dawson VL, Dawson TM. Recent Advances in the Genetics of Parkinson's Disease. Annual review of genomics and human genetics. 2011;12;301-325. doi:10.1146/annurev-genom-082410-101440

75. Del Prete D, Checler F, Chami M. Ryanodine receptors: physiological function and deregulation in Alzheimer disease. Molecular Neurodegeneration. 2014;9;21. doi:10.1186/1750-1326-9-21

76. Johansen KK, Wang L, Aasly JO, et al. Metabolomic profiling in LRRK2-related Parkinson's disease. PLoS One. 2009;4(10);e7551.

77. LeWitt PA, Li J, Lu M, Guo L, Auinger P; Parkinson Study Group–DATATOP Investigators. Metabolomic biomarkers as strong correlates of Parkinson disease progression. Neurology. 2017 Feb 8 pii: 10.1212/WNL.0000000000003663.

78. Sakakibara R, Tateno F, Kishi M, Tsuyusaki Y, Terada H, Inaoka T. MIBG myocardial scintigraphy in pre-motor Parkinson's disease: a review. Parkinsonism Relat Disord. 2014;20(3);267-73.

79. Pellecchia MT, Pivonello R, Salvatore E, Faggiano A, Barone P, De Michele G,Lombardi G, Colao A, Filla A. Growth hormone response to arginine test distinguishes multiple system atrophy from Parkinson's disease and idiopathic late-onset cerebellar ataxia. Clin Endocrinol (Oxf). 2005;62(4);428-33.

80. Lim NS, Swanson CR, Cherng H, et al. Plasma EGF and cognitive decline in Parkinson's disease and Alzheimer's disease. Annals of Clinical and Translational Neurology. 2016;3(5);346-355. doi:10.1002/acn3.299

81. Costa A, Peppe A, Carlesimo GA, et al. Brain-derived neurotrophic factor serum levels correlate with cognitive performance in Parkinson's disease patients with mild cognitive impairment. Front Behav Neurosci. 2015;9;253. doi:10.3389/fnbeh.2015.00253

82. Negida A, Menshawy A, El Ashal G, et al. Coenzyme Q10 for Patients with Parkinson's Disease: A Systematic Review and Meta-Analysis. CNS Neurol Disord Drug Targets.2016;15(1);45-53. Review.

83. Lolekha P, Wongwan P, Kulkantrakorn K. Association between serum uric acid and motor subtypes of Parkinson's disease. J Clin Neurosci. 2015;22(8);1264-7.doi: 10.1016/j.jocn.2015.02.015

84. Ragnhild E, Skogseth, Kolbjorn Bronnick, et al. Associations between Cerebrospinal Fluid Biomarkers and Cognition in Early Untreated Parkinson's Disease. Journal of Parkinson's Disease, 2015;5(4);783 DOI: 10.3233/JPD-150682

Chapter 6

Computers in Classification of Parkinson's Disease from Healthy Subjects Using Magnetic Resonance Imaging

Bharti Rana, Akanksha Juneja, RK Agrawal, S Senthil Kumaran and Madhuri Behari

Introduction

With the advent of neuroimaging techniques, differential diagnosis of individual subjects is becoming an emerging field. This has attracted attention of machine learning research communities to provide computer-aided diagnosis (CAD). Recently, a few CAD models have been proposed for Parkinson's disease (PD), which utilizes T1-weighted Magnetic Resonance Imaging (T1w-MRI).

Parkinson's disease (PD), a disorder related to movement that influences elderly persons. The progressive reduction in dopamine-producing cells in substantia nigra region is the main reason of PD. Dopamine is a neurotransmitter that sends messages from one part to another part of brain and from brain to other parts of the body. It plays a major role in controlling the movements. Thus, in case of inadequate dopamine, the functioning of the nerve cells is impaired and nerve cells are incapable to forward the brain messages. The main features of PD[1], related to motor signs, are resting tremor, bradykinesia, postural instability and rigidity.

Although, PD patients can be diagnosed easily in later phase, it is hard to discriminate PD patients from its mimickers in early stage. However, effective medication in later phase is still problematic for clinicians. Furthermore, treatments in later phase leaves patient in a disabled position in the early stage. Therefore, timely diagnosis of PD is necessary. This will help the clinicians to develop right treatment strategies and may provide a good quality of life to the patients. Despite significant efforts by researchers,[2] till date there is no biochemical or non-invasive pathological test for effective diagnosis of PD, especially at the early stage[3]. Currently, clinicians carry out preliminary diagnosis of PD using neurological examination, like Hoehn and Yahr[4] staging scale (H&Y), mini-mental state examination[5] and Unified Parkinson's Disease Rating Scale (UPDRS)[6] and medical records[1]. However, these tests

DOI: 10.1201/9781003505488-6

and criteria produce subjective decisions and are highly dependent on experience of the clinicians[6,7]. Also, significant time and support from patients as well as clinicians are required to collect all the necessary information[8].

Recently, neuroimaging techniques (or modalities) are becoming a boon for clinicians to understand brain anatomy, inner working of the brain, investigational studies, diagnosis, disease progression and surgeries. It provides an objective assessment[9] and generates images of the nervous system to quantify the atrophy in the various regions of brain in any disease. Among numerous neuroimaging modalities, structural MRI (sMRI) is the most preferred neuroimaging modality because it is non-invasive and generates high resolution structural images of brain. However, inherent complexity of brain anatomy introduces complication to identify neurodegeneration in brain. Additionally, acquisition of sMRI has been utilized to discard other pathologies. Furthermore, in PD, sMRI is utilized in surgical treatments like stem cell implantation and deep brain stimulation.

Nowadays, image processing and machine learning research communities are attempting to build automated CAD tools using sMRI that could be a valuable help for clinicians in PD diagnosis and its stage/progression. This chapter presents a comparative study of the existing CAD models for classification of PD and healthy subjects using T1w-MRI.

Key components of a CAD method

In general, key components of a CAD method are: pre-processing, dimensionality reduction using feature extraction and selection, scheme used for training-testing the decision model and the classifier to learn a decision model. These components are shown in Figure 1.

The pre-processing steps include: (i) skull-stripping (to remove external parts of the brain); (ii) normalization (to deform MRI to a standard template); (iii) segmentation (to obtain cerebrospinal fluid (CSF), white matter (WM) and gray matter (GM)), and (iv) smoothing. Mostly, SPM8 toolbox[10] is used to carry out pre-processing.

After pre-processing, dimension reduction is performed using feature extraction and feature selection. Feature extraction transforms original features (or attributes) in to another set of features which are more meaningful representation for classification and analysis. Voxel-based morphometry (VBM), independent component analysis and principal component analysis are commonly employed feature extraction techniques for analysis of MRI data. In feature selection, a subset of significant and uncorrelated attributes are chosen from a given set of features, which helps in dimensionality reduction and may result into better classification results. Some commonly employed feature selection methods are correlation coefficient, T-test based ranking, information gain, mutual information (MI) and minimum redundancy maximum relevance (mRMR).

The whole data is split into training and testing data. Training data is used to train a decision model and testing data is used to evaluate it. Training data is used to develop a decision model, a prototype to classify the test data. Test data is used to check the efficacy of the decision model. The most widely and commonly employed supervised classification method in the medical domain is support vector machine (SVM).

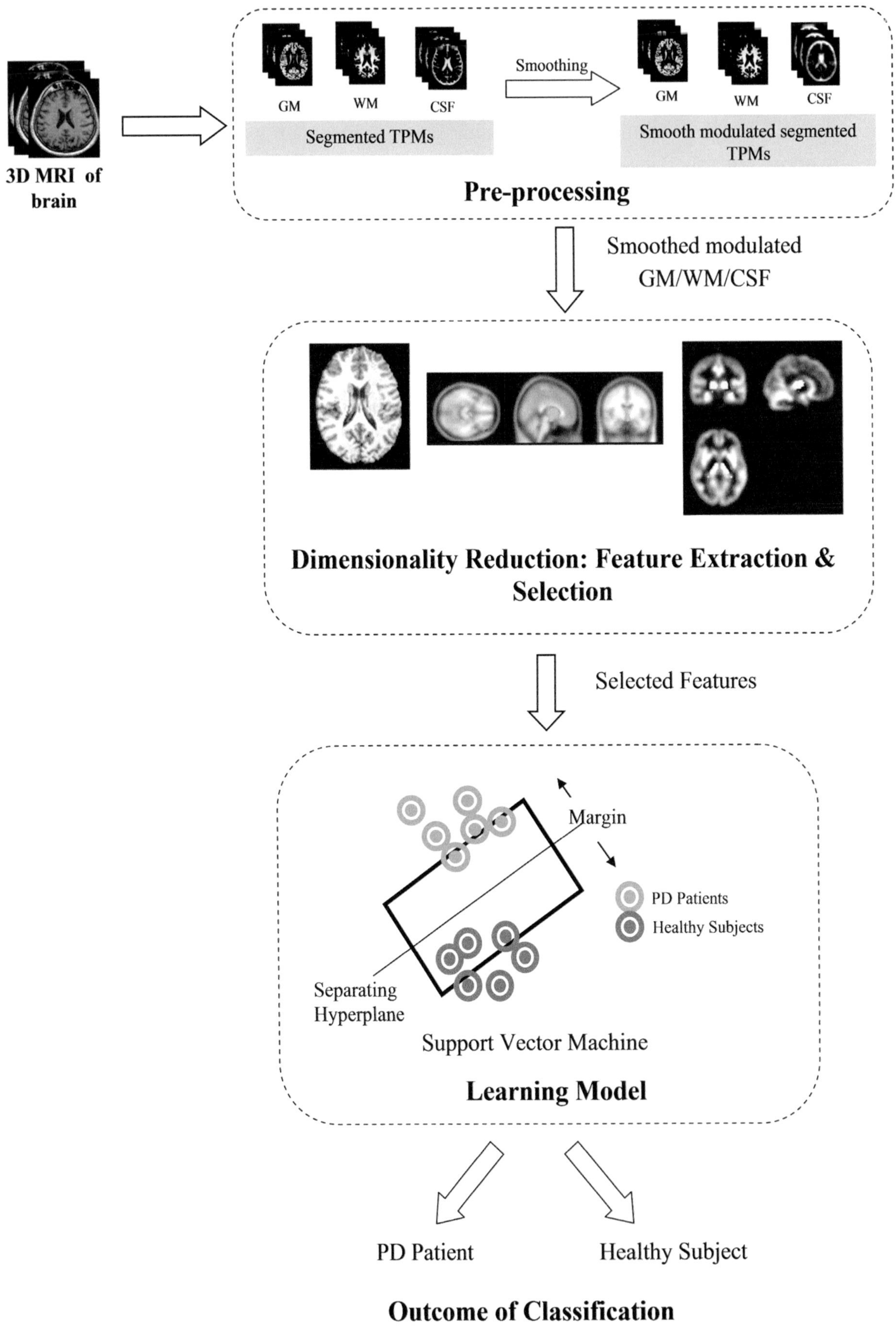

Figure 1: *Key components of CAD tool*

Literature review

This section presents the existing research works that proposed fully automatic CAD methods using T1-weighted MRI data. CAD methods using T1-weighted MRI are limited.[11-17]

Focke et al. (2011)[11] performed differential diagnosis of PD on individuals. They constructed features from GM and WM individually and used SVM to build-up a decision model. From GM features, they reported maximum classification accuracy of 39.53% and from WM features, 41.86% classification accuracy was achieved. However, the classification performance is not considerable.

Babu et al. (2014)[12] proposed CAD of PD wherein VBM was utilized to extract features from GM only. Thereafter, projection based learning and meta-cognitive radial basis function network (PBL-McRBFN) was used for classification. They reported 82.32% classification accuracy. Further, they used independent component analysis for obtaining a reduced set of features and noticed 10-11% decrease in the classification accuracy. Additionally, recursive feature elimination (RFE) in PBL-McRBFN was employed on VBM identified features to improve the classification results and classification accuracy improved to 87.21%. However, the proposed CAD method was highly complex computationally.

Another research group[13] attempted classification of PD patients from healthy subjects. MRI brain volumes were pre-processed using skull-stripping followed by normalization. After pre-processing, principal component analysis (PCA) was used to construct features. Thereafter, PCA components were ranked using Fisher discriminant ratio to find the most significant features. They also utilized SVM classifier and achieved 85.8% classification accuracy. However, PCA being an unsupervised approach does not consider class label and may increase within class variance, which may deteriorate the decision model's performance.

Rana et al. (2015)[14] proposed region(s)-of-interest (ROIs) based CAD method using GM, WM and CSF individually to differentiate PD patients from healthy subjects. This is the first research work wherein analysis of CSF was suggested for constructing CAD tool. Also, they highlighted the automatic delineation of ROIs instead of manual delineation. In this work,[14] five ROIs which are mentioned to be affected in PD were considered individually as well as their various combinations for feature constructed. Feature selection was carried out by employing MI in combination with t-test (MIFST). Classification was performed using SVM with leave-one-out cross-validation (LOOCV) method. For GM, 86.67% classification accuracy was achieved with SN region only while for WM, combination of SN and frontal lobe (FL) achieved 86.67% classification accuracy. However, their method included prior knowledge about ROIs and considered features from five ROIs only and hence may have ignored other affected regions. Besides, ROIs were demarcated with a well-known brain atlas that is not proposed to analyze data of patients.

Rana et al. (2015)[15] proposed to extract features from whole brain instead of pre-defined ROIs. In their CAD method, dimensionality reduction is performed using spectral feature selection approach; a group of discerning attributes from skull-stripped normalized volumes was chosen. They also employed SVM classifier with LOOCV scheme. They reported maximum classification accuracy of 86.67% and also identified the affected brain region(s) from the whole brain.

In other research work,[16] VBM is utilized at first stage on GM, WM and CSF individually. Thereafter, clusters of spatially contiguous voxels were created and three statistical features (entropy, mean and variance) were extracted from each cluster. Thereafter, mRMR feature selection method with t-test was employed to find-out a subset of significant and uncorrelated features. They also employed SVM for classification. They proposed the fusion of features extracted from GM, WM and CSF. They[16] achieved a classification accuracy of 88.33%. Their model was also capable to determine the affected brain regions.

In[17] a fused feature descriptor (FFD), a function of GM, WM and CSF was proposed. They extended 2D local binary pattern (LBP) to 3D LBP and utilized it for feature extraction from 3D FFD. Then, mRMR feature selection method and SVM classifier was utilized. They reported 95% classification accuracy and also mentioned the affected brain regions in PD.

A brief detail regarding the key components (pre-processing, dimensionality reduction covering feature extraction and selection, dataset size, data division scheme for training and testing, and classifier) employed by the existing methods is provided in Table 1.

Table 1: *The key components used in various research works for classification of healthy subjects and PD patients*

Research work	Pre-processing	Dimensionality reduction	Dataset	Division scheme*	Classifier
Focke et al. 2011[11]	Segmentation, Modulation and smoothing to obtain GM and WM	All Voxels as features and feature reduction using weights obtained with f-contrast map	22HS 21PD	LOOCV	SVM
Salvatore et al. 2014[13]	NSSV	PCA followed by FDR	28HS 28PD	Hold-out (50% data as training and 50% as testing) and LOOCV	SVM
Babu et al. 2014[12]	Segmentation and smoothing to obtain GM	Value of voxels which were identified using VBM	112HS 127PD	Hold-out (3:1 for training and testing) with 10 random trials	PBL-McRBFN with RFE
Rana et al. 2015[14]	Segmentation, modulation and smoothing to obtain GM, WM and CSF volumes	Voxels' values of five pre-defined ROIs independently, in pairs and triplets followed by MIFST	30HS 30PD	LOOCV	SVM
Rana et al. 2015[15]	NSSV	All voxels' value followed by spectral feature selection	D1: 30HS, 30PD and D2: 27HS, 27PD	LOOCV	SVM
Rana et al. 2016[16]	Segmentation, modulation and smoothing to obtain GM, WM and CSF volumes	Statistical features from each spatial cluster of VBM statistical map and mRMR	30HS 30PD	LOOCV	SVM
Rana et al. 2017[17]	Theta obtained from smooth modulated GM, WM and CSF	3D Local Binary Pattern and mRMR	30HS 30PD	LOOCV	SVM

*Division scheme represents how the data is split into training and testing datasets. †HS: Healthy subjects, NSSV: Normalization and skull-stripping

In Table 2, the best classification outcome obtained by the existing research works are reported. All the existing research works have captured the performance using specificity, sensitivity and classification accuracy. Table 2 also mentions 3D volume with which maximum performance was achieved.

Table 2: *Best classification performance achieved by the existing methods for discriminating PD patients from healthy subjects*

Existing research work	Volume used	Performance measures (in percentage)		
		Specificity	Sensitivity	Classification accuracy
Focke et al. 2011[11]	GM	50.00	28.57	39.53
	WM	45.45	38.10	41.86
Salvatore et al. 2014[13]	NSSV	86.00	86.00	85.80
Babu et al. 2014[12]	GM	87.00	87.39	87.21
Rana et al. 2015[14]	GM	83.33	90.00	86.67
	WM	86.67	86.67	86.67
	CSF	90.00	76.67	83.33
Rana et al. 2015[15]	NSSV (D1: 30HS, 30PD)	83.33	90.00	86.67
	NSSV (D2: 27HS, 27PD)	81.48	96.29	88.89
Rana et al. 2016[16]	GM+WM+CSF	86.67	90.00	88.33
Rana et al. 2017[17]	FFD	96.67	93.33	95.00

* NSSV: Normalization and skull-stripping, HS: Healthy subjects, and FFD: fused feature descriptor

Also, existing methods attempted to identify the brain regions that were used in PD and healthy subjects' classification. These brain regions are mentioned in Table 3. Also, most of the existing work studied the alteration in volume of the affected regions in PD in contrast to controls. This is also highlighted in Table 3.

Table 3: *Identified affected brain regions in PD reported in the existing research works*

Research work	Utilized volume	Identified brain regions
Focke et al. 2011[11]	GM	Right precuneus (↓)
	WM	Right superior longitudinal (↑) and fasciculus/right angular gyrus (↑)
Babu et al. 2014[12]	GM	Superior temporal gyrus (↓)
Salvatore et al. 2014[13]	NSSV	The caudal part of the pons and medial part of the midbrain (containing SN) *
Rana et al. 2015†[14]	GM	SN (↓)
	WM	Combination of SN and FL (↓)
	CSF	Combination of FL and mid-brain (↑)
Rana et al. 2015[15]	NSSV (D1) NSSV (D2)	Cuneus (↑), caudate (belongs to basal ganglia) (↓), extra nuclear (↓), thalamus(↓), lingual gyrus (↑) and lateral ventricle (↓)
Rana et al. 2016[16]	GM+WM+CSF	Superior, medial, middle, and inferior frontal gyrus (↓), orbital gyrus (↓), superior temporal gyrus (↓), anterior cingulate (↓), rectal gyrus (↓), hypothalamus (↓), caudate (↓), insula (↓), thalamus (↓)and frontal sub-gyral (↓)
Rana et al. 2017[17]	FFD	Superior frontal gyrus (↓), precuneus (↓), lentiform nucleus (↓), and caudate regions (↓)

↓: Reduction in volume of brain regions in PD patients, ↑: Increase in volume of brain regions in PD patients, *: No comments on volume reduction and †: Used prior knowledge and studied 5 ROIs

From Tables 1-3, we can observe the following:

(a) Existing CAD methods are able to classify PD and healthy subjects with good classification results except in 1 study[11]. Maximum classification accuracy of 95% is obtained by Rana et al. (2017).[17]

(b) CSF should also be analyzed for developing CAD of PD.

(c) Further, Rana et al. (2017)[17] pointed out that instead of studying CSF, WM and GM individually, FFD (a function of GM, WM and CSF) should be utilized.

(d) All the methods attempted to identify affected brain region(s) that might be considered as promising imaging-marker(s) in PD.

(e) Also, decrease in volume of WM and GM, and increase in volume of CSF was noted.

(f) A small sized dataset are utilized in all except the study by Babu et al.[12] that may lead to poor generalization performance.

(g) Classification performance is also influenced by unbalanced dataset. Thus, while learning a decision model, a balanced dataset should be utilized as was done in some studies.[13-17]

Further, research work[15,16] attempted to establish a relationship between UPDRS/H&Y scores and volume of identified regions. However, no significant correlation was observed in these studies. A significant value of correlation coefficient may help the clinicians in understanding the effect on brain regions. In addition, most of the researchers' extracted features at voxel level except the research works by Rana et al.[16,17] However, degeneration occur in group of voxels belonging to some regions instead of individual voxels. Therefore, while developing a decision model, features should be extracted at group level instead of individual voxel level.

Conclusion

In summary, the existing CAD methods for PD diagnosis achieved good classification performance. The classification performance of the existing methods encourages research community to develop CAD methods for PD diagnosis using sMRI. These CAD tools would be helpful for clinicians to diagnose PD easily. However, these methods suffer from a few limitations such as small and (or) unbalanced dataset, analysis of individual GM/WM/CSF and extracted features at voxel level. Thus, in future, there is a need to develop efficient methods that overcome the drawbacks of the existing methods. Also, features should be extracted at local regional level and should not involve prior knowledge while developing a CAD method. In addition, multi-modal study should be carried out to strengthen the findings.

References

1. Jankovic J. Parkinson's Disease: Clinical Features and Diagnosis. Journal of Neurology, Neurosurgery and Psychiatry. 2008;79(4);368–376.
2. Schwarz ST, Timothy R, Vamsi G, Morgan PS, Bajaj N, Auer DP. T1-Weighted MRI Shows Stage-Dependent Substantia Nigra Signal Loss in Parkinson's Disease. Movement Disorders. 2011;26(9);1633-1638.
3. National Institute of Neurological Disorders and Stroke. [Online].; 2015. Available from: http://www.ninds.nih.gov/disorders/parkinsons_disease/parkinsons_disease.htm.

4. Hoehn MM, Yahr MD. Parkinsonism: Onset, Progression, and Mortality. Neurology. 1967;17(5);427–442.
5. Folstein MF, Folstein SE, McHugh PR. "Mini-Mental State". A Practical Method for Grading the Cognitive State of Patients for the Clinician. Journal of Psychiatric Research. 1975;12(3);189-198.
6. Fahn S, Elton RL. Members of the UPDRS Development Committee. Unified Parkinson's Disease Rating Scale. In Recent developments in Parkinson's disease. Florham Park: MacMillan Healthcare Information; 1987;153–164, 293–304.
7. Machado A, Rezai AR, Kopell BH, Gross RE, Sharan AD, Benabid AL. Deep Brain Stimulation for Parkinson's Disease: Surgical Technique and Perioperative Management. Movement Disorders. 2006;Suppl 14;S247-58.
8. Post B, Merkus MP, de Bie RM, de Haan RJ, Speelman JD. Unified Parkinson's Disease Rating Scale Motor Examination: Are Ratings of Nurses, Residents in Neurology, and Movement Disorders Specialists Interchangeable? Movement disorders. 2005;20(12);1577–1584.
9. Piccini P, Whone A. Functional Brain Imaging in the Differential Diagnosis of Parkinson's Disease. Lancet Neurology. 2004;4;284–290.
10. SPM by members & collaborators of the Welcome Trust Centre for Neuroimaging. [Online]. Available from: http://www.fil.ion.ucl.ac.uk/spm/software/spm8/.
11. Focke NK, Helms G, Scheewe et al. Individual Voxel-Based Subtype Prediction can Differentiate Progressive Supranuclear Palsy from Idiopathic Parkinson Syndrome and Healthy Controls. Human Brain Mapping. 2011;32(11);1905–1915.
12. Babu Gidutkuri Sateesh, Suresh Sundaram, Mahanand Belathur Suresh. A Novel PBL-McRBFN-RFE Approach for Identification of Critical Brain Regions Responsible for Parkinson's Disease. Expert Systems with Applications. 2014; 41(2);478-488.
13. Salvatore C, Cerasa A, Castiglioni I, et al. Machine Learning on Brain MRI Data for Differential Diagnosis of Parkinson's Disease and Progressive Supranuclear Palsy. Journal of Neuroscience Methods. 2014;222;230-237.
14. Rana Bharti, Juneja Akanksha, Saxena Mohit, et al. Regions-of-Interest based Automated Diagnosis of Parkinson's Disease using T1-weighted MRI. Expert Systems with Applications. 2015;42(9);4506–4516.
15. Rana Bharti, Juneja Akanksha, Saxena Mohit, et al. Graph-theory-based spectral feature selection for computer aided diagnosis of Parkinson's disease using T1-weighted MRI. International Journal of Imaging Systems and Technology. 2015; 25(3);245–255.
16. Rana B, Juneja A, Saxena M, Gudwani S, Kumaran SS, Agrawal RK, et al. Voxel-based morphometry and minimum redundancy maximum relevance method for classification of Parkinson's disease and controls from T1-weighted MRI. In Tenth Indian Conference on Computer Vision, Graphics and Image Processing; 2016; Guwahati: ACM.22.
17. Rana B, Juneja A, Saxena M, Gudwani S, Kumaran SS, Behari M, et al. Relevant 3D local binary pattern based features from fused feature descriptor for differential diagnosis of Parkinson's disease using structural MRI. Biomedical Signal Processing and Control. 2017;34;134-143.

Chapter 7

Imaging in Parkinson's Disease

Nirendra K Rai

Introduction

Idiopathic Parkinson's disease (PD) is one of the commonest neurodegenerative diseases. Although, it is considered to be a Lewy body pathology and neuronal loss are not always associated with PD, suggesting other factors responsible for neuronal degeneration. Degenerative processes do not involve only dopaminergic system but are more widespread and involve other central neurotransmitter system including serotoninergic, cholinergic, noradrenergic and other neurotransmitters of central nervous system. Neuronal network involving basal ganglia including substantia nigra (SN), subthalamic nucleus (STN) and thalamus have important role in pathogenesis of PD and related disorders. Healthy substantia nigra (SN) possesses cluster of few important dopaminergic cells (nigrosomes) having special immune-histochemical properties. There are five types of nigrosomes, but nigrosome-1, the most important and largest one, are present in posterior third of SN. It can be visualized by special sequences of magnetic resonance imaging (MRI) in patients affected with PD. Dysfunctional nigrostriatal projections result in striatal dopamine deficiency (SDD) causing symptoms of PD. Other atypical parkinsonian syndromes associated with SDD include multiple system atrophy (MSA), progressive supranuclear palsy (PSP), corticobasal degeneration (CBD), vascular Parkinsonism and Wilson's disease. Non-striatal dopamine deficiency (SDD) disorders mimicking PD include drug induced Parkinsonism; Parkinsonism due to toxins (manganese or CO poisoning); essential tremors; or psychogenic Parkinsonism. Detecting functional status of dopaminergic network is one of the potential targets to differentiate various related disorders. Because of multiple pathophysiological changes in PD including gliosis, neuronal loss, neuronal dysfunction, Lewy body accumulation and other related changes; it is practically difficult to have a single imaging modality to diagnose it. Hence, a good biomarker (imaging as well as biochemical) is needed to differentiate idiopathic PD from other parkinsonian syndrome especially in early stage of disease. In this chapter, imaging biomarker would be discussed in detail, which may help in differentiating one from other Parkinsonism syndromes. They are as follows:

DOI: 10.1201/9781003505488-7

1. Magnetic Resonance Imaging

Various MRI sequences being used in neurodegenerative disorders including PD are shown in Table 1.

Table 1: *Magnetic resonance imaging sequences/techniques used in PD and related disorders*

	Conventional
1	Structural
2	Iron: relaxometry T2*
3	Susceptibility-weighted
4	Neuromelanin
	Newer methods
5	Magnetization transfer imaging
6	Diffusion tensor imaging or MR Tractography
7	Functional MRI
8	Perfusion MRI and pharmacologic MRI
9	Neurochemical or metabolic MRI

1.1 Structural imaging:

Neurodegenerative processes are characterized by neuronal cell loss, by astrogliosis or microglial proliferation or accumulation of paramagnetic substances like iron, resulting in characteristic signal alteration in affected brain areas on MRI.[1] These signal changes are based on a variety of factors, including proton density (free and mobile), relaxation time, diffusion, magnetic susceptibility and field strength. SN is the key structure involved in PD and one of the most important imaging targets. Anatomy of SN needs to be reviewed briefly for better understanding of imaging changes. SN is a small structure, which constitutes only 0.5% of midbrain volume and runs in the antero-superolateral aspect to the postero-inferomedial aspect, between red nucleus and crural fibers through the entire midbrain.[2] Due to peculiar tilted orientation of SN, there is partial volume loss in conventional axial sequences of MRI. To overcome this problem and better resolution, slices of 3-T MRI sequences at 45-degree angulations to commissural line (AC-PC) (i.e. perpendicular to the midbrain axis) may be used.[3] Further, conventional MRI may not differentiate substantia nigra pars compacta (SNc) from substantia nigra pars reticulata (SNr), as the ventral parts of the SNr extend and intermix with crural fibers. Imaging characteristics for red nucleus and SNr may be used to recognize ill-defined SNc. Usually, high iron content of red nucleus and SNr produces hypointense signal on T2-weighted images. The hyperintense region between hypointense SNr and red nucleus may be considered as SNc.[4,5] Due to small size and ill-defined imaging boundaries, volumetric analysis of SNc is difficult to assess by conventional T2 weighted MRI, resulting its low sensitivity despite a high specificity.[6] Although, conventional MRI is one of the most frequently used imaging modalities for work-up of a patient with parkinsonian features, it is used to exclude other parkinsonism syndrome like strategic space occupying lesions, small vessel disease, altered basal ganglia signal, hydrocephalus or rarely other degenerative parkinsonism like MSA or PSP.[7] Dedicated imaging study to target SNc is still lacking in common clinical setting.

As already discussed above, Nigrosome-1 produces hyperintense signal in a "linear/comma/wedge" shaped structure in susceptibility weighted imaging (SWI) sequences of MRI at posterior one-third part of SN. It is outlined anterolaterlly by a hypointense signal of SNc and medially by medial leminiscus. In healthy individuals, the nigrosome-1 and its nearby structures resemble like a "tail of a swallow" on axial High Resolution-SWI imaging, hence known as "Swallow-tail appearance". HR-SWI sequences using ultrahigh magnetic field strengths of 7T MRI was successfully used to detect PD related radiological changes in nigrosome-1 as well as healthy nigrosome-1.[8] Considering these facts, 3T MRI was also tried to detect these signal changes and could picked-up a "swallow tail" appearance of the dorsolateral SN in healthy nigrosome-1. These changes were absent in PD (absent swallow tail sign). Interestingly, this sign had very high sensitivity, specificity and accuracy.[9] The "absent swallow tail" has been suggested to be used as useful diagnostic biomarker for PD. However, a recent study reported abnormal "swallow-tail appearance" in levy body dementia, questioning the usefulness of this sign.[10]

1.2 Imaging of iron deposition

Iron deposition in patients with PD was considered as possible imaging biomarker and various longitudinal studies exploring use of relaxometry technique (T2-weighted imaging technique, T2* and inverse R2*) did not correlate extent of iron deposition with duration of PD, but with motor severity. At present, relaxometry is probably not a useful biomarker for PD, but it may be a promising method for future. Other iron detecting important MRI sequences are SWI and quantitative susceptibility mapping (QSM). Probably in patients with PD, QSM have similar or higher sensitivity than relaxometry, but still it could not differentiate PD from various parkinsonian disorders.[11,12] In future, either uses of high magnetic field like 7 T MRI or additional imaging methods to detect iron load may help in better understanding of PD.

1.3 Neuromelanin imaging

Neuromelanin is biochemically similar to melanin and is accumulated in the locus ceruleus and the substantia nigra. Probably it has protective role against oxidative stress and it is gradually reduced in patients with PD. T1 weighted MRI is a sensitive method to detect level of neuromelanin and has a possible role to understand its change in PD.[13] Altered neuromelanin in PD may be a potential diagnostic test in future. However, it is not clear whether this approach is better than other available modalities.

2. Newer modalities of MRI

Newer MRI modalities as shown in Table-1 have increased the utility of MRI in better diagnosis of PD and to differentiate it from atypical Parkinsonism.

2.1 Magnetization transfer imaging (MTI)

Magnetization transfer imaging (MTI) is another modality, in which imaging sequences are based on the differential signal produced by free protons of water and bound protons within various cell structures like myelin or cell membranes in high magnetic field. It is primarily expressed as a ratio i.e. magnetization transfer ratio (MTR). The MTR in SN is reported to be less in PD than in controls. However, there is considerable overlap between two groups, resulting limited utility of these sequences. Quantitative magnetization transfer protocol rather than MTRs is probably more sensitive and has a potential role in differentiating PD from other Parkinson syndrome. MTI may be useful in neurodegenerative disorders, as there are multiple types of pathological processes, resulting in reduced imaging signal.[14,15]

2.2 Diffusion imaging

Diffusion imaging is based on changes in water diffusion in presence of a magnetic field. Diffusion tensor imaging (DTI)/tractography, a diffusion-based technique, provides better understanding of axons and myelin. It detects quantitative loss of anisotropy (directionality) or change in amplitude of diffusion of water. The diffusion data can be used to detect any alteration between fibre tracts connecting two regions; hence any change in structural connectivity especially between SN and striatum/thalamus can be demonstrated.[16] Along with DTI, many other parameters such as apparent diffusion coefficient (ADC), fractional anisotropy (FA), mean diffusivity (MD) and directionality of diffusivity (radial or transverse; and axial or longitudinal) may be used.[17] Ability of DTI to detect local fractional anisotropy (FA) within SN increases its possible utility in idiopathic PD.[18] To explore possible role of DTI, Schwarz et al. explored potential diagnostic role of DTI by using 3-T MRI by imaging SN in patients with PD. They outlined boundaries of SN and substructures manually for analysis as well as did a voxel-based analysis. Authors reported an increase in MD in SN of PD (when SN and other structures were selected manually), but no difference on voxel-based analysis. FA was also non-contributory by nigral regional selected manually or voxel-based analysis in their study. By the meta-analysis of 11 studies, authors did not find any change in mean diffusivity in substantia nigral, but a significant decrease in FA.[19] The inconsistent results of DTI raise questions regarding validity or usefulness of this modality in PD. However, recently published data evaluating newer diffusion techniques for imaging SN highlights possible role of newer method for diagnosis and staging of PD.[20]

2.3 Magnetic resonance spectroscopy (MRS)

Magnetic resonance spectroscopy (MRS) is another non-invasive method to evaluate metabolic changes in brain. It acquires signal commonly from hydrogen (1H) proton, but information regarding other metabolites like phosphorus, carbon or nitrogen can also be acquired. MRS frequently acquires signals from localized small region known as voxel. Use of MRS was not very impressive in PD, till one recent study, which demonstrated elevated NAA/Cr ratio in the SN of early PD and the authors highlighted its importance as possible biomarker.[21] However, sample size of this study was small with only 23 patients and six controls, with short follow-up period of only three months. Post-mortem studies in PD reported reduced glutathione (a prevalent brain antioxidant) concentration in nigral tissue. However, high magnetic field MRS with better signal-noise ratios failed to demonstrate its reduced concentration in nigral tissue in vivo.[22] Various other MRS biomarkers are being explored for early diagnosis and 31-P MRS may be used to determine important pathophysiological data, especially energy metabolites in PD patients in future studies.

2.4 Functional magnetic resonance imaging (fMRI)

Functional MRI (fMRI) is a newer technique based on blood oxygen level dependent (BOLD) imaging to generate images and relies on regional differences in cerebral blood flow. fMRI detects change in the oxyhaemoglobin and deoxyhaemoglobin in specific brain site, which changes with activity of different neurons of that region. fMRI can also be used to detect neuronal integrity between various parts of brain in healthy as well as diseased persons. Resting state functional connectivity MRI (rsfMRI) is a newer application to evaluate functional networks of brain.[23] Studies with rsfMRI reported that in PD patients, there is increased resting state neuronal integrity between STN and cortical motor areas.[24]

rsfMRI has been also used to evaluate synucleinopathy related disorders like Lewy body dementia (LBD), PD and REM sleep behaviour disorder (RBD) and to differentiate them from healthy subjects. It is important to highlight that intake of medications for PD normalizes the rsfMRI abnormalities. Hence, to differentiate PD from healthy subjects, rsfMRI should be done when subjects are off medication. Reproducibility is another most important challenge of rsfMRI and standardization of the process is also required.

3. Techniques to evaluate the dopamine dysfunction in PD

3.1 Presynaptic assessment

Presynaptic assessment of dopaminergic system is one of the most important features in evaluation of patients with PD. PET and SPECT are widely used to detect dopamine function of striatum in PD and other parkinsonian syndromes.[25] There are many important targets used to assess dopaminergic function at nerve ending. In the process of intraneural sysnthesis of dopamine in a healthy person, the enzyme dopa decorboxylase (aromatic amino-acid decarboxylase (AADC)) plays an important role. [18F] flurodopa (a common tracer for PET), is converted to 18[F] fluorodopamine by AADC, hence, [18F] flurodopa PET can be used to measure striatal AADC activity. Putamen uptake of [18F] flurodopa is reduced to 50% in PD when a patient develops motor symptoms. Many processes are involved in uptake of [18F] flurodopa, but the principal determinants are axonal terminal plexus density and AADC activity. With disease progression, there is reduction in nerve density, with a compensatory upregulation of AADC level in remaining terminals. Hence, in early PD, [18F] flurodopa may not be able to estimate severity of the degeneration.[26,27] Despite the fact that striatal [18F]flurodopa uptake does not provide true status of density of functional nerve terminals, it is commonly used as important tool to monitor disease progression objectively. Increased [18F] flurodopa uptake in other part of brain such as dorsolateral prefrontal area, anterior cingulate, and globus pallidus interna is also reported in early PD. It is probably due to compensatory up regulation of AADC and uptake of [18F] flurodopa into noradrenergic and serotonergic nerve endings. The putaminal [18F] flurodopa uptake declines by about 10% of its baseline level annually, along with gradual loss of compensatory mechanisms with advancement of disease. It is important to highlight that there is an inverse correlation between the severity of motor abnormality (rigidity and bradykinesia) and [18F] flurodopa uptake but not with severity of tremor suggesting different mechanism for various motor abnormality in PD.

[18F] fluorodopamine is further transported to intraneuronal storage vesicles from cytoplasm by vesicular monoamine transporter 2 (VMAT2) at nerve ending. 11C- or 18F-dihydrotetrabenazine (DTBZ) PET can measure density of VMAT2. It is a promising marker and probably a most valid measurement of dopamine terminal density.[28,29] In PD, striatal 11C-DTBZ is reported to be reduced significantly.

Third important target is dopamine transporter (DAT). Extracellular dopamine released by nerve terminals at synaptic cleft need to be removed rapidly to stop continuous signalling. DAT is an important component for this action. DAT function in PD can be measured by various PET ligands (11C-CFT, 18F-FP-CIT, 11CRTI-32 or 11C-methylphenidate) and SPECT ligands (123I-b-CIT, 123I-FP-CIT, 123I-IPT, 123I-altropane and 99mTc-TRODAT-1). These DAT ligands are non-selective and can bind to any of dopamine, serotonin and noradrenaline transporters. As dopaminergic transmitters are major neurotransmitter within striatum, and with the reduction in its level at nerve ending, its uptake is also reduced in PD patients. Although DAT imaging probably does not estimate actual nerve terminal density due to its down regulation in remaining functional nerve terminals, DAT loss is considered as a useful marker in PD. More than 90% of clinically probable PD have reduced DAT binding in stratum.[30]

As explained above, three targets AADC, DAT and VMAT2 are being used commonly by PET/SPECT for dopaminergic striatal presynaptic evaluation. These modalities are now being used clinically to determine functional status of presynaptic nigrostriatal system and to differentiate various neurodegenerative Parkinsonism (PD; MSA; PSP; and CBS), from other causes of Parkinsonism. However, it is difficult to differentiate various neurodegenerative Parkinsonism by these methods.[31] Functional imaging may have additional prognostication role. Studies suggested that patients with early possible Parkinsonism without presynaptic dopamine deficiency have better prognosis and only 3% develop PD after two years of follow-up.[32] Functional dopaminergic imaging also guides regarding inappropriate use of anti-parkinsonian medication. Clinically suspected PD with normal dopaminergic imaging may have alternative diagnosis and anti-parkinsonian medication may have limited role in these patients. Another important role of SPECT/PET is to detect subjects at risk, before developing clinical symptoms. Preclinical detection will be more relevant with availability of neuroprotective agents. PET and SPECT ligands can also be used to evaluate functional status of non-dopaminergic systems like serotonergic, noradrenergic, cholinergic neurotransmitters in PD. Detection of changes in these non-dopaminergic transmitters may be used to evaluate non-motor symptoms such as depression, psychosis, dementia and sleep abnormalities. In future, it may help to classify PD in cognitive PD, depression PD, sleep PD and autonomic PD.

3.2 Post synaptic assessment

PET and SPECT ligands can be used to study availability of postsynaptic dopaminergic receptor. Many studies used 11C-raclopride, a D2/3 receptor ligand in PD patients. These studies revealed higher binding of 11C-raclopride in the putamen of new PD patients. This was probably due to increase in postsynaptic D2 receptor level in early PD. It was inversely related with presynaptic reduction of 18Fdopa uptake.[33,34] Increased putaminal 11C-raclopride binding in early PD may be due to a compensatory upregulation of postsynaptic receptors or due to decreased competition for its binding at receptors (low levels of synaptic dopamine resulting in reduced competing for these reagents). In advanced PD, there is significant reduction in D2 binding of these reagents, suggests progression of disease or down regulation of receptor due to chronic exposure to dopaminergic treatment. One SPECT study reported severely reduced striatal 123I-IBZM binding in MSA than PD, suggesting its possible differentiating role in two conditions.[35]

3.3 Synaptic assessment

PET/SPECT can be used to assess changes in dopamine level at synaptic clefts. It may provide further insightful information regarding various motor complications in advanced PD.

4. Perfusion imaging and pharmacologic imaging

Various PET methods have been used to evaluate neuronal integrity in PD. For example, $^{15}O\text{-}H_2O$ is used to study the cerebral blood flow (CBF) whereas 18F-fluorodeoxyglucose (FDG) to evaluate resting brain metabolic changes. FDG PET reveals a specific PD-related pattern of involvement with abnormalities restricted mainly to a specific region of brain, for example, basal ganglia, thalamus and supplementary motor and premotor areas. These changes are probably due to abnormality in their functional neuronal integrity rather than their degeneration. 18F-fluorodopa PET imaging is used to assess functional integration of various dopaminergic network systems. Along with a post-synaptic dopamine receptor tracer (raclopride), 18F-fluorodopa PET can evaluate release of dopamine in response to various specific tasks.[36]

MRI can also be used for pharmacodynamic imaging to evaluate drug effects and to evaluate clinical features of PD. Although there are limited studies evaluating perfusion MRI in PD, pharmacodynamic/pharmacologic MRI (phMRI) study may be a promising tool to evaluate effect of medicine on perfusion changes, with negligible radiation hazard. Blood flow change in different brain region after administrating antiparkinsonian therapy has been reported by phMRI.[37] BOLD signal changes and arterial spin labelling (ASL) are two commonly used methods to study phMRI. BOLD is already discussed earlier in this chapter and ASL, probably a better method is discussed below.

ASL is another fMRI modality which provides details of regional blood flow without any nuclear radiotracer, which makes it a potential useful tool. It is now validated in PD patients. There is good correlation between images of regional perfusion changes (ASL images) with images of altered glucose metabolism (FDG PET images) in the same subjects.[38] By assessing loss of caudate perfusion laterality index, ASL images may be used to stage advanced PD.[39] However, better study is required to detect utility of this modality in routine clinical settings.

5. Trans-cranial sonography (TCS)

B-mode ultrasound imaging is safe, cost effective and easily available; rapid diagnostic modality is used extensively in the field of medicine. However, B-mode TCS is a novel technique to evaluate mid-brain region in patients with PD. TCS can be used to detect increased echogenicity in the midbrain. In one study, up to 90% of PD had bilaterally increased echogenicity in their lateral midbrain.[40] Interestingly, this signal was static despite clinical progression of disease and it was further suggested that iron deposition, rather than dopaminergic cell loss is responsible for signal changes.[41] A recently published prospective study evaluated both TCS and SPECT in various neurologic disorders.[42] Clinical diagnosis of PD after follow-up of two years was used as gold standard. Sensitivity of TCS was only 40% and specificity was 61% for PD and for SPECT, it was 66% and 88% respectively. Another prospective blinded study assessed 60 patients of PD and followed them for 12 months to evaluate sensitivity and specificity of TCS in PD.[43] Compared to clinical diagnosis as gold standard; TCS had high sensitivity and specificity. The sensitivity and specificity of TCS for detecting early PD at time of enrolment was 91% and 82% respectively. In another study, TCS was used to differentiate between idiopathic PD, MSA and PSP.[44] Substantia nigra had normal echogenicity in MSA-P but not in PD (90% sensitivity; 98% specificity; 86% positive predictive value). In patients with PSP, third-ventricle were dilated by > 10 mm along with increased ecogenicity in lentiform nucleus (84% sensitivity; 98% specificity; with 89% positive predictive value). In this study, TCS imaging with normal substantia nigra and hyperechogenic lentiform nucleus was suggestive of MSA-P or PSP (59% sensitivity; 100% specificity; with 100% positive predictive value). Young onset Parkinsonism (age <60 years) and normal echogenicity in

SN were most important feature to exclude PD (75% sensitivity; 100% specificity; 100% positive predictive value).

TCS appears to be a promising modality but wide range of sensitivity (40 - 95%) and specificity (60 - 96 %) of TCS has been reported in PD. Poor echo window of skull bone and ill-defined boundaries of substantia nigra, raphe nucleus, etc. in mid brain are important limitations for routine use of TCS in PD or other neurodegenerative disorders. Also, there is limited and variable spatial resolution as well as undefined diagnostic cut-off value for various PD populations, making its routine clinical use in PD patients difficult and challenging. Due to high observer dependent variability, an automated computer-assisted program is needed to evaluate an area of interest of brain to reduce observer dependent variability. Few TCS studies performed computer assisted program to study various parts of the brain, but results were variable.[45,46,47] Despite these limitations, TCS is now considered as an important imaging tool by European Federation of Neurological Society and Movement Disorder Society's European Section for early diagnosis, differential diagnosis and screening for PD.[48,49]

Conclusion

Neuroimaging has important place in management of PD. Conventional MRI is being used for long time in PD, especially to differentiate it from related disorders. However, MRI with higher magnetic strength and newer sequences of MRI have promising role in understanding PD pathogenenesis as well as early diagnosis and follow up. PET and SPECT are another important modalities being used for long time and various ligands are used to understand presynaptic, synaptic and postsynaptic changes. It is approved to be used to differentiate degenerative parkinsonian syndrome from other secondary Parkinsonism, but not approved presently to differentiate various degenerative parkinsonian syndrome like PD, PSP, MSA and CBS. Transcranial sonography, one of the interesting and safe modalities of imaging is now a recommended tool in PD patients. However, there is lack of standardized protocol, non-availability of normative data for PD patients in various population and inter-observer variability, limiting widespread use of these modalities in routine clinical setting.

References

1. Gerlach M, Double KL, Youdim MB, Riederer P. Potential sources of increased iron in the substantia nigra of parkinsonian patients. J Neural Transm. 2006;70(Suppl.);133–142.
2. Oikawa H, Sasaki M, Tamakawa Y, Ehara S, Tohyama K. The substantia nigra in Parkinson Disease, Proton density-weighted spin-echo and fast short inversion time inversion-recovery MR findings. AJNR Am J Neuroradiol. 2002;23;1747–1756.
3. Niehaus L, Boelmans K. Diagnosis of Parkinson's disease--transcranial sonography in relation to MRI. Int Rev Neurobiol. 2010;90;63-79.
4. Sommer U, Hummel T, Cormann K, Mueller A, Frasnelli J, Kropp J, Reichmann H. Detection of presymptomatic Parkinson's disease, Combining smell tests, transcranial sonography, and SPECT. Mov Disord. 2004;19;1196–1202.
5. Antonini A, Leenders KL, Meier D, Oertel WH, Boesiger P, Anliker M. T2 relaxation time in patients with Parkinson's disease. Neurology. 1993;43;697–700.

6. Meijer FJ, Goraj B. Brain MRI in Parkinson's disease. Front Biosci (Elite Ed). 2014;6;360–9.
7. Massey LA, Micallef C, Paviour DC, et al. Conventional magnetic resonance imaging in confirmed progressive supranuclear palsy and multiple system atrophy. Mov Disord. 2012;27;1754–62.
8. Blazejewska AI, Schwarz ST, Pitiot A, Stephenson MC, Lowe J, et al. (2013) Visualization of nigrosome 1 and its loss in PD: pathoanatomical correlation and in vivo 7 T MRI. Neurology. 81; 534–540.
9. Schwarz ST, Afzal M, Morgan PS, Bajaj N, Gowland PA, et al. The 'Swallow Tail' Appearance of the Healthy Nigrosome – A New Accurate Test of Parkinson's Disease: A Case-Control and Retrospective Cross-Sectional MRI Study at 3T. PLoS ONE. 2014;9(4);e93814.
10. Shams S, Fällmar D, Schwarz S, et al. MRI of the Swallow Tail Sign: A Useful Marker in the Diagnosis of Lewy Body Dementia? AJNR Am J Neuroradiol. 2017;38;1737-1741.
11. He N, Ling H, Ding B, et al. Region-specific disturbed iron distribution in early idiopathic Parkinson's disease measured by quantitative susceptibility mapping. Hum Brain Mapp. 2015; 36;4407–20.
12. Barbosa JH, Santos AC, Tumas V, et al. Quantifying brain iron deposition in patients with Parkinson's disease using quantitative susceptibility mapping, R2 and R2. Magn Reson Imaging. 2015;33;559–65.
13. Castellanos G, Fernandez-Seara MA, Lorenzo-Betancor O, et al. Automated neuromelanin imaging as a diagnostic biomarker for Parkinson's disease. Mov Disord. 2015;30;945–52.
14. Tambasco N, Belcastro V, Sarchielli P, et al. A magnetization transfer study of mild and advanced Parkinson's disease. Eur J Neurol. 2011;18;471–7.
15. Morgen K, Sammer G, Weber L, et al. Structural brain abnormalities in patients with Parkinson disease: a comparative voxel-based analysis using T1-weighted MR imaging and magnetization transfer imaging. AJNR Am J Neuroradiol. 2011;32;2080–6.
16. Menke RA, Scholz J, Miller KL, et al. MRI characteristics of the substantia nigra in Parkinson's disease: a combined quantitative T1 and DTI study. Neuroimage. 2009;47;435–41.
17. Le Bihan D. Looking into the functional architecture of the brain with diffusion MRI. Nat Rev Neurosci. 2003;4;469–80.
18. Zhan W, Kang GA, Glass GA, et al. Regional alterations of brain microstructure in Parkinson's disease using diffusion tensor imaging. Mov Disord. 2012;27;90–7.
19. Schwarz ST, Abaei M, Gontu V, Morgan PS, Bajaj N, Auer DP. Diffusion tensor imaging of nigral degeneration in Parkinson's disease: a region-of-interest and voxel-based study at 3 T and systematic review with meta-analysis. Neuroimage Clin. 2013;3;481–8.
20. Ofori E, Pasternak O, Planetta PJ, et al. Longitudinal changes in free-water within the substantia nigra of Parkinson's disease. Brain. 2015;138;2322–31.
21. Seraji-Bozorgzad N, Bao F, George E, et al. Longitudinal study of the substantia nigra in Parkinson disease: a high-field (1) HMR spectroscopy imaging study. Mov Disord. 2015;30;1400–4.
22. Emir UE, Tuite PJ, Oz G. Elevated pontine and putamenal GABA levels in mild-moderate Parkinson disease detected by 7 tesla proton MRS. PLoS One. 2012;7;e30918.
23. Zhang D, Raichle ME. Disease and the brain's dark energy. Nat Rev Neurol. 2010;6;15–28.

24. Baudrexel S, Witte T, Seifried C, et al. Resting state fMRI reveals increased subthalamic nucleus-motor cortex connectivity in Parkinson's disease. Neuroimage. 2011;55;1728–38.
25. Brooks DJ. Technology insight: imaging neurodegeneration in Parkinson's disease. Nat Clin Pract Neurol. 2008;4;267–277.
26. Ribeiro MJ, Vidailhet M, Loc'h C, et al.Dopaminergic function and dopamine transporter binding assessed with positron emission tomography in Parkinson disease. Arch Neurol. 2002;59;580–586.
27. Morrish PK, Rakshi JS, Sawle GV, Brooks DJ. Measuring the rate of progression and estimating the preclinical period of Parkinson's disease with [18F]dopa PET. J Neurol Neurosurg Psychiat. 1998; 64;314–319.
28. Bohnen NI, Albin RL, Koeppe RA, et al. Positron emission tomography of monoaminergic vesicular binding in aging and Parkinson disease. J Cereb Blood Flow Metab. 2006;26;1198–1212.
29. Okamura N, Villemagne V, Drago J, et al. In vivo measurement of vesicular monoamine transporter type 2 density in Parkinson disease with (18)F-AV-133. J Nucl Med. 2010;51;223–228.
30. Benamer TS, Patterson J, Grosset DG, et al. Accurate differentiation of parkinsonism and essential tremor using visual assessment of [123I]-FP-CIT imaging: the [123I]-FP-CIT Study Group. Mov Disord 2000;15;503–510.
31. Ba F, Martin WR. Dopamine transporter imaging as a diagnostic tool for parkinsonism and related disorders in clinical practice, Parkinsonism Relat Disord. 2015;21;87-94.
32. Marshall VL, Patterson J, Hadley DM, Grosset KA, Grosset DG. Two year follow-up in 150 consecutive cases with normal dopamine transporter imaging. Nucl Med Commun. 2006;27;933–937.
33. Rinne JO, Laihinen A, Rinne UK, Nagren K, Bergman J, Ruotsalainen U. PET study on striatal dopamine D2 receptor changes during the progression of early Parkinson's disease. Mov Disord. 1993;8;134–138.
34. Turjanski N, Lees AJ, Brooks DJ. PET studies on striatal dopaminergic receptor binding in drug naive and L-dopa treated Parkinson's disease patients with and without dyskinesia. Neurology. 1997; 49;717–723.
35. Plotkin M, Amthauer H, Klaffke S, et al. Combined 123I-FP-CIT and 123I-IBZM SPECT for the diagnosis of parkinsonian syndromes: study on 72 patients. J Neural Transm. 2005;112;677–692.
36. Strafella AP, Bohnen NI, Perlmutter JS, et al. Molecular imaging to track Parkinson's disease and atypical parkinsonisms: New imaging frontiers. Mov Disord. 2017;32;181–192.
37. Andersen AH, Hardy PA, Forman E, et al. Pharmacologic MRI (phMRI) as a tool to differentiate Parkinson's disease-related from age-related changes in basal ganglia function. Neurobiol Aging. 2015;36;1174–82.
38. Teune LK, Renken RJ, de Jong BM, et al. Parkinson's disease-related perfusion and glucose metabolic brain patterns identified with PCASL-MRI and FDG-PET imaging. NeuroImage Clin. 2014;5;240–244.
39. Yamashita K, Hiwatashi A, Togao O, et al. Cerebral blood flow laterality derived from arterial spin labeling as a biomarker for assessing the disease severity of parkinson's disease. J Magn Reson Imaging. 2017;45;1821–1826.

40. Berg D, Siefker C, Becker G. Echogenicity of the substantia nigra in Parkinson's disease and its relation to clinical findings. J Neurol. 2001;248;684–689.

41. Berg D, Merz B, Reiners K, Naumann M, Becker G. Five-year follow-up study of hyperechogenicity of the substantia nigra in Parkinson's disease. Mov Disord. 2005;20;383–385.

42. Bouwmans AEP, Vlaar AMM, Mess WH, Kessels A, Weber WEJ. Specificity and sensitivity of transcranial sonography of the substantia nigra in the diagnosis of Parkinson's disease: Prospective cohort study in 196 patients. BMJ Open. 2013;3;1–8.

43. Gaenslen A, Unmuth B, Godau J, et al. The specificity and sensitivity of transcranial ultrasound in the differential diagnosis of Parkinson's disease: a prospective blinded study. Lancet Neurol 2008;7; 417–424.

44. Walter U, Dressler D, Probst T, et al. Transcranial brain sonography findings in discriminating between parkinsonism and idiopathic Parkinson disease. Arch Neurol. 2007;64;1635–1640.

45. Ahmadi SA, Baust M, Karamalis A, et al. Midbrain segmentation in transcranial 3D ultrasound for Parkinson diagnosis. Med Image Comput Assist Interv. 2011;14;362–369.

46. Sakalauskas A, Laučkaitė K, Lukoševičius A, Rastenytė D. Computer-Aided Segmentation of the Mid-Brain in Trans-Cranial Ultrasound Images. Ultrasound Med Biol. 2016;42;322-32.

47. Sakalauskas A, Lukoševičius A, Laučkaitė K, Jegelevičius D, Rutkauskas S. Automated segmentation of transcranial sonographic images in the diagnostics of Parkinson's disease. Ultrasonics. 2013;53;111-21.

48. Lauckaite K, Rastenyt_e D, Surkiene D, et al. Specificity of transcranial sonography in parkinson spectrum disorders in comparison to degenerative cognitive syndromes. BMC Neurology. 2012;12;12.

49. Berardelli A, Wenning GK, Antonini A, et al. EFNS/MDS-ES recommendations for the diagnosis of Parkinson's disease. Eur J Neurol. 2013;20;16–34.

Chapter 8

Functional Magnetic Resonance Imaging

M Saxena, Vaishna Narang, S Senthilkumaran and Madhuri Behari

Parkinsonism exhibits a combination of six important features namely tremor at rest, rigidity, gait disturbances, bradykinesia, stooped posture and freezing. Parkinsonism is represented by neurological disorders such as Parkinson's disease (PD), multiple system atrophy (MSA) and progressive supranuclear palsy (PSP). Investigation of neurological correlates for motor, memory and speech becomes essential as patients exhibit tremor, memory deficits and hampered speech. Postmortem is still considered to be the golden standard for differentiation of PD from MSA and PSP, however this provides information of end stage patients, and does not help in assessment of brain function across various stages of disease. Functional magnetic resonance imaging (fMRI) is a recently developed neuroimaging technique which has enabled the better understanding of the brain structure and function at various stages of disease on the basis of structural and/or physiological parameters. Using carefully designed task based experiments; it is now possible to assign certain brain areas to specific movements (motor), sensory perceptions (somatic, visual and auditory) and functions such as memory, mood and attention (cognitive); using haemoglobin as an endogenous contrast agent. fMRI signal also called blood oxygen level dependent (BOLD), relies on the changes in concentration of deoxyhaemoglobin associated with an increase in blood flow to active regions of the brain during an active task.

Brain activation pattern in PD has been extensively studied with fMRI to report activation in left hemispheric primary motor cortex (M1), inferior parietal cortex (IPC), superior parietal lobule (SPL) and pre-supplementary motor area (pSMA), together with right putamen, in response to right hand motor tasks such as pressing a key (finger/thumb), simple or complex finger tapping (FT) task or controlling a joystick.[1, 2] Studies report either hypo[3] or hyper[1,4] activations of SMA (Supplementary motor area) and M1 regions on dopaminergic intake but may lead to severe dyskinesias in some PD, possibly a result of excessive increase in putamen bilaterally and in pSMA.[2,5] While most studies were restricted to early PD with minimal tremors during 'off-state' (meso-cortical region being less affected),[4] studies with action tremors or patient that could hardly perform a motor task (MSA and PSP) are rare, including studies that compare PD, MSA, and PSP.[6]

DOI: 10.1201/9781003505488-8

PD present impaired immediate and delayed memory recalls, deficits together with executive dysfunction making efficient learning and recognition difficult. Declarative memory constitutes semantic memory (knowledge of objects and meaning of words). This is referred to "working memory" which involves regions in medial and lateral, parietal and frontal cortex. Information about the properties of an object like its appearance and its utility etc. is a function of sensory and motor perception that is active at the time information is acquired.[7] Previous functional neuroimaging studies relate left posterior temporal cortex towards processing pictures and words.[7] Therefore, damage to sensory or motor areas of brain, as in Parkinsonism, may affect short term memory as reported by studies investigating the neuroanatomical correlates of memory impairment.[8] Due to paucity of literature in MSA and PSP, short term memory deficits were assumed to be similarly associated with degeneration of fronto-striatal connections as in PSP.[9] Previous studies reporting L-Dopa effects on memory function (using psychological and verbal memory tests) also present conflicting results reporting a decline, improvement, or no effect on the short term memory.[10-13]

Human speech is a very complex communication effort that requires accurate control of anatomical structures of vocal tract including the use of tongue to manipulate the oral/nasal resonances. Speech dysfunction, common in parkinsonism manifests as difficulty in initiating speech, low volume (hypophonia), voice tremors, festination, mutism, articulatory apraxia and akinesia that lead to breathless voice, variable dysarthria, and imprecise consonants, in PD and MSA, and dysarthria with variable degrees of spasticity, and palilalia and stuttering in PSP.[14] Previous language dependent fMRI studies involved word pronunciation, word production, reading sentences, or were focused on aspects of verbal fluency, such as volume, the thoracic effort in sound production, in addition to resting state brain networks in speech production[15] to report more medial frontal and left hemispheric frontal, parieto-occipital and cerebellar activation during 'off-state' which improved upon dopa replacement.[16]

No consistent findings are reported, though, some report reduced motor activation in left M1, left parietal cortex and pre SMA, while increased activation in others, as is with effects of dopaminergic therapy, but increased activation of left putamen has been reported in most studies. These differences are attributed to different motor tasks (self-paced/cued) performed by differently staged patients and also different imaging protocols. The interpretation of these findings is also confounded by inability of PD in performing the motor tasks efficiently and so resting state fMRI helps in establishing the sensory-motor network connectivity to show disrupted functional integration in cortico-striatal loops and cerebellar activation compensating for the disruption in the SMA-basal ganglia network with disease progression.[17] Similarly, memory studies directly mapping the brain function is lacking as all the previous studies depended on motor skills (for at least the responses) of the subjects.[11, 13] Also, previous speech being language dependent there is a need to understand the neurological control of vocal musculature.

Our fMRI study with PD, MSA and PSP patients (Table 1) evaluated motor, memory, and speech during "off-state" and "on-state" using 1.5 Tesla Clinical MR Scanner (Avanto: M/s. Siemens). We showed that Finger tapping (FT) could be substituted with Fist clenching (FC) motor task especially in MSA and PSP patients that can barely perform efficient noticeable FT. Similarly, with encoding and recollection in the same state i.e. during the "off-state" or during the "on-state", our study with

Table Ia. The clinical profile of all the subjects recruited for the study.

Group	N	*Subjects (M /F)*	*Age (Years)*	*Duration (Years)*	*LEDD (mg/ day)*	*Stage*	*MMSE*	*UPDRS III*
Control	*40*	*24 / 16*	*51 ± 7.4*	-	-	-	*29.0 ± 1.0*	-
PD	*29*	*22 / 7*	*59.5 ± 12.0*	*5.3 ± 3.4*	*348.6 ± 129.9*	*H & Y 1.8 ± 0.6*	*27.14 ± 2.4*	*22.7 ± 11.2*
MSA	*20*	*13 / 7*	*61.5 ± 6.9*	*4.7 ± 3.9*	*279.3 ±115.2*	*UMSARS 2. 8 ± 1.0*	*28.1 ± 2.2*	*23.8 ± 14.7*
PSP	*19*	*16 / 3*	*63.0 ± 6.9*	*3.08 ± 2.3*	*288.89 ± 68.0*	*PSP rating 1.92 ± 1.7*	*26.9 ± 2.8*	*20.8 ± 6.0*

Table Ib. The clinical profile of controls and PD recruited for the comparative FT and FC motor tasks.

Group	N	*Subjects (M /F)*	*Age (Years)*	*Duration (Years)*	*LEDD (mg/ day)*	*Stage*	*MMSE*	*UPDRS III*
Control	*10*	*6 / 4*	*57.0 ± 7.4*	-	-	-	*29.0 ± 1.0*	-
PD	*13*	*8 / 5*	*59.6 ± 12.0*	*5.3 ± 3.4*	*348.6 ± 29.9*	*1.8 ± 0.6*	*27.1 ± 2.4*	*22.7± 11.16*

Table Ic. The clinical profile of all the subjects actually analyzed for the study and Multivariate Correlation statistics between the BOLD activation and duration in the left hemisphere during the 'off-state' in Parkinsonism.

Group	N	*Subjects (M /F)*	*Age (Years)*	*Duration (Years)*	*LEDD (mg/ day)*	*Stage*	*MMSE*	*UPDRS III*	*LEDD & Motor Score (MS)*	*BA (235) & duration (D)*	*BA(4) & duration (D)*	*BA(6) & duration (D)*
Control	*30*	*19/11*	*51.7 ± 7.5*	-	-	-	-	-	-	-	-	-
PD	*16*	*12/4*	*60.8 ±13.2*	*5.2 ± 3.3*	*484.0 ± 218.1*	*1.8 ± 0.5*	*28.1 ± 1.7*	*22.3 ± 11.9*	*LEDD= 26.04 - 0.02x MS (0.04)*	*BA 235= 29.75 - 3.24 *D (0.25)*	*BA 4= 2.08 + 3.30 *D (0.20)*	*BA6= 5.28 + 0.36 *D (0.02)*
MSA	*15*	*10/5*	*61.9 ± 6.9*	*5.1 ± 6.9*	*293.3 ± 117.9*	*3.1 ± 1.0*	*25.5 ± 7.7*	*23.1 ± 13.7*	*LEDD= 7.23 + 0.04 x MS (0.52)**	*BA 235= 49.32 - 5.86 *D (0.31)*	*BA 4= 1.33 + 5.28 *D (0.35)**	*BA6= 10.98 +1.11 *D (0.06)*
PSP	*17*	*14/3*	*63.9 ± 7.6*	*2.9 ± 2.2*	*294.8 ± 62.0*	*2.5 ± 1.4*	*27.5 ± 2.5*	*30.6 ± 7.2*	*LEDD= 8.49 + 0.03 x MS (0.06)*	*BA 235= 6.18 +1.14 *D (0.03)*	*BA 4= 18.22 + 3.27 *D (0.07)*	*BA6= 11.32 + 0.60 *D (0.02)**

Table Id. The clinical profile of all the subjects actually completing the follow-up study.

Group	N	*Subjects (M /F)*	*Age (Years)*	*Duration (Years)*	*LEDD (mg/ day)*	*Stage*	*MMSE*	*UPDRS III*	*months*
PD Baseline	*11*	*9/2*	*60.6 ± 13.0*	*5.6 ± 3.6*	*436.4 ±160.5*	*1.6 ± 0.5*	*28.4 ± 1.8*	*18.3 ± 10.6*	*13.18 ± 2.6*
PD Fup	*11*	*9/2*	*61.6 ± 13.0*	*6.6 ± 3.6*	*508.2 ± 152.5*	*2.1 ± 0.6*	*27.4 ± 1.7*	*25.5 ± 13.6*	
MSA Baseline	*06*	*3/3*	*63.2 ± 4.2*	*3.7 ± 1.9*	*346.3 ± 177.9*	*3.0 ± 0.9*	*27.0 ± 3.7*	*27.2 ±11.5*	*9.8 ± 3.0*
MSA Fup	*06*	*3/3*	*63.3 ± 3.9*	*3.7 ± 1.9*	*369.6 ± 175.3*	*3.3 ± 0.8*	*25.2 ± 3.6*	*31.0 ± 10.2*	
PSP Baseline	*06*	*5/1*	*63.8 ± 8.1*	*2.1 ± 1.2*	*245.0 ± 30.0*	*2.3 ± 0.8*	*27.0 ± 3.3*	*26.5 ± 8.6*	*7.6 ± 3.0*
PSP Fup			*64.5 ± 7.9*	*2.1 ± 1.2*	*293.0 ± 31.6*	*2.7 ± 0.8*	*26.0 ± 4.2*	*26.0 ± 4.2*	

*(Significant P≤0.05)

N: number of subjects; PD: Parkinson's disease; MSA: Multiple System Atrophy; PSP: Progressive Supranuclear Palsy, M: Male; F: Female; BA: Brodmann Area; LEDD : Levodopa equivalent daily dose; MMSE: Mini Mental State Examination score; UPDRS: Unified Parkinson's Disease Rating Scale; UMSARS: Unified Multiple System Atrophy Rating Scale; Y&H: Hoehn and Yahr staging system for PD.

short term memory made memory retrieval more efficient.[11] Also, speech study is exclusive being non-language dependent and observes movement of articulatory structures (sustained phonation and articulatory dysarthria) in relation to the neural correlates both during the "on" and "off-state".[14,18]

A definite difference in activation pattern was observed among the three patient groups in contrast to that observed in controls. Controls recruited contralateral medial frontal gyrus (MeFg BA6), postcentral gyrus (PoCg) and ipsilateral cerebellum for FC. The "off-state" revealed more SMA recruitment in PD, reduced recruitment in MSA and no recruitment in PSP as compared to controls. Also, a reduced activation in PD and MSA, and greater sensory-motor activation in PSP was observed as compared to controls. Regarding the ipsilateral cerebellar activation, MSA presented no activation, while PD and PSP presented a reduced recruitment. During the "on-state" however, no SMA activation was observed in any of the patient groups. A reduced PoCg activation was observed only in PD, while MSA presented an increased M1 and right parietal cortex (BA 40) activation and ipsilateral cerebellar activation similar to controls. PSP on the other hand expressed a loss of cerebellar activation, with increased activations in the parietal, middle occipital (MOg, BA 19) cortex and right fusiform gyrus (fg). With reference to the baseline state, results of the paired t-test in the three patients' groups indicate that patients of PD, MSA and PSP present drastic reduction in contralateral SMA, sensory motor cortex (SMC) and ipsilateral cerebellum for either "off" or "on-state". However, follow up study during the "on-state" presented some aberrant activation in ipsilateral cerebellum for MSA and SMC for PSP (Figures 1, 2). It appears that M1 activation is complemented by SMA in PD and MSA, which may be a consequence of a functional and anatomical connectivity of SMA to the M1 and the potential compensation of M1 by SMA as it communicates with M1, pre-motor cortex, primary somato-sensory cortex including striatum, and thalamus to execute and initiate simple voluntary movements, motor planning, motor learning, with its involvement directly proportional to the complexity and cueing of movements.[19,20] A reduced SMC recruitment, a main executive locus for simple voluntary movements, reflects abnormal sensory processing.[21] Cerebellar reduction in PD and PSP upon dopaminergic replacement suggests its role in compensating the deficits of basal ganglia.[22] Degenerative disruptions of the striatal–frontal connections lead to aberrant parietal and premotor recruitment.[23] However, during motor task in MSA, fronto-parietal pathway being part of the cortico-striatal may compensate for deficits in M1, posterior parietal cortex and cerebellar dysfunction.[24] The motor system is differentially affected in PD, MSA, and PSP; in MSA and PSP more symmetric functional loss is observed in putamen than in PD.[25]

The memory task involved a visual recollection of familiar pictures (common household items and fruits/vegetables) and unfamiliar pictures (landscapes and scenes of water bodies) shown during the encoding phase within the MR environment, through a binocular camera. Significant low short term memory response was observed in PD for both familiar objects and scenes during "off-state" which did not improve during "on-state" as reported in previous studies, suggesting that memory loss is not due to dopaminergic deficiency and hence dopaminergic therapy has no effect.[26] Fusiform gyrus has been shown to be involved in memory encoding.[27] Altered volumes of gray matter (GM) and white matter (WM) in five ROIs cuneus and precuneus (ROI_CP), medial, middle and inferior frontal lobe (ROI_MMI), thalamus (ROI_TH), fusiform gyrus (ROI_FuG), and superior frontal gyrus (ROI_SFG)[28] and surface based analyses in cortical volumes for both hemispheres (FreeSurfer, Table 2). As

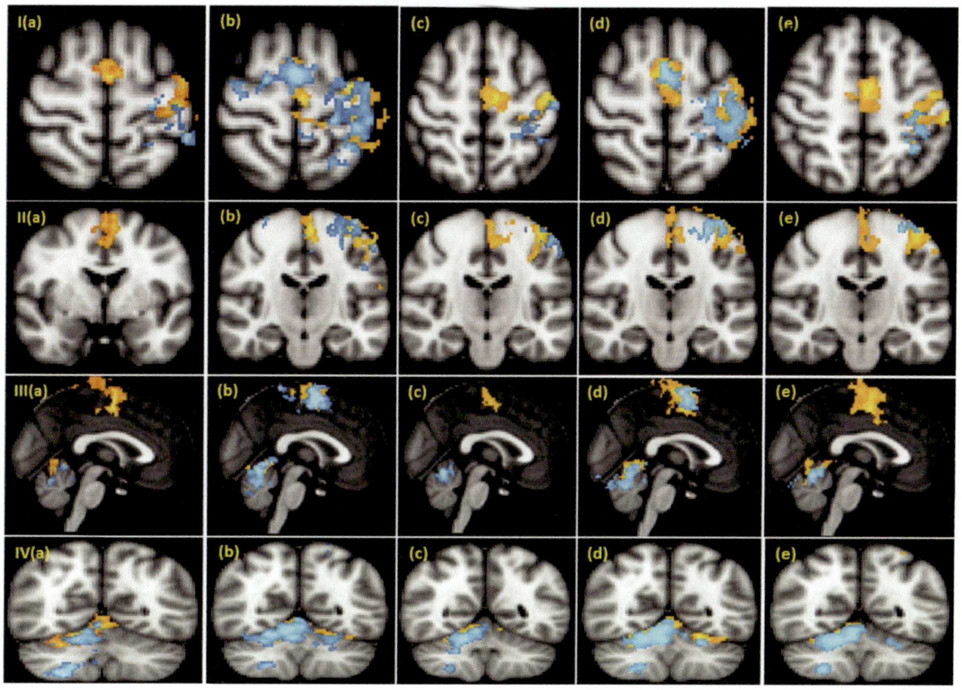

Figure 1: Group activation maps of higher level group analyses (P≤0.01) single sample t-test (a) Controls (FT Vs FC) (b) PD-Off (FT Vs FC) (c) PD-On (FT Vs FC); two sample t-test (d) Controls Vs PD-Off (FT Vs FC) and (e) Controls Vs PD-On (FT Vs FC). The Yellow-Red color for finger tapping (FT) and Blue-light blue for fist clenching (FC). Rows I-III present vies in axial coronal and sagittal planes and row IV presents the cerebellar activations.

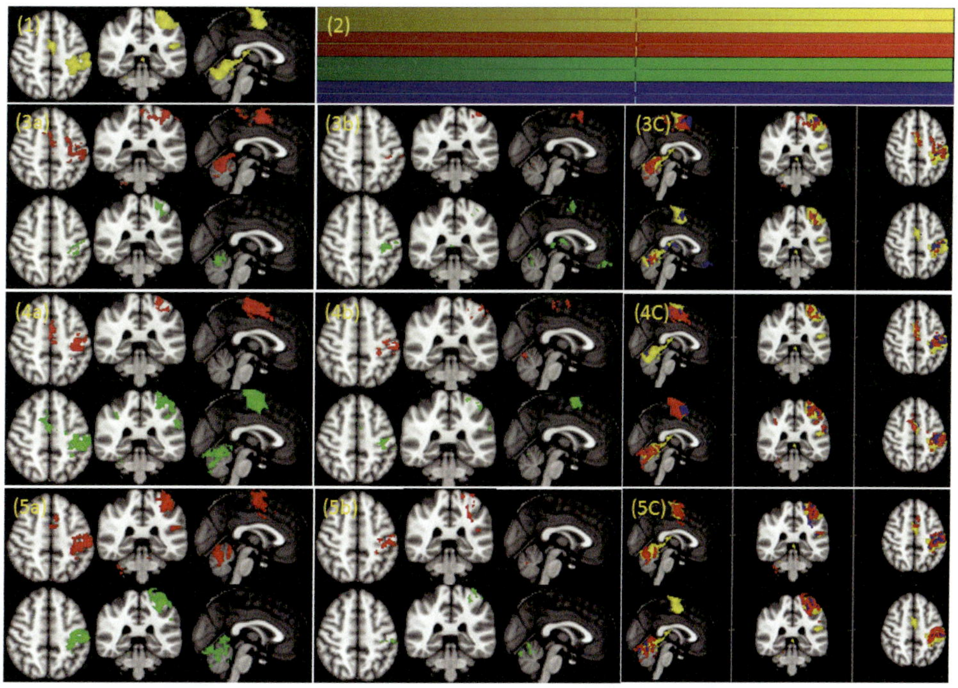

Figure 2: Results of the group average (one sample t-test) illustrating significant BOLD activations in Controls (1), PD (3), MSA (4), and PSP (5) for baseline(a) and follow-up (b) during 'off-state' (upper row) and 'on-state' (lower row). The color bar (2) indicates threshold. Yellow-Controls; Red- 'off-state'; green- 'on-state'. Simultaneous overlay in PD (3c), MSA (4c), and PSP (5c) for baseline(Red) and follow-up (blue) during 'off-state' (upper row) and the 'on-state' (lower row) as compared to Controls (yellow).

compared to controls we observed significant volume reduction in bilateral superior frontal gyrus in each of the three groups, bilateral fusiform gyrus in MSA and PSP through right hemispheric in PD and bilateral precentral gyrus in PD while left precentral gyrus in MSA. Additionally, left precuneus in PD and right cuneus in PSP also showed significantly reduced volumes (Tables 2; Figure 3, 4).

Table 2. Mean ($\bar{X}$) grey matter (GM) and white matter (WM) values and the statistical analysis using t-test indicating significant differences between normal and Parkinsonism patients i.e. Parkinson's disease (PD), Multiple System Atrophy (MSA) and Progressive Supranuclear Palsy (PSP). Region of interests (ROIs): cuneus and precuneus (ROI_CP), medial, middle and inferior frontal lobe (ROI_MMI), thalamus (ROI_TH), fusiform gyrus (ROI_FuG), and superior frontal gyrus (ROI_SFG).

ROIs (GM)	$\bar{X}_{Normal}$	$\bar{X}_{PD}$	$t_{Normal, PD}$ $T_0 = \pm 2.0211$	$\bar{X}_{PSP}$	$t_{Normal, PSP}$ $T_0 = \pm 2.0281$	$\bar{X}_{MSA}$	$t_{Normal, MSA}$ $T_0 = \pm 2.0196$
ROI_CP	4375.93	4105.06	**2.52**	3948.49	**3.29**	4023.83	**2.82**
ROI_MMI	10141.51	9391.42	**2.43**	8875.26	**3.76**	9052.1	**3.26**
ROI_TH	552.89	562.01	-0.52	514.26	**2.05**	548.05	0.17
ROI_FuG	1549.64	1475.53	1.5	1442.24	1.98	1387.68	**2.97**
ROI_SFG	3447.11	3173.6	**2.34**	2960.18	**3.91**	3022.94	**3.16**
ROIs(WM)	$\bar{X}_{Normal}$	$\bar{X}_{PD}$	$t_{Normal, PD}$ $T_0 = \pm 2.0211$	$\bar{X}_{PSP}$	$t_{Normal, PSP}$ $T_0 = \pm 2.0281$	$\bar{X}_{MSA}$	$t_{Normal, MSA}$ $T_0 = \pm 2.0196$
ROI_CP	3046.52	2863.51	1.48	2924.25	1.07	3343.1	-0.84
ROI_MMI	5872.39	5636.82	0.96	5525.97	1.3	6534.11	-0.83
ROI_TH	695.54	699.21	-0.1	679.94	0.38	747.79	-0.84
ROI_FuG	474.42	471.43	0.13	499.05	-1.05	558.07	-1.33
ROI_SFG	1655.45	1548.83	1.38	1519.65	1.74	1832.2	-0.64

* bold values represent the significantly difference and T_0 is standard t-test value for n-1 degree

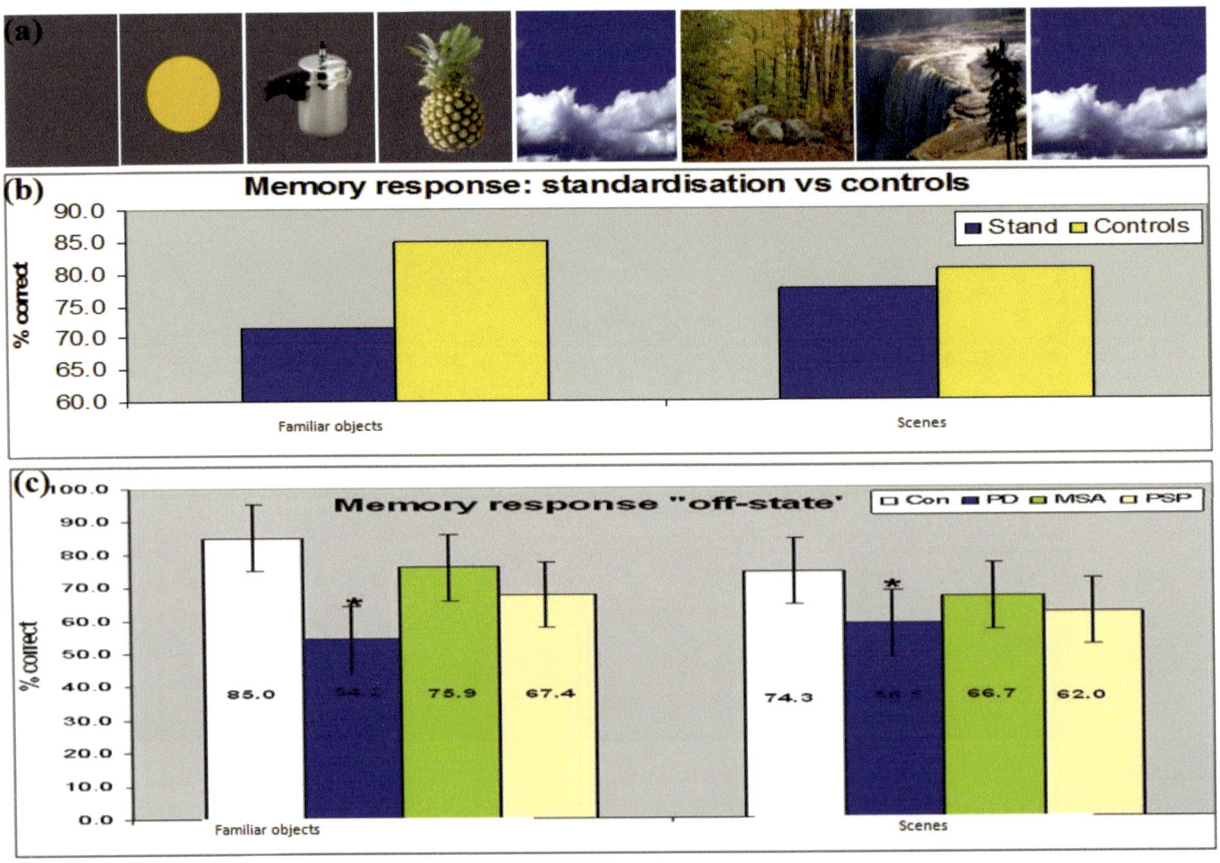

Figure 3: (a) The design for memory task. (b) Memory recall responses of controls during the standardization (outside the MR scanner) and (c) that during the scan for memory task in patient groups obtained from Lumina responses.

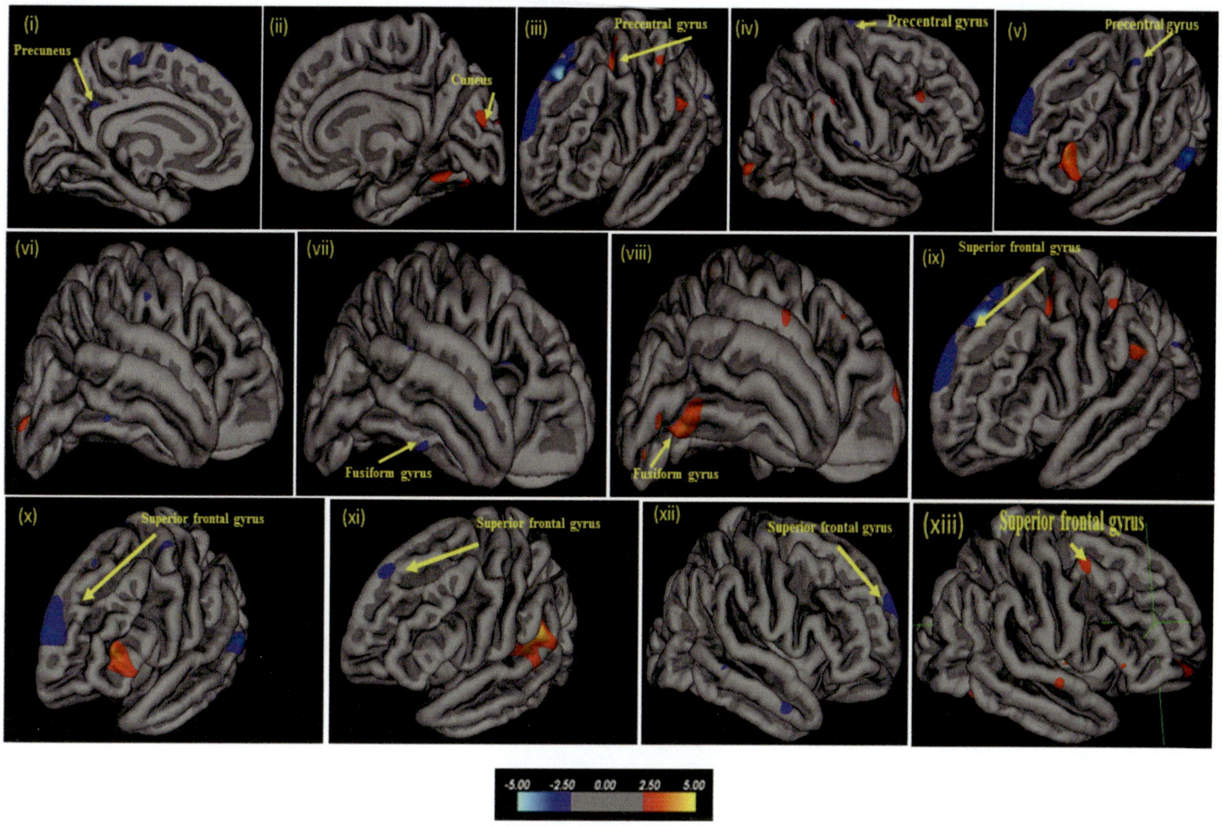

Figure 4: Results of DODS analyses in FreeSurfer's Qdec. The results shown represent a comparison of each patient group as with controls with respect to the volume loss in various cortical regions. (i) lh-Precuneus (controls-PD), (ii) rh-cuneus (controls-PSP),(iii) lh-precentral gyrus (controls-PD), (iv) rh-precentral gyrus (controls-PD), (v) lh-precentral gyrus (controls-MSA), (vi) rh-fusiform gyrus (controls-PD), (vii) rh-fusiform gyrus (controls-MSA), (viii) rh-fusiform gyrus (controls-PSP); lh-superior frontal gyrus: (ix) controls-PD, (x) controls-MSA, (xi) controls-PSP,rh-superior frontal gyrus: (xii) control-MSA, and (xiii) control-PSP.

The mechanism of speech production is very complicated and involves neural co-ordination of vocal structures i.e. lips, tongue, larynx and the respiratory system.[18] The air forced from lungs is occluded at various points in the oral chamber to make a resonant chamber. In absence of any occlusion, vowel sounds are produced, while tongue shape and position is the key in producing different consonant sounds. Also, velum (extension of soft palate) allows air to flow into the nasal cavity to produce nasal sounds. A precise interaction of these vocal structures enables a clear speech and requires a coordinated neural integration of motor regions such as SMA, basal ganglia, thalamus and cerebellum.[18] Our study involved loud reading of simple bi-syllabic (Consonant-Vowel) Hindi words chosen from five stop consonant categories i.e. velars, palatals, retroflexes, dentals, bilabials and nasals, while fMRI scans were acquired.[18] Up to 56% reduced vital lung capacity and a significant increase in mean intensities for all consonant categories indicates that patients force more air out of their lungs to achieve audible speech.[18] Also, delayed plosion (release of air during speech production) and increased voice onset times suggests reduced movements of oral articulators, thereby affecting speech in these patients (Table 3; Figure 5).

Table 3: Summary of Spectrogram Analysis using PRAAT

	Intensities	VOT	F1	F2	F2 Transition	F3	F4	Pitch
Velars	↑MSA	↑PD	*	↑PD/MSA	↑PSP	*	*	*
Palatals	*	↑PSP	↓PD	*	*	*	*	↓PD/MSA
Retroflexes	↑MSA/PSP	↑PD/MSA/PSP	*	*	*	↑PD	*	*
Dentals	↑PD/MSA/PSP	*	*	*	↓PD	*	*	*
Bilabials	↑PD/MSA/PSP	↑PSP	↑MSA	*	↓MSA	*	*	*
Nasals	*	↑PSP	↑PD/MSA	↑PD/MSA	*	↑MSA	↑MSA	*

* No significant difference between the controls and the patient group. ↑ Significant increase ↓ significant decrease

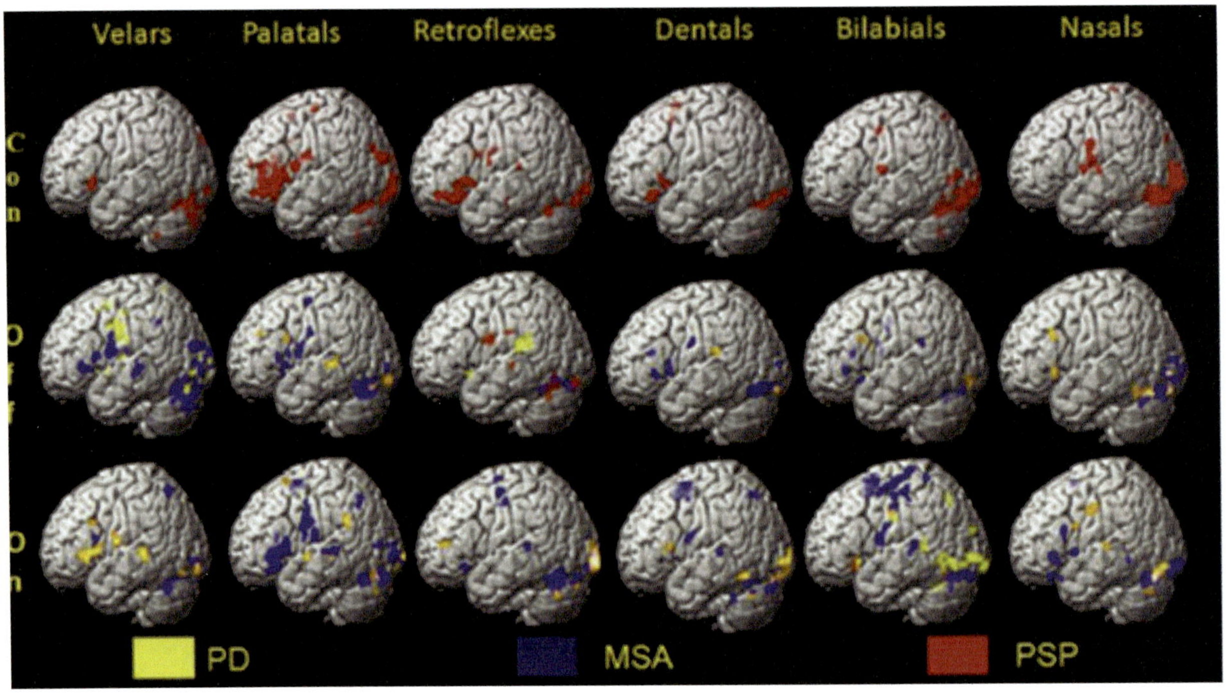

Figure 5: Brain activation for various stop consonants ($p \leq 0.001$, cluster threshold 10). First row: controls; Second row: Patients during the 'off-state'; Bottom row: 'on-state'

Conclusion

In general, FC presented a more focused recruitment of motor predominant areas and can substitute FT, in patients manifesting action tremors or in movement disorders like MSA, PSP. Comparison of FC in patient groups to controls confirmed an overall reduction of activated clusters in the ipsilateral cerebellum, and SMA and somatosensory cortex. The PSP presented neither M1 nor SMA activity. PD and PSP presented similar trends with respect to activations in cerebellum SMA and somatosensory cortex with the parietal cortex having emerged as a component in the compensatory pathway. Grey matter volume reductions in cuneus, precuneus, medial-middle and inferior frontal lobe and fusiform gyrus and thalamus in Parkinsonism, probably explains the poor memory responses. Parkinsonism exhibits decline in short term memory as compared to healthy controls; more significant in PD as compared to MSA and PSP. Dopaminergic administration, although restores the motor functions, has been observed to have no significant effect on improving the memory functions. Patients expel more air out of their lungs to initiate speech in these patients. With more organized future studies, differentiation among PD, MSA and PSP, on the basis of fMRI studies might help improve clinical management.

References

1. Haslinger B, Erhard P, Kämpfe N, et al. Event related functional magnetic resonance imaging in Parkinson's disease before and after levadopa. Brain. 2001;124;558-570.
2. Herz DM, Eickhoff SB, Lokkegaard A, Siebner HR. Functional Neuroimaging of Motor Control in Parkinson's Disease: A Meta-Analysis. Hum Brain Mapp. 2014 35;3227-3237.
3. Playford ED, Jenkins IH, Passingham RE, Nutt J, Frackowiak RS, Brooks DJ. Impaired mesial frontal and putamen activation in Parkinson's disease: a positron emission tomography study. Ann Neurol. 1992;32;151-161.
4. Sabatini U, Boulanouar K, Fabre N, Martin F, Carel C, Colonnese C. Cortical motor reorganization in akinetic patients witk Parkinson's disease: a functional MRI study. Brain. 2000;123;394-403.
5. Herz DM, Haagensen BN, Christensen MS, et al. The Acute Brain Response to Levodopa Heralds Dyskinesias in Parkinson Disease. Ann Neurol. 2014;75;829-836.
6. Worker A, Blain C, Jarosz J, et al. Cortical Thickness, Surface Area and Volume Measures in Parkinson's Disease, Multiple System Atrophy and Progressive Supranuclear Palsy. PLoS ONE. 2014;9.
7. Kao AW, Racine CA, Quitania LC, Kramer JH, Christine CW, Miller BL. Cognitive and Neuropsychiatric Profile of the Synucleinopathies:Parkinson's Disease, Dementia with Lewy Bodies and Multiple System Atrophy. Alzheimer Dis Assoc Disord. 2009;23;365-370.
8. Lucas-Jimenez O, Dıez-Cirarda M, Ojeda N, Pena J, Cabrera-Zubizarreta A, Ibarretxe-Bilbao N. Verbal Memory in Parkinson's Disease: A Combined DTI and fMRI Study. Journal of Parkinson's Disease. 2015;5;793-804.
9. Ghosh BCP, Rowe JB, Calder AJ, Hodges JR, Bak TH. Emotion Recognition in Progressive Supranuclear Palsy. J Neurol Neurosurg Psychiatry. 2009;80;1143-1145.
10. Kulisevsky J, Garcı´a-Sa´nchez C, Berthier ML, et al. Chronic Effects of Dopaminergic Replacement on Cognitive Function in Parkinson's Disease: A Two-Year Follow-Up Study of Previously Untreated Patients. Movement Disorders. 2000;15;613-626.
11. Mohr E, Fabbrini G, Williams J, et al. Dopamine and Memory Function in Parkinson's Disease. Movement Disorders. 1989;4;113-120.
12. Sawamoto N, Piccini P, Hotton G, Pavese N, KrisThielemans, Brooks DJ. Cognitive deficits and striato-frontal dopamine release in Parkinson's disease. Brain. 2008;131;1294-1302.
13. Singh S, Behari M. Verbal and visual memory in patients with early Parkinson's disease: Effect of levodopa. Neurology India. 2006;54;33-37.
14. Sachin S, Kumaran SS, S S, et al. Functional mapping in PD and PSP for sustained phonation and phoneme tasks. J Neurol Sci. 2008;273;51-56.
15. New AB, Robin DA, Parkinson AL, et al. The intrinsic resting state voice network in Parkinson's disease. Hum Brain Mapp. 2015;36;1951-1962.
16. Arantes PR, Gobato HH, Davoglio BB, et al. Levodopa versus non-levodopa brain language fMRI in Parkinson's disease. Einstein. 2012;10;171-179.

17. Tessitore A, Giordano A, Micco RD, Russo A, GioacchinoTedeschi. Sensory motor connectivity in Parkinson's disease: the role of functional neuroimaging. Front Neurol. 2014;5.

18. Saxena M, Behari M, Kumaran SS, Goyal V, Narang V. Assessing speech dysfunction using BOLD and acoustic analysis in parkinsonism. Parkinsonism and related disorders. 2014;20;855-861.

19. Witt ST, Meyerand ME, Laird AR. Functional neuroimaging correlates of finger tapping task variations: An ALE meta-analysis. NeuroImage. 2008;42;343-356.

20. Yu R, Liu B, Wang L, Chen J, Liu X. Enhanced Functional Connectivity between Putamen and Supplementary Motor Area in Parkinson's Disease Patients. PLoS ONE. 2013;8;e59717.

21. Boecker H, Ceballos-Baumann A, Bartenstein P, et al. Sensory processing in Parkinson's and Huntington's diseaseInvestigations with 3D H215O-PET. Brain. 1999;122;1651-1665.

22. Yu H, Sternad D, Corcos DM, Vaillancourt DE. Role of Hyperactive Cerebellum and Motor Cortex in Parkinson's Disease. NeuroImage. 2007;35;222-233.

23. Samuel M, Ceballos-Baumann AO, Blin J, et al. Evidence for lateral premotor and parietal overactivity in Parkinson's disease during sequential and bimanual movements: A PET study. Brain. 1997;120;963-976.

24. Payoux P, Brefel-Courbon C, Ory-Magne F, et al. Motor activation in multiple system atrophy and Parkinson disease. A PET study. Neurology. 2010;75;1174-1180.

25. Burciu RG, Chung JW, Shukla P, et al. Functional MRI of disease progression in Parkinson's disease and atypical parkinsonian syndromes. Neurology. 2016;87;709-717.

26. Press D, Mechanic D, Manoach D. Working memory deficits in Parkinson's disease resolve after consolidation. Mov Disord. 2001;16;S25.

27. Kelly C, Zubicaray Gd, Martino AD, et al. L-Dopa Modulates Functional Connectivity in Striatal Cognitive and Motor Networks: A Double-Blind Placebo-Controlled Study. J Neurosci. 2009 29;7364-7378.

28. Rana B, Juneja A, Saxena M, et al. Regions-of-interest based automated diagnosis of Parkinson's disease using T1-weighted MRI. Expert Systems with Applications. 2015;42;4506–4516.

Chapter 9

Non-Motor Features of Parkinson's Disease

Deepti Vibha, Muthukani Sankaranarayanan

Introduction

Parkinson' disease (PD) is a clinically heterogeneous progressive disorder of the nervous system with various motor and non-motor features. Non-motor symptoms (NMS) are considered to be integral to the disease and define the premotor stage.[1] These non-motor symptoms (NMS) are often not recognized and pose important health care related issues in PD. Various NMS and their prevalence are listed below:

Table 1: Non-motor symptoms of PD

Non-motor symptoms	Prevalence[2-7]
Cognitive dysfunction	41-78 %
Psychosis, hallucinations	20-40 %
Mood disorders	~50 %
Sleep disturbances	55-80 %
Fatigue	33-58 %
Autonomic system dysfunction	58%
Olfactory deficits	74-96 %
Gastro-intestinal dysfunction	40-82%
Painful sensory symptoms	46%
Skin findings (seborrhea)	18.6-59 %

Psychiatric features are the most prevalent and disabling non-motor symptoms in PD. Some of the patients might present with early non-motor features (such as olfactory deficits, constipation, depression, and REM sleep related behavior disorder) and precede the clinical onset of motor symptoms by several years.[8]

The main non-motor features will be reviewed further individually.

DOI: 10.1201/9781003505488-9

Non motor symptoms:

The non-motor symptoms (NMS) have a huge toll on the quality of life (QoL) of patients with PD. There are several scores on the measurement and quantification of these symptoms which have been translated in Hindi and validated in Indian population[9] (unpublished material) like NMS questionnaire (NMSQuest),[10] EuroQoL5Dimension (EQ-5D)][11] and PD disease- specific QoL questionnaire (PDQ39).[12]

1. Cognitive dysfunction:

Cognitive deficits are one of the common non-motor features in PD and its prevalence is reported up to 78% in observational studies.[13] It has a huge impact on instrumental activities of daily living and its presence is associated with rapid worsening of motor symptoms, increased admissions at nursing homes. It is also an important predictor of mortality in PD.[4] Older age at onset, longer disease duration, neuropsychiatric features such as depression and positive family history of cognitive dysfunction are risk factors of dementia.[3,4] Patients who develop dementia after five years of parkinsonian motor symptoms are referred to as PD dementia (PDD).

Dementia in PD is typically subcortical dementia where patients might present with mental slowing, psychomotor retardation with behavior and personality changes. They usually present with executive deficits such as decision making, problem solving, multitasking, difficulty in set shifting and visuo-spatial deficits, retrieval of memory, fluctuations in cognition which are in contrast with dementia associated with Alzheimer's disease.

Though the exact anatomical and pathological correlates are not very clear, the neuropathological studies have shown presence of cortical and limbic neuronal Lewy bodies which are made up of alpha synuclein and it is inversely associated with cognitive performance in patients with PDD. Some of these patients have higher cortical amyloid plaques and neurofibrillary tangles (NFT) which probably implies the coexistence of AD pathology and PD. But the presence of Lewy bodies seems to correlate better with dementia than amyloid plaques.[5]

2. Psychosis and hallucinations:

Psychosis and hallucinations are one of the most challenging symptoms in the natural history of PD. Psychosis develops in about 20-40% of treated patients with PD. Visual hallucinations and delusions are the commonest psychotic symptoms in PD patients.[6] They usually develop in patients with established PD with clear sensorium and preserved insight. They usually develop constellation of features such as paranoid delusions, visual hallucinations and other illusions. Though these visual hallucinations can happen at any time, they are more common during the evening hours, probably due to low environmental stimulation and these are usually vivid, colourful and rich in detail.[7] The common delusions in PD are delusions of infidelity, though grandiose delusions, delusion of broadcasting do also occur. Tactile, gustatory and auditory hallucinations are less common in PD.

The mechanisms behind these psychotic features are either the primary disease itself or due to anti-parkinsonian medicines or both. Among the anti-parkinsonian drugs, the propensity of causing hallucinations are more with anticholinergic agents, followed by Amantadine, MAO-B inhibitors,

COMT inhibitors, dopaminergic agonists and finally Levodopa in the descending order and they should be stopped in the same order if the patient develops these issues.

Psychotic features are the most common reason for nursing home placement as compared to motor symptoms in PD. In addition, various risk factors have now been identified, including advanced age, cognitive dysfunction, depression, severe and longer disease course and sleep disorders.[7]

3. Mood disturbances:

Depression, anxiety, apathy/abulia are the most common mood disorders noted in Parkinson disease. These are troublesome symptoms in both early as well as late PD.[14] Indeed apathy is much more common and often missed.

a. Depression:

Around 50% of the PD patients have symptoms of depression and occasionally as a presenting complaint. The prevalence of major depressive disorder vary from 10-20% among various studies.[14] Though the depressive symptoms are mild to moderate, they usually produce significant impairment in QoL and they are usually undertreated too. In a study by Behari et al depression was observed in 38.5% of Indian PD patients before the motor symptoms were identified.[11]

Recognition of depressive symptoms is difficult as the psychomotor retardation and loss of interest, which is associated with depression, might mimic the typical bradykinesia and decreased facial expression in PD. But depression usually presents with sadness, anhedonia, loss of interest in daily activities in PD.

b. Anxiety:

The second common psychiatric symptom after depression in PD is anxiety and it occurs in around 30 % of patients with PD.[12] Generalized anxiety disorders (GAD) and social phobia are among the most common anxiety syndromes in PD. Depression and anxiety was usually co-morbid conditions in PD. Like motor fluctuations in PD, these mood disturbances are also improve during the "ON" state and worsen during the "OFF" state.

c. Apathy:

Apathy has been increasingly recognized as a significant non-motor symptom with a prevalence of 17-62% of PD patients. Though it usually occurs in association with cognitive dysfunction and depression, it had been reported in early stages of PD independently. A meta-analysis showed that the prevalence of apathy was less in patients with higher doses of dopaminergic agonists implying that dopaminergic agonists might be useful for symptoms of apathy.[13] It is also associated with negative impact on patient's QoL.

4. Sleep disturbances:

Sleep disorders including insomnia, excessive daytime somnolence (EDS), restless legs syndrome (RLS), Rapid Eye Movement sleep behavior disorder (RBD) are prevalent in 55-80% of patients with PD. The commonest sleep disorders are sleep fragmentation and early morning awakening.

Sleep fragmentation is because of multiple awakenings after sleep and these might be due to rigidity interfering with turning in the bed, nocturia, tremors, leg cramps, unpleasant dreams, nocturnal confusion and hallucinations. Some of the PD patients have early morning painful dystonia and that produces early awakening. The psychotic symptoms such as depression also contribute to poor sleep quality and early morning awakening.

Symptoms such as RLS (urge to move the limbs) and periodic limb movement during sleep (PLMS) are also frequently noted in PD and their individual prevalence was difficult to interpret due to overlapping symptoms in PD.

RBD is prevalent in around 15-47% of patients with PD.[14] This parasomnia is characterized by excessive motor activity in the dream state with retention of muscle tone during REM sleep. Various longitudinal studies showed that 30-80% of patients with idiopathic RBD developed neurodegenerative diseases such as synucleinopathy over many years after onset depending upon the follow-up period.[15,16]

Excessive day-time somnolence has also been described in PD and it can be due to various sleep disturbances which are mentioned above, anti-parkinsonian drug induced, depression or multifactorial.

5. Fatigue:

Fatigue is another common problem in PD patients for which exact pathophysiology is not known. It is reported in 33-58% of patients with PD in observational studies.[21] The presence and severity of fatigue is neither associated with severity of motor dysfunction nor with duration of illness and it is distinct from apathy. It can be physical or mental. Unlike fatigue in other diseases, fatigue associated with PD improves with exercise. It is associated with depression, autonomic dysfunction, sleep disorders in some patients and it usually worsens with disease progression.

6. Autonomic dysfunction:

Autonomic dysfunction is reported in 50-80% patients with PD.[22] It may manifest as dysphagia, sudomotor disturbances and cardiovascular, urogenital, thermoregulatory and gastrointestinal dysfunction. It may be due to primary degenerative process affecting the autonomic ganglia, brainstem nuclei or hypothalamus. The autonomic dysfunction in PD occurs late in the course of disease and is less severe as compared to patients with multisystem atrophy. In addition, the response to levodopa is usually rather poor.

Orthostatic hypotension (OH) is a marker of sympathetic dysfunction in PD and it is worsened by anti-parkinsonian medicines. It is often associated with supine hypertension which is shown to be causing target end organ damage. Urogenital dysfunction produces urinary urgency, hesitancy; urge incontinence, sexual dysfunction (hypo/hyper sexuality). Urodynamic studies reveal occurrence of involuntary detrusor contraction at the early phase of filling in patients with PD.

Males with sexual dysfunction may complain of decreased libido, inability to maintain erection and females may complain of vaginal dryness, inability to achieve orgasm and reduced sexual drive. These symptoms are complex and multifactorial and the various etiologies are being autonomic involvement, coexistent depression, motor symptoms (bradykinesia, rigidity) or drug induced.

Gastrointestinal dysfunction varies from swallowing difficulty, drooling, delayed transit in stomach and intestine, increased Helicobacter pylori infection, bacterial overgrowth in small intestine, constipation.

7. Olfactory dysfunction:

Olfactory dysfunction in the form of hyposmia or anosmia is one of the earliest NMS in PD and it usually precedes the motor features by years. It may present with difficulty in identifying the odours or its discrimination. It is reported in upto 90% of patients with PD.[23] The exact pathophysiology is not known. It is postulated that deficits in serotonergic, noradrenergic, cholinergic systems in brain might be the mechanisms involved (See the chapter by Pathak at al. in this volume). Most of the patients with olfactory dysfunction in PD are not aware of it. In a study by Behari et al. using Indian smell identification test (INSIT), the authors concluded that INSIT, an indigenously prepared test battery for identification of smell has strong agreement with commercially available Sniffin-Stic Test.[24] They also found that sense of smell is impaired in 54.7% PD patients.

8. Pain:

Pain is another frequently reported symptom in Parkinson disease and at times it can be quite disabling. Pain and other sensory disturbances are reported in 30-83% of patients with PD.[21] Kings PD pain scale classified pain into seven domains including nocturnal pain, oro-facial and central pain. There are both dopaminergic and non-dopaminergic pathways implicated in the pathophysiology of pain in PD. This pain can be of lancinating, burning or tingling and it can be either localized or generalized.

Painful sensory symptoms may be related to motor fluctuations (on-off phenomenon), painful dyskinesias or painful dystonia.

9. Dermatologic findings:

Seborrheic dermatitis and hyperhidrosis are the most prevalent skin conditions reported in Parkinson disease and it is considered to be a premotor symptom but often overlooked. It is postulated that seborrhea (increased sebum production) is an indication of autonomic dysfunction (parasympathetic over-activity) though it is not proven. Malignant melanoma and bullous pemphigoid are also comparatively more common in PD patients.

Treatment

Dementia in PD can be treated with acetylcholine esterase inhibitors (Rivastigmine, Donepezil, Galantamine) and among these Rivastigmine is the only drug approved by European medical agency for the same. In a multicenter randomized blinded study by Emre M et al.,[26] Rivastigmine improved all primary and secondary outcomes significantly in patients with PD dementia and interestingly it delayed the symptoms onset by 1.5-2 years.

Psychosis and hallucinations can be treated with atypical antipsychotic agents. Quetiapine and clozapine are equally effective. Recently, Pimavanserin, a 5-HT 2A receptor inverse agonist was found to be efficacious in control of PD related psychosis and has been approved by US-FDA in 2016 for hallucinations and delusions.[16] Anti-parkinsonian drug induced psychosis and hallucinations also should be kept in mind and the same should be treated according to the probability of drug causing it.

Depression can be treated with selective Serotonin Reuptake Inhibitors (SSRI) or Tricyclic antidepressants (TCA). SSRI is usually preferred due to safer side effect profile. Rare aggravation of motor symptoms with SSRI and occurrence of serotonin syndrome should be kept in mind in appropriate clinical settings while treating with this group of drugs. TCA can worsen postural hypotension, cognitive decline and urogenital dysfunction, particularly in the elderly.

The treatment of excessive daytime somnolence should be started with sleep hygiene and treating the causes of poor night time sleep. In those who do not respond effectively, stimulants like modafinil and methylphenidate can be tried. The combination of levodopa-carbidopa reduces the spontaneous leg movements during sleep and can be considered in the treatment of PLMS.

REM sleep behavior disorder can be treated with either clonazepam or melatonin.

Orthostatic hypotension can be treated by stopping the concerned anti-parkinsonian drugs if feasible, increasing salt intake, avoiding sudden change in the posture and various counter physical maneuvers. In recalcitrant cases, drugs like midodrine, fludrocortisone, desmopressin, pyridostigmine, droxidopa can be tried but randomized controlled trials are lacking in this respect. Antihypotensive drugs (midodrine) should not be given after 5 pm as most PD patients develop rise in BP during night time.

Irritative urinary symptoms like urgency, urge incontinence can be treated with anticholinergics like imipramine, oxybutynin, trospium, tolterodine. Sildenafil can be tried for patients with erectile dysfunction in PD. Nocturia can be treated with desmopressin nasal spray at night time.

Constipation can be treated with high fluid intake, stool bulking agents and in non-responders with Lubprostone or polyethylene glycol

Siallorhea and drooling can be treated with anticholinergics such as glycopyrrolate, sublingual atropine, ipratropium bromide. Botulinum toxin injection into the salivary glands can be tried in refractory cases.

There is no effective treatment for fatigue and the causes like depression and excessive daytime somnolence should be dealt with first. In refractory cases, methylphenidate, pemoline, amantadine can be tried.

Seborrheic dermatitis usually improves with dopaminergic therapy (levodopa) and it can be treated with topical or systemic antifungals in refractory cases.

Conclusion

Focused research on non-motor features of PD has expanded our understanding of the pathophysiology of PD and its role in affecting the quality of life in PD. As some of non-motor features such as RBD, Olfactory dysfunction, autonomic dysfunction could precede the onset of motor symptoms of PD, they could served as potential biomarkers and should aid in drug trials for prevention of further neurodegeneration. It is the duty of the treating neurologists to educate the patients and their family regarding these symptoms and treat them to reduce its burden.

Take home message:

- Non-motor symptoms may precede the onset of motor symptoms and add to the morbidity of the disease.
- Non-motor symptoms include cognitive dysfunction, neuropsychiatric features, sleep disturbances, fatigue, autonomic system dysfunction, olfactory dysfunction, gastrointestinal dysfunction, painful sensory symptoms, skin findings (seborrhea).
- The optimal recognition and treatment of non-motor symptoms in Parkinson's disease may improve quality of life (QoL).
- The identification of non-motor symptoms in pre motor stage might provide opportunity for early diagnosis and drug trials.

Reference

1. Lim S-Y, Fox SH, Lang AE. Overview of the Extranigral Aspects of Parkinson Disease. Arch Neurol. 2009;66(2);167–72.
2. Langston JW. The Parkinson's complex: parkinsonism is just the tip of the iceberg. Ann Neurol. 2006;59(4);591–6.
3. Emre M. Dementia in Parkinson's disease: cause and treatment. Curr Opin Neurol. 2004;17(4);399–404.
4. Tsuboi Y, Uchikado H, Dickson DW. Neuropathology of Parkinson's disease dementia and dementia with Lewy bodies with reference to striatal pathology. Parkinsonism Relat Disord. 2007;13(3);S221-224.
5. Forsaa EB, Larsen JP, Wentzel-Larsen T, Alves G. What predicts mortality in Parkinson disease?: a prospective population-based long-term study. Neurology. 2010;75(14);1270–6.
6. Kraybill ML, Larson EB, Tsuang DW, Teri L, McCormick WC, Bowen JD, et al. Cognitive differences in dementia patients with autopsy-verified AD, Lewy body pathology, or both. Neurology. 2005;64(12);2069–73.
7. Fénelon G, Mahieux F, Huon R, Ziégler M. Hallucinations in Parkinson's disease: prevalence, phenomenology and risk factors. Brain J Neurol. 2000;123(4);733–45.
8. Langston JW. The parkinson's complex: Parkinsonism is just the tip of the iceberg. Ann Neurol. 2006;59(4);591–6.
9. Behari M, Malhotra. To assess the severity of non motor symptoms and quality of life of Indian Parkinson's disease patients after translating and validating the Hindi version of NMSS, PDQ- 39, and EQ5-D.
10. Chaudhuri KR, Martinez-Martin P, Schapira AHV, Stocchi F, Sethi K, Odin P, et al. International multicenter pilot study of the first comprehensive self-completed nonmotor symptoms questionnaire for Parkinson's disease: the NMSQuest study. Mov Disord Off J Mov Disord Soc. 2006;21(7);916–23.
11. EuroQol Group. EuroQol--a new facility for the measurement of health-related quality of life. Health Policy Amst Neth. 1990;16(3);199–208.

12. Jenkinson C, Fitzpatrick R, Peto V, Greenhall R, Hyman N. The Parkinson's Disease Questionnaire (PDQ-39): development and validation of a Parkinson's disease summary index score. Age Ageing. 1997;26(5),353–7.
13. Emre M. Dementia in Parkinson's disease: cause and treatment. Curr Opin Neurol. 2004;17(4);399–404.
14. Weintraub D, Moberg PJ, Duda JE, Katz IR, Stern MB. Effect of psychiatric and other nonmotor symptoms on disability in Parkinson's disease. J Am Geriatr Soc. 2004;52(5);784–8.
15. Behari M, Srivastava AK, Pandey RM. Quality of life in patients with Parkinson's disease. Parkinsonism Relat Disord. 2005;11(4);221–6.
16. Broen MPG, Narayen NE, Kuijf ML, Dissanayaka NNW, Leentjens AFG. Prevalence of anxiety in Parkinson's disease: A systematic review and meta-analysis. Mov Disord Off J Mov Disord Soc. 2016;31(8);1125–33.
17. den Brok MGHE, van Dalen JW, van Gool WA, Moll van Charante EP, de Bie RMA, Richard E. Apathy in Parkinson's disease: A systematic review and meta-analysis. Mov Disord Off J Mov Disord Soc. 2015;30(6);759–69.
18. Comella CL, Nardine TM, Diederich NJ, Stebbins GT. Sleep-related violence, injury, and REM sleep behavior disorder in Parkinson's disease. Neurology. 1998;51(2);526–9.
19. Chendo I, Ferreira JJ. Pimavanserin for the treatment of Parkinson's disease psychosis. Expert Opin Pharmacother. 2016;17(15);2115–24.
20. Schenck CH, Bundlie SR, Mahowald MW. Delayed emergence of a parkinsonian disorder in 38% of 29 older men initially diagnosed with idiopathic rapid eye movement sleep behaviour disorder. Neurology. 1996;46(2);388–93.
21. Alves G, Wentzel-Larsen T, Larsen JP. Is fatigue an independent and persistent symptom in patients with Parkinson disease? Neurology. 2004;63(10);1908–11.
22. Asahina M, Vichayanrat E, Low DA, Iodice V, Mathias CJ. Autonomic dysfunction in parkinsonian disorders: assessment and pathophysiology. J Neurol Neurosurg Psychiatry. 2013;84(6);674–80.
23. Ponsen MM, Stoffers D, Booij J, van Eck-Smit BLF, Wolters EC, Berendse HW. Idiopathic hyposmia as a preclinical sign of Parkinson's disease. Ann Neurol. 2004;56(2);173–81.
24. George J, Jose T, Behari M. Use of Indian smell identification test for evaluating olfaction in idiopathic Parkinson's disease patients in India. Neurol India. 2013;61(4;365–70.
25. Beiske AG, Loge JH, Rønningen A, Svensson E. Pain in Parkinson's disease: Prevalence and characteristics. Pain. 2009;141(1–2);173–7.
26. Emre M, Aarsland D, Albanese A, Byrne EJ, Deuschl G, De Deyn PP, et al. Rivastigmine for dementia associated with Parkinson's disease. N Engl J Med. 2004;351(24);2509–18.

Chapter 10

Sleep Disorders in Parkinson's Disease

Vibhor Pardasani

Disorders of sleep Parkinsons disease patients commonly have sleep related disorders. 60-98% of PD patients are known to report sleep disorders.[1] In a few patients of PD, sleep related problems may predate motor findings and provide the potential window period for introducing neuro-protective agents in the future.

Excessive daytime sleepiness is known to occur more commonly in subjects who go on to develop PD than in those who do not.[2] The evidence that REM Sleep Behaviour Disorder (RBD) predates PD in several patients and that idiopathic RBD is also a synucleinopathy like idiopathic PD is now unequivocal.

Two recent questionnaire studies further quantify the significance of sleep issues in PD. The PRIAMO study completed on 1,072 patients with PD found that 15% patients were found to have Restless Leg Syndrome (RLS), 21% Excessive Daytime Somnolence (EDS), 30% RBD, 37% insomnia, while 8% PD patients were reported to have fatigue that limited daytime activities.[3] The study by Martinez-Martin et al. demonstrated that of the 411 patients with PD, RLS was found in 32%, daytime sleepiness in 47%, difficulty falling asleep in 50%, fatigue in 66%, while nocturia was reported in 68% patients.[4]

Pathophysiology of PD involves neuronal cell loss in substantia nigra, brainstem nuclei and in cerebral cortex. Synuclein accumulation in RBD follows the topography and the same temporal sequence as PD in the traditional Braak staging, where α-synuclein aggregates are first present in the peripheral autonomic nervous system and in the lower brainstem before advancing to the midbrain and reaching the substantia nigra. There is Lewy body accumulation in the pedunculopontine tegmental nucleus which is the center for REM sleep. As a result, RBD can at times present in the premotor phase of PD.[5] Patients with different motor problems of PD can have sleep disturbances because he sleep disturbances observed in PD patients may be due to motor problem like slowness, rigidity, dystonia with decrease ability to turn or dyskinesias.[6, 7, 8]

DOI: 10.1201/9781003505488-10

In 2002, Kumar S et al. first reported the prevalence of sleep disturbances among our PD patients. 149 consecutive PD patients and 115 age-matched healthy controls were interviewed; their sleep was assessed with the help of a 23-question validated sleep questionnaire.[9] Sleep disturbances were reported in 42% PD patients as against 12% of the controls. About 32% patients reported insomnia (5% among controls), nightmares were reported in 32% patients (5% among controls), while 15% PD patients reported EDS (6% among controls) ($P < 0.025$). Prevalence of sleep disturbances co-related with Unified Parkinson's Disease Rating Scale (UPDRS) Part III, Schwab and England score, the dose of levodopa, the scores on the rigidity and the bradykinesia scores. When we assessed EDS using the Epworth Sleepiness Scale (ESS), we found that the mean ESS scores in PD patients was significantly higher than the ESS score among matched controls ($P<0.05$).[10] We found higher ESS scores to be significantly associated with higher Hoehn and Yahr (H&Y) stage of disease and higher UPDRS part I, III and total scores.

Later, we looked at the prevalence of different sleep disorders in familial PD patients and compared to those among sporadic PD patients. Over a two-year period, we evaluated 134 consecutive patients with sporadic (104 patients) and familial PD (30 patients). We did an in-depth clinical history, performed a detailed physical examination and administered a questionnaire on sleep related problems.[11] The disease severity scores as well as the disease duration were comparable in the two groups. We found that insomnia was significantly more common in patients with sporadic PD than in familial PD patients (26.9% vs. 3.3%; $p=0.006$). While 25% patients with sporadic PD were found to have RBD, none of the familial PD patients had RBD ($p=0.002$). In 34.6% of these, clinical RBD was found to be present even before parkinsonian symptoms were reported. RLS was detected in 11.54% patients with sporadic PD as against 3.33% patients with familial PD; however, the difference was not statistically significant ($p= 0.181$).

The prevalence of insomnia in our two studies was quite similar (30.2% and 32%).[9] The prevalence among PD patients of clinical RBD (25%) and RLS (14.8%) was also comparable to previous studies.[3,12,13]

RBD has been found to have a strong association with Lewy body pathology (similar to other synucleinopathies like PD, diffuse Lewy body dementia and multiple system atrophy). Several longitudinal studies have reported that a large number of RBD patients go on to develop clinical Parkinsonism.[14,15,16] In the longest followed series by Schenck et al., 81% of the original cohort of RBD patients, developed features of parkinsonism, around 14 years after they reported RBD.[17] Among PD patients, 50% report RBD; the prevalence being higher in males. Patients with RBD symptoms have worse prognosis with higher rates of dementia and hallucinations, more frequent falls and incidents of autonomic dysfunction.[18,19]

We had systematically studied the characteristics of PD patients, who had RBD and compared them to those not reporting RBD. We found that those with had RBD reported insomnia significantly more frequently early morning as well as nocturnal awakenings. In addition, they had more snoring and excessive daytime sleepiness when compared to patients without RBD.[20]

Unlike the findings of Postuma et al. we did not see any higher incidence of falls in patients with RBD.[21] Gjerstad found that RBD was common in males and co-related it with the use of higher dopaminergic dosage.[22] These findings were neither seen in or study nor by Wetter's group.[23] These differences suggest that the clinical characteristics and progression of RBD in PD may be different in different populations.

We found our patients with RBD to be having significantly more hallucinations similar to the findings of Pacchetti et al.[24] The finding that RBD and REM-related hallucinations co-occur in a substantial number of patients suggests that the causation of dopaminergic drugs associated hallucinations may involve altered REM sleep regulation.[25]

Sleep related injuries (SRI) have been reported in up to 70% PD patients.[26] Our patients did not report any SRI. PD patients had a milder form of RBD with little or no progression and low frequency of sleep-associated motor behaviour.

Insomnia in PD is usually a defect in sleep maintenance rather than sleep onset. Almost 80% with PD have been found to report sleep fragmentation as well as early morning awakenings. The causes may include overnight emergence of negative motor symptoms, co-occurrence of other primary sleep disorders, side effects of dopaminergic medications, psychiatric disorders including depression and autonomic dysfunction in addition to the influence of primary neurodegenerative process on centers regulating sleep–wake cycles.

Unlike RBD, which predates motor manifestations of PD in a substantial number of patients, there are no reports to show that RLS predisposes to PD. Patients with PD/RLS develop features of PD before developing RLS.[27] Iatrogenic RLS sometimes follows treatment of PD. Dopaminergic medications, which are also useful in low doses in idiopathic RLS, may at times cause the RLS symptoms.[28] This phenomenon is similar to augmentation noted in some RLS patients, who showed worsening of RLS symptoms with dopaminergic drugs.[29] In idiopathic RLS, this frequently responds to withdrawal of dopaminergic agents despite immediate withdrawal-related worsening.

Interestingly, RLS has also been reported to occur after deep brain stimulation (DBS) of subthalamic nucleus (STN) in 11 of 195 patients after reduction in antiparkinsonian medication.[30] On the other hand, some studies have reported improvement in RLS symptoms following both ablative as well as DBS procedures on the globus pallidus internus and STN.[31] At times RLS appears following initiation of other non-dopaminergic drugs used in PD patients. These include serotonin re-uptake inhibitors (SSRIs), anticholinergics/anti-histaminergics and anti-psychotics.[27]

Periodic limb movements of sleep, as defined by the American Academy of Sleep Medicine as "at least four periodic episodes of repetitive and highly stereotyped limb movements that occur during sleep," is associated with PD in some studies.[32,33] However, this has not been replicated by some later reports.[34,35] Some studies have also reported PLMS to be correlating with the severity of PD .[36] Sleep disordered breathing occurs in PD patients at the same frequency as in general population. The earlier reports of co-relation between the two, suffered from methodological errors.[37,38] In most patients who develop OSA in the presence of PD, the classical risk factors of OSA are absent. Pharyngeal narrowing as a result of degeneration of pharynx-innervating brainstem neurons probably contributes to OSA.[39]

Garima Shukla et al. continue to look carefully at sleep disorders occurring in association with Parkinsonism. In 2015, they described two patients, in whom circadian rhythm sleep disorder (CRSD) – advanced sleep phase (ASP) type was one of the presenting manifestations of Parkinsonian plus syndromes, one with Progressive Supranuclear Palsy and the other with Multiple System Atrophy.[40] ASP type of CRSD, is a condition in which early in the evening, patients start feeling sleepy and go to bed. They then wake up very early in the morning. This association had not been previously described.

Conclusion

Sleep disorders are extremely common among patients with Parkinson's disease and affect their quality of life adversely. There needs to be greater awareness among sleep specialists and neurologists regarding sleep disorders affecting patients with PD and Parkinson's plus syndrome. RBD would at time occur several years before the onset of PD and in future may provide the window of opportunity for the action of potential neuro-protective agents.

Take home message

1. Patients with Parkinson's disease frequently complain of sleep related symptoms.
2. Sleep related symptoms may predate onset of PD.
3. RBD, RLS, EDS and insomnia are the common sleep disorders in PD patients.
4. The occurrence and type of sleep disorders may vary between different Parkinsonian syndromes as well as between sporadic and familial PD.

Recognition and management of sleep disorders may add tremendously to the quality of life in patients with Parkinsonism.

References

1. Adler CH, Thorpy MJ. Sleep issues in Parkinson's disease. Neurol Clin. 2005;64;S12–20.
2. Abbott RD, Ross GW, White LR, et al. Excessive daytime sleepiness and subsequent development of Parkinson disease. Neurology. 2005;65;1442–6.
3. Barone P, Antonini A, Colosimo C, et al. The PRIAMO study: a multicenter assessment of non-motor symptoms and their impact on quality of life in Parkinson's disease. Mov Disord. 2009;24;1641–9.
4. Martinez-Martin P, Rodriguez-Blazquez C, Kurtis MM, et al. The impact of nonmotor symptoms on health-related quality of life of patients with Parkinson's disease. Mov Disord. 2011;26;399–406.
5. Adler C, Hentz J, Shill H, et al. Probable RBD is increased in Parkinson's disease but not in essential tremor or restless legs syndrome. Parkinsonism Relat Disord. 2011;17;456–8.
6. Claassen DO, Josephs KA, Ahlskog JE, et al. REM sleep behavior disorder preceding other aspects of synucleinopathies by up to half a century. Neurology. 2010;75;494–9.

7. Boeve BF, Silber MH, Saper CB, et al. Pathophysiology of REM sleep behavior disorder and relevance to neurodegenerative disease. Brain. 2007;130;2770–88.
8. Driver-Dunckley ED, Adler CH. Movement disorders and sleep. Neurol Clin. 2012;30(4);1345-58.
9. Kumar S, Bhatia M, Behari M. Sleep disorders in Parkinson's disease. Mov Disord. 2002;17(4);775-81.
10. Kumar S, Bhatia M, Behari M. Excessive daytime sleepiness in Parkinson's disease as assessed by Epworth Sleepiness Scale (ESS). Sleep Med. 2003;4(4);339-42.
11. Vibha D, Shukla G, Singh S, Goyal V, Srivastava AK, Behari M. Lower prevalence of sleep disturbances in familial versus sporadic Parkinson's disease: a questionnaire based study. J Neurol Sci. 2010;295(1-2);27-30.
12. Comella CL, Nardine TM, Diederich NJ, Stebbins GT. Sleep-related violence, injury, and REM sleep behavior disorder in Parkinson's disease. Neurology. 1998;51;526–9.
13. Nomura T, Inoue Y, Miyake M, Yasui K, Nakashima K. Prevalence and clinical characteristics of restless legs syndrome in Japanese patients with Parkinson's disease. Mov Disord. 2006;2(3);380–4.
14. Iranzo A, Molinuevo JL, Santamaria J, et al. Rapid-eye-movement sleep behaviour disorder as an early marker for a neurodegenerative disorder: a descriptive study. Lancet Neurol. 2006;5;572–577.
15. Britton TC, Chaudhuri KR. REM sleep behavior disorder and the risk of developing Parkinson disease or dementia. Neurology. 2009;72(15);1294-5.
16. Postuma RB, Gagnon JF, Vendette M, Montplaisir JY. Idiopathic REM sleep behavior disorder in the transition to degenerative disease. Mov Disord. 2009;24;2225–2232;72;1294–1295.
17. Schenck CH, Boeve BF, Mahowald MW. Delayed emergence of a parkinsonian disorder or dementia in 81% of older men initially diagnosed with idiopathic rapid eye movement sleep behavior disorder: a 16-year update on a previously reported series. Sleep Med. 2013;14;744–748.
18. Vendette M, Gagnon JF, Decary A, et al. REM sleep behavior disorder predicts cognitive impairment in Parkinson disease without dementia. Neurology. 2007;69;1843–1849.
19. Postuma RB, Bertrand JA, Montplaisir J, et al. Rapid eye movement sleep behavior disorder and risk of dementia in Parkinson's disease: a prospective study. Mov Disord. 2012;27(6);720-726.
20. Vibha D, Shukla G, Goyal V, Singh S, Srivastava AK, Behari M. RBD in Parkinson's disease: a clinical case control study from North India. Clin Neurol Neurosurg. 2011;113(6);472-6.
21. Postuma RB, Gagnon JF, Vendette M, Charland K, Montplaisir J. REM sleep behavior disorder in Parkinson's disease is associated with specific motor features. J Neurol Neurosurg Psychiatry. 2008;79;1117–21.
22. Gjerstad MD, Boeve B, Wentzel-Larsen T, Aarsland D, Larsen JP. Occurrence and clinical correlates of REM sleep behaviour disorder in patients with Parkinson's disease over time. J Neurol Neurosurg Psychiatry. 2008;79;387–91.
23. Wetter TC, Collado-Seidel V, Pollmächer T, Yassouridis A, Trenkwalder C. Sleep and periodic leg movement patterns in drug free patients with Parkinson's disease and multiple system atrophy. Sleep. 2000;23;361–7.
24. Pacchetti C, Manni R, Zangaglia R, et al. Relationship between hallucinations, delusions, and rapid eye movement sleep behavior disorder in Parkinson's disease. Mov Disord. 2005;20;1439–48.

25. Comella CL. Sleep disorders in Parkinson's disease: an overview. Mov Disord. 2007;22(S17);S367–73.
26. Scaglione C, Vignatelli L, Plazzi G, et al. REM sleep behaviour disorder in Parkinson's disease: a questionnaire-based study. Neurol Sci. 2005;25;316–21.
27. Ondo WG. Sleep/wake problems in Parkinson's disease: pathophysiology and clinico-pathologic correlations. J Neural Transm (Vienna). 2014;121(1);S3-13.
28. Calzetti S, Angelini M, Negrotti A, Marchesi E, Goldoni M. A long-term prospective follow-up study of incident RLS in the course of chronic DAergic therapy in newly diagnosed untreated patients with Parkinson's disease. J Neural Transm. 2014;121;499–506.
29. Garcia-Borreguero D, Allen RP, Kohnen R, et al. Diagnostic standards for dopaminergic augmentation of restless legs syndrome: report from a World Association of Sleep Medicine-International Restless Legs Syndrome Study Group consensus conference at the Max Planck Institute. Sleep Med. 2007;8;520–530.
30. Kedia S, Moro E, Tagliati M, Lang AE, Kumar R. Emergence of restless legs syndrome during subthalamic stimulation for Parkinson disease. Neurology. 2004;63;2410–2412.
31. Chahine LM, Ahmed A, Sun Z. Effects of STN DBS for Parkinson's disease on restless legs syndrome and other sleep related measures. Parkinsonism Relat Disord. 2011;17;208–211.
32. Trenkwalder C. Sleep dysfunction in Parkinson's disease. Clin Neurosci. 1998;5;107–114.
33. Wetter TC, Collado-Seidel V, Pollmacher T, Yassouridis A, Trenkwalder C. Sleep and periodic leg movement patterns in drug-free patients with Parkinson's disease and multiple system atrophy. Sleep. 2000;23;361–367.
34. Yong MH, Fook-Chong S, Pavanni R, Lim LL, Tan EK. Case control polysomnographic studies of sleep disorders in Parkinson's disease. PLoS One. 2011;6(7);e22511.
35. Peeraully T, Yong MH, Chokroverty S, Tan EK. Sleep and Parkinson's disease: a review of case-control polysomnography studies. Mov Disord. 2012;27;1729–1737.
36. Covassin N, Neikrug AB, Liu L, et al. Clinical correlates of periodic limb movements in sleep in Parkinson's disease. J Neurol Sci. 2012;316;131–136.
37. Cochen De Cock V, Abouda M, Leu S, et al. Is obstructive sleep apnea a problem in Parkinson's disease? Sleep Med. 2010;11;247- 252.
38. Trotti LM, Bliwise DL. No increased risk of obstructive sleep apnea in Parkinson's disease. Mov Disord. 2010;25;2246-2249.
39. Friedman JH, Millman RP. Sleep disturbances and Parkinson's disease. CNS Spectr. 2008;13;12–17.
40. Shukla G, Kaul B, Gupta A, Goyal V, Behari M. Parkinsonian syndromes presenting with circadian rhythm sleep disorder- advanced sleep-phase type. Natl Med J India. 2015;28(5);233-5.

Chapter 11

Neuropsychiatric Symptoms in Parkinson's Disease

Ashish Sharma, Vinay Goyal

Neuropsychiatric symptoms are common in Parkinson's disease and may surpass motor symptoms as the major factor impacting patient's quality of life. The symptoms may be broadly separated into those associated with the disease process and those that represent adverse effects of treatment. Symptoms attributed to the disease arise from pathologic changes within multiple brain regions and are not restricted to dysfunction in the dopaminergic system. Mood symptoms such as depression, anxiety and apathy are common and may precede the development of motor symptoms by years, while other neuropsychiatric symptoms such as cognitive impairment, dementia and psychosis are more common in later stages of the disease. Neuropsychiatric symptoms attributed to treatment include impulse control disorders, pathologic use of dopaminergic medications and psychosis.

Mood disorders

Mood disorders are major psychiatric disorders in early as well as late stages PD. Common mood disorders in PD are depression, anxiety and abulia or apathy. [1]

Depression

Most common psychiatric disturbance in PD is depression.[2,3] Depression has negative impact on motor disability and decreased quality of life (QoL).[4-6]

Depression may be early symptom of PD or even predate motor symptoms and is present in about 50% patients[2, 3, 6-8] and it varies from 10-20% in many community based and clinic based studies.[7,9,10-11] Despite the high prevalence, however, depression in PD remains undertreated.[7,10,12]

Detecting depression in PD is rewarding, psychomotor slowing, blunted affect and masked facial expression may confuse with bradykinesia in PD, while associated somatic features (decreased appetite, difficulty with concentration and sleep disturbances) help to differentiate.

Common depressive symptoms are sadness, anhedonia and decreased interest in activities. Less common are guilt and feelings of worthlessness.[13] Incidence of suicide is higher than general

DOI: 10.1201/9781003505488-11

population and it remains same even after deep brain stimulation (DBS) (contrary to common belief that DBS increases suicide).[14]

Anxiety

Anxiety is another common psychiatric disturbance in PD,[2] occurring in about 30% patients.[15] Common are generalized anxiety disorder and social phobia.[15-17]

Depression and anxiety are often co-morbid conditions in PD[18] and may be associated with "on-off" fluctuations, with worsened mood and anxiety during "off" periods and with improvement when in the "on" state.[19-21]

Apathy and abulia

Apathy is primary loss of motivation, suggested by diminished speech, motor activity and emotional expression.[22] Abulia is loss of the impulse/motivation to think, speak and act. Diminished motivation is important factor in apathy and abulia. Localization for apathy and abulia is medial frontal lobe and limbic system dysfunction.[23-25] Apathy may be associated with depression.[26,27] It is important to remember that mood is neutral in apathy and negative in depression.[22] In a systematic review and meta-analysis of 23 studies with over 5,300 subjects with PD, the following observations were made:[28]

- Pooled prevalence of apathy was 40 percent, but ranged from 17 to 62 percent due to heterogeneity of studies
- Apathy was associated with older age, cognitive impairment, depression, increased motor symptoms and more severe disability
- Among patients with apathy, co-existing depression was present in 58 percent
- Prevalence of apathy among those without depression or cognitive impairment was 23 percent; this finding supports the view that apathy is a separate clinical feature of PD
- Apathy was less prevalent in patients with higher levodopa equivalent dosage (mainly dopamine agonists), suggesting that dopamine agonists is a possible therapy.

Presence of apathy has a negative impact on QoL in patients with PD.[29]

Cognitive dysfunction and dementia

Cognitive involvement of various degrees is common in PD and it is independent predictor of mortality in PD.[30] Prevalence of dementia in PD is About 40% in community based studies, while its as high as about 80% in longitudinal studies.[31] Predictors of dementia in PD are older age and motor severity.[32] Other risk factors are later age of onset of PD, presence of hallucinations, depressive symptoms and a family history of dementia.[33]

Psychomotor retardation, memory difficulty and altered personality are suggestive of subcortical dementia in PD, mainly affecting executive function (decision making or multi-tasking), memory retrieval and visuospatial misperception. Early executive dysfunction differentiates it from Alzheimer disease (AD).

Pathophysiologic basis of Parkinson disease dementia (PDD) is not fully understood. Neuropathological changes are mixture of cortical neuronal Lewy inclusion bodies filled with alpha synuclein, amyloid plaques and neurofibrillary tangles of AD. Because of the overlapping clinical as well as pathological similarities (Lewy bodies in the limbic cortex and neocortex) between DLB and PD with dementia, many experts believe the two conditions may be two ends of the same disorder.

Clinically, patients are diagnosed with PDD if their illness begins with PD and they develop dementia at least a year after the onset of parkinsonian motor symptoms.

When dementia begins before or concurrently with parkinsonian motor symptoms, DLB is the more likely diagnosis.

Psychosis and hallucinations

Psychosis is common and occurs in one third of drug-treated patients with PD. While the most common is visual hallucinations, other hallucinations (auditory, olfactory and tactile hallucinations) may also be present.[34-37]The prevalence and severity of hallucinations increase over time in patients with PD.[38]

Psychosis is attributable to underlying pathology of DLB, antiparkinsonian drugs, or both. All antiparkinsonian medications have been reported to induce psychosis,[39,40] but some may cause it like DOPA-agonist[41,42] and resolve on discontinuation of drugs.[43] However, duration or dose of anti-parkinsonian drug therapy correlates poorly with the severity of psychosis.[44]

Delusions can also occur in PD, which is mainly paranoid type. Common delusions include spousal infidelity, people stealing money, intruders living in the house, or nurses planning harmful plots; though some patients may retain insight.

Psychosis is also one single predictor for nursing home placement,[45,46] increased caregiver burden[47] and higher mortality.[30,48] Risk factors for psychosis are high doses of antiparkinsonian drugs, dementia, old age, impaired vision, depression, sleep disorders, comorbid disease burden and longer disease duration.[35,44,49,50]

Impulse control disorders (ICD)

ICD is associated with use of DA which includes pathologic gambling, compulsive sexual behaviour or compulsive buying.[51-55] This issue is illustrated by the following reports:

- In a Canadian case series of 297 patients with PD, the lifetime prevalence of pathologic gambling, hypersexuality, or compulsive shopping was 13.7 percent in patients on DAs.[56] Lifetime prevalence of pathologic gambling for all patients and for those receiving any DA was 3.4 and 7.2 percent, respectively, compared with a lifetime prevalence of 1.0 percent in the Ontario population.[57] There was a statistically significant association of pathologic gambling and hyper-sexuality with earlier PD onset and DA therapy.[56,57] In line with earlier retrospective studies,[58, 59] pathologic gambling did not develop in patients receiving levodopa monotherapy.[57]
- Another series of 272 patients with PD found that criteria for ICD, either anytime during the course of PD or currently active, was met by 6.6 and 4.0 percent of patients, respectively.[60] Frequency

of pathologic gambling, compulsive sexual behaviour and compulsive buying anytime during PD were 2.6, 2.6 and 1.5 percent respectively. On multivariate analysis, significant predictors of an active ICD were use of DA and a history of ICD symptoms before the onset of PD.

- Findings from a study that compared recently diagnosed, untreated patients with PD (n=168) with healthy controls (n=143) suggest that untreated PD itself is not associated with increased risk of developing ICD.[61]

Study by Dodd ML et al. suggested that tapering DA resolved ICD in 8 of 11 cases. One had Dopaminergic dysregulation syndrome (DDS) while 6 patients simultaneously developed other inappropriate behaviours such as hyper sexuality.[59]

In a study, amantadine (target dose 100 mg twice daily) resolved pathologic gambling, [62] though few dropped out due to adverse effects, suggesting that it is poorly tolerated in patients with DA induced ICD.

Dopaminergic dysregulation syndrome (DDS) is compulsive use of dopaminergic drugs in spite having good effect or even drug induced dyskinesia. [63]

DDS is more common in males, in early stage of PD, especially when taking higher doses of DA, in spite of severe drug-induced dyskinesia.[63,64] DDS is associated with mood disorder (hypomania or manic psychosis). ICD including hyper sexuality and pathologic gambling may accompany DDS.[63]A form of complex, prolonged, purposeless and stereotyped behaviour called punding may also be associated with DDS.[65]

DDS appears to be uncommon. Incidence of DDS was 3.4 % in a series of 202 patients with PD.[66] DDS is more frequent with DA than levodopa.[66] Risk factors for DDS are younger age at disease onset, higher novelty seeking personality traits, depressive symptoms and alcohol intake.[67]

Management of DDS is simple, but not well established. Preferably use lower doses of DA. Withdrawal of DA in DDS is controlled. Continuous subcutaneous apomorphine infusions may suppress off-period dysphoria where low doses of clozapine or quetiapine is also another alternate option.[67]

Cognitive and behavioural complications of neuromodulation

Most studies have reported only mild or no significant deleterious effects of STN DBS in terms of long-term neuropsychological function and mood, with the exception of declines in tests of verbal fluency.[68-74]However, neuropsychological and psychiatric complications in the immediate postoperative period were not uncommon.[75]

Other studies suggest that STN stimulation may be associated with adverse cognitive, behavioural or psychiatric problems, despite clear improvements in motor disability, activities of daily living and quality of life.[76-79] In one of these reports, deficits in frontal executive function were seen primarily in patients' aged 70 or older.[78] In addition, cognitive and mood disturbances may be more common with STN than GPi stimulation,[80, 81] though here too the evidence is inconsistent. The NSTAPS trial (n=128), found no differences at one year in cognitive, mood and behavioural side effects between

STN DBS and GPi DBS.[74, 88] In contrast, a small trial found evidence of persistent cognitive changes in two of ten patients assigned to STN DBS at 12 months and none of 10 patients assigned to GPi DBS.[80] Other perioperative complications in patients assigned to STN included transient mild delirium in three, transient anxiety in two and hallucinations in one that responded to reduction of the levodopa dose. No perioperative complications in mood or cognition were seen in the patients assigned to GPi stimulation.

The occasional psychiatric reactions seen in patients after STN stimulation have included transient psychosis, hypomania, pathologic gambling and sexual disinhibition.[76,77] These reactions are similar to those seen in dopamine dysregulation syndrome associated with dopamine agonist medications.[81]

The cause of cognitive and behavioural problems in STN stimulation is unclear, but given the small size of the STN nucleus, one plausible mechanism is spread of electrical current from the sensorimotor region of STN into the nearby limbic region of STN.[82]

Risk of suicide

There is serious concern about increasing incidence of suicide after DBS (especially after STN-DBS).[83-86] In an international survey of 55 centers (n= 5311 patients STN-DBS),[86] suicide rate in first postoperative year after DBS was 0.26 percent per year, which was 13 times higher than suicide rate of general population controls (with matched for age, gender and country of origin). The general population comparator was considered reasonable because of evidence that patients with PD may have reduced baseline suicide rate compared with the general population.[87] Risk factors for post-operative suicides were postoperative depression, single marital status and previous history of ICD or DDS.[86] These accounted for 51 percent of the variance for attempted suicide risk. Suicides were also associated with younger age, younger Parkinson disease onset and a previous suicide attempt.

Although an increased risk of suicide associated with DBS treatment is not firmly established,[91] it seems prudent to screen candidates for DBS surgery and exclude those at high risk for suicide.[85,89]

Take home message

- Neuropsychiatric symptoms affect nearly all patients with PD and optimal management of the disease requires careful attention to both motor and non-motor symptoms.
- Patients should be screened for mood disorders, cognitive impairment, psychosis and compulsive behaviours and detailed counselling about the risks of medical and surgical treatments should be provided.
- Screening and counselling for impulsive and compulsive behaviours are particularly important, as many patients minimize these behaviours or may not identify them as adverse affects of their medical treatment.
- Careful adjustment of medications, addition of pharmacologic treatments when needed, counselling and appropriate surgical intervention will enable patients with PD to attain optimal control of a challenging and multifaceted illness.

References

1. Politis M, Wu K, Molloy S, et al. Parkinson's disease symptoms: the patient's perspective. Mov Disord. 2010;25;1646.
2. Aarsland D, Larsen JP, Lim NG, et al. Range of neuropsychiatric disturbances in patients with Parkinson's disease. J Neurol Neurosurg Psychiatry. 1999;67;492.
3. Rai NK, Goyal V, Kumar N, et al. Neuropsychiatric co-morbidities in nondemented Parkinson's disease. Ann Indian Acad Neurol. 2015;18;33.
4. Weintraub D, Moberg PJ, Duda JE, et al. Effect of psychiatric and other non motor symptoms on disability in Parkinson's disease. J Am Geriatr Soc. 2004;52;784.
5. Schrag A, Jahanshahi M, Quinn N. What contributes to quality of life in patients with Parkinson's disease? J Neurol Neurosurg Psychiatry. 2000;69;308.
6. Ravina B, Camicioli R, Como PG, et al. The impact of depressive symptoms in early Parkinson disease. Neurology. 2007;69;342.
7. Tandberg E, Larsen JP, Aarsland D, Cummings JL. The occurrence of depression in Parkinson's disease. A community-based study. Arch Neurol. 1996;53;175.
8. Global Parkinson's Disease Survey Steering Committee. Factors impacting on quality of life in Parkinson's disease: results from an international survey. Mov Disord. 2002;17:60.
9. Leentjens AF, Marinus J, Van Hilten JJ, et al. The contribution of somatic symptoms to the diagnosis of depressive disorder in Parkinson's disease: a discriminant analytic approach. J Neuropsychiatry Clin Neurosci. 2003;15;74.
10. Weintraub D, Moberg PJ, Duda JE, et al. Recognition and treatment of depression in Parkinson's disease. J Geriatr Psychiatry Neurol. 2003;16;178.
11. Starkstein SE, Preziosi TJ, Bolduc PL, Robinson RG. Depression in Parkinson's disease. J Nerv Ment Dis. 1990;178;27.
12. Meara J, Mitchelmore E, Hobson P. Use of the GDS-15 geriatric depression scale as a screening instrument for depressive symptomatology in patients with Parkinson's disease and their carers in the community. Age Ageing. 1999; 28;35.
13. Gotham AM, Brown RG, Marsden CD. Depression in Parkinson's disease: a quantitative and qualitative analysis. J Neurol Neurosurg Psychiatry. 1986;49;381.
14. Stenager EN, Wermuth L, Stenager E, Boldsen J. Suicide in patients with Parkinson's disease. An epidemiological study. Acta Psychiatr Scand. 1994;90;70.
15. Broen MP, Narayen NE, Kuijf ML, et al. Prevalence of anxiety in Parkinson's disease: A systematic review and meta-analysis. Mov Disord. 2016;31;1125.
16. Richard IH. Anxiety disorders in Parkinson's disease. Adv Neurol. 2005;96;42.
17. Nuti A, Ceravolo R, Piccinni A, et al. Psychiatric comorbidity in a population of Parkinson's disease patients. Eur J Neurol. 2004;11;315.
18. Wee N, Kandiah N, Acharyya S, et al. Depression and anxiety are co-morbid but dissociable in mild Parkinson's disease: A prospective longitudinal study of patterns and predictors. Parkinsonism Relat Disord. 2016;23;50.

19. Maricle RA, Nutt JG, Carter JH. Mood and anxiety fluctuation in Parkinson's disease associated with levodopa infusion: preliminary findings. Mov Disord. 1995;10;329.
20. Maricle RA, Nutt JG, Valentine RJ, Carter JH. Dose-response relationship of levodopa with mood and anxiety in fluctuating Parkinson's disease: a double-blind, placebo-controlled study. Neurology. 1995;45;1757.
21. Menza MA, Sage J, Marshall E, et al. Mood changes and "on-off" phenomena in Parkinson's disease. Mov Disord. 1990;5;148.
22. Richard IH. Apathy does not equal depression in Parkinson disease: why we should care. Neurology. 2006;67;10.
23. Reijnders JS, Scholtissen B, Weber WE, et al. Neuroanatomical correlates of apathy in Parkinson's disease: A magnetic resonance imaging study using voxel-based morphometry. Mov Disord. 2010;25;2318.
24. Skidmore FM, Yang M, Baxter L, et al. Apathy, depression, and motor symptoms have distinct and separable resting activity patterns in idiopathic Parkinson disease. Neuroimage. 2013;81;484.
25. Robert G, Le Jeune F, Lozachmeur C, et al. Apathy in patients with Parkinson disease without dementia or depression: a PET study. Neurology. 2012;79;1155.
26. Pedersen KF, Larsen JP, Alves G, Aarsland D. Prevalence and clinical correlates of apathy in Parkinson's disease: a community-based study. Parkinsonism Relat Disord. 2009; 15;295.
27. Dujardin K, Langlois C, Plomhause L, et al. Apathy in untreated early-stage Parkinson disease: relationship with other non-motor symptoms. Mov Disord. 2014;29;1796.
28. Den Brok MG, van Dalen JW, van Gool WA, et al. Apathy in Parkinson's disease: A systematic review and meta-analysis. Mov Disord. 2015;30;759.
29. Barone P, Antonini A, Colosimo C, et al. The PRIAMO study: A multicenter assessment of nonmotor symptoms and their impact on quality of life in Parkinson's disease. Mov Disord. 2009; 24;1641.
30. Forsaa EB, Larsen JP, Wentzel-Larsen T, Alves G. What predicts mortality in Parkinson disease?: a prospective population-based long-term study. Neurology. 2010;75;1270.
31. Aarsland D, Zaccai J, Brayne C. A systematic review of prevalence studies of dementia in Parkinson's disease. Mov Disord. 2005;20;1255.
32. Hobson P, Meara J. Risk and incidence of dementia in a cohort of older subjects with Parkinson's disease in the United Kingdom. Mov Disord. 2004;19;1043.
33. Aarsland D, Andersen K, Larsen JP, et al. Prevalence and characteristics of dementia in Parkinson disease: an 8-year prospective study. Arch Neurol. 2003;60;387.
34. Fénelon G, Mahieux F, Huon R, Ziégler M. Hallucinations in Parkinson's disease: prevalence, phenomenology and risk factors. Brain. 2000;123(4);733.
35. Sanchez-Ramos JR, Ortoll R, Paulson GW. Visual hallucinations associated with Parkinson disease. Arch Neurol. 1996;53;1265.
36. Lee AH, Weintraub D. Psychosis in Parkinson's disease without dementia: common and comorbid with other non-motor symptoms. Mov Disord. 2012;27;858.
37. Chou KL, Messing S, Oakes D, et al. Drug-induced psychosis in Parkinson disease: phenomenology and correlations among psychosis rating instruments. Clin Neuropharmacol. 2005;28;215.

38. Goetz CG, Wuu J, Curgian LM, Leurgans S. Hallucinations and sleep disorders in PD: six-year prospective longitudinal study. Neurology. 2005;64;81.
39. Goetz CG, Tanner CM, Klawans HL. Pharmacology of hallucinations induced by long-term drug therapy. Am J Psychiatry .1982;139;494.
40. Cummings JL. Behavioral complications of drug treatment of Parkinson's disease. J Am Geriatr Soc. 1991;39;708.
41. Parkinson Study Group. Pramipexole vs levodopa as initial treatment for Parkinson disease: A randomized controlled trial. Parkinson Study Group. JAMA. 2000; 284:1931.
42. Rascol O, Brooks DJ, Korczyn AD, et al. A five-year study of the incidence of dyskinesia in patients with early Parkinson's disease who were treated with ropinirole or levodopa. N Engl J Med. 2000;342;1484.
43. Friedman JH. 'Drug holidays' in the treatment of Parkinson's disease. A brief review. Arch Intern Med. 1985;145;913.
44. Aarsland D, Larsen JP, Cummins JL, Laake K. Prevalence and clinical correlates of psychotic symptoms in Parkinson disease: a community-based study. Arch Neurol. 1999; 56;595.
45. Aarsland D, Larsen JP, Tandberg E, Laake K. Predictors of nursing home placement in Parkinson's disease: a population-based, prospective study. J Am Geriatr Soc. 2000;48;938.
46. Goetz CG, Stebbins GT. Risk factors for nursing home placement in advanced Parkinson's disease. Neurology. 1993;43;2227.
47. Aarsland D, Brønnick K, Ehrt U, et al. Neuropsychiatric symptoms in patients with Parkinson's disease and dementia: frequency, profile and associated care giver stress. J Neurol Neurosurg Psychiatry. 2007;78;36.
48. Factor SA, Feustel PJ, Friedman JH, et al. Longitudinal outcome of Parkinson's disease patients with psychosis. Neurology. 2003;60;1756.
49. Arnulf I, Bonnet AM, Damier P, et al. Hallucinations, REM sleep, and Parkinson's disease: a medical hypothesis. Neurology. 2000;55;281.
50. Biglan KM, Holloway RG Jr, McDermott MP, et al. Risk factors for somnolence, edema, and hallucinations in early Parkinson disease. Neurology. 2007;69;187.
51. Voon V, Fox SH. Medication-related impulse control and repetitive behaviors in Parkinson disease. Arch Neurol. 2007; 64;1089.
52. Weintraub D, Koester J, Potenza MN, et al. Impulse control disorders in Parkinson disease: a cross-sectional study of 3090 patients. Arch Neurol. 2010;67;589.
53. Moore TJ, Glenmullen J, Mattison DR. Reports of pathological gambling, hypersexuality, and compulsive shopping associated with dopamine receptor agonist drugs. JAMA Intern Med. 2014;174;1930.
54. Smith KM, Xie SX, Weintraub D. Incident impulse control disorder symptoms and dopamine transporter imaging in Parkinson disease. J Neurol Neurosurg Psychiatry. 2016; 87;864.
55. Sharma A, Goyal V, Behari M, et al. Impulse control disorders and related behaviours (ICD-RBs) in Parkinson's disease patients: Assessment using "Questionnaire for impulsive-compulsive disorders in Parkinson's disease" (QUIP). Ann Indian Acad Neurol. 2015;18;49.

56. Voon V, Hassan K, Zurowski M, et al. Prevalence of repetitive and reward-seeking behaviors in Parkinson disease. Neurology. 2006;67;1254.
57. Voon V, Hassan K, Zurowski M, et al. Prospective prevalence of pathologic gambling and medication association in Parkinson disease. Neurology. 2006;66;1750.
58. Driver-Dunckley E, Samanta J, Stacy M. Pathological gambling associated with dopamine agonist therapy in Parkinson's disease. Neurology. 2003;61;422.
59. Dodd ML, Klos KJ, Bower JH, et al. Pathological gambling caused by drugs used to treat Parkinson disease. Arch Neurol. 2005;62;1377.
60. Weintraub D, Siderowf AD, Potenza MN, et al. Association of dopamine agonist use with impulse control disorders in Parkinson disease. Arch Neurol. 2006;63;969.
61. Weintraub D, Papay K, Siderowf A, Parkinson's Progression Markers Initiative. Screening for impulse control symptoms in patients with de novo Parkinson disease: a case-control study. Neurology. 2013; 80;176.
62. Thomas A, Bonanni L, Gambi F, et al. Pathological gambling in Parkinson disease is reduced by amantadine. Ann Neurol. 2010;68;400.
63. Giovannoni G, O'Sullivan JD, Turner K, et al. Hedonistic homeostatic dysregulation in patients with Parkinson's disease on dopamine replacement therapies. J Neurol Neurosurg Psychiatry. 2000;68;423.
64. Lawrence AD, Evans AH, Lees AJ. Compulsive use of dopamine replacement therapy in Parkinson's disease: reward systems gone awry? Lancet Neurol. 2003;2;595.
65. Evans AH, Katzenschlager R, Paviour D, et al. Punding in Parkinson's disease: its relation to the dopamine dysregulation syndrome. Mov Disord. 2004;19;97.
66. Pezzella FR, Colosimo C, Vanacore N, et al. Prevalence and clinical features of hedonistic homeostatic dysregulation in Parkinson's disease. Mov Disord. 2005;20:77.
67. Evans AH, Lawrence AD, Potts J, et al. Factors influencing susceptibility to compulsive dopaminergic drug use in Parkinson disease. Neurology. 2005;65;1570.
68. Ardouin C, Pillon B, Peiffer E, et al. Bilateral subthalamic or pallidal stimulation for Parkinson's disease affects neither memory nor executive functions: a consecutive series of 62 patients. Ann Neurol. 1999;46;217.
69. Woods SP, Fields JA, Tröster AI. Neuropsychological sequelae of subthalamic nucleus deep brain stimulation in Parkinson's disease: a critical review. Neuropsychol Rev. 2002;12;111.
70. Daniele A, Albanese A, Contarino MF, et al. Cognitive and behavioural effects of chronic stimulation of the subthalamic nucleus in patients with Parkinson's disease. J Neurol Neurosurg Psychiatry. 2003;74;175.
71. Funkiewiez A, Ardouin C, Caputo E, et al. Long term effects of bilateral subthalamic nucleus stimulation on cognitive function, mood, and behaviour in Parkinson's disease. J Neurol Neurosurg Psychiatry. 2004;75;834.
72. Parsons TD, Rogers SA, Braaten AJ, et al. Cognitive sequelae of subthalamic nucleus deep brain stimulation in Parkinson's disease: a meta-analysis. Lancet Neurol. 2006;5;578.
73. Witt K, Daniels C, Reiff J, et al. Neuropsychological and psychiatric changes after deep brain stimulation for Parkinson's disease: a randomised, multicentre study. Lancet Neurol. 2008;7;605.

74. Odekerken VJ, Boel JA, Geurtsen GJ, et al. Neuropsychological outcome after deep brain stimulation for Parkinson disease. Neurology. 2015;84;1355.
75. Morrison CE, Borod JC, Perrine K, et al. Neuropsychological functioning following bilateral subthalamic nucleus stimulation in Parkinson's disease. Arch Clin Neuropsychol. 2004;19;165.
76. Smeding HM, Speelman JD, Koning-Haanstra M, et al. Neuropsychological effects of bilateral STN stimulation in Parkinson disease: a controlled study. Neurology. 2006;66;1830.
77. Schüpbach M, Gargiulo M, Welter ML, et al. Neurosurgery in Parkinson disease: a distressed mind in a repaired body? Neurology. 2006;66;1811.
78. Saint-Cyr JA, Trépanier LL, Kumar R, et al. Neuropsychological consequences of chronic bilateral stimulation of the subthalamic nucleus in Parkinson's disease. Brain. 2000; 123(10);2091.
79. Rothlind JC, York MK, Carlson K, et al. Neuropsychological changes following deep brain stimulation surgery for Parkinson's disease: comparisons of treatment at pallidal and subthalamic targets versus best medical therapy. J Neurol Neurosurg Psychiatry. 2015; 86:622.
80. Anderson VC, Burchiel KJ, Hogarth P, et al. Pallidal vs subthalamic nucleus deep brain stimulation in Parkinson disease. Arch Neurol. 2005;62;554.
81. Saint-Cyr JA, Albanese A. STN DBS in PD: selection criteria for surgery should include cognitive and psychiatric factors. Neurology. 2006;66;1799.
82. Okun MS, Foote KD. Subthalamic nucleus vs globus pallidus interna deep brain stimulation, the rematch: will pallidal deep brain stimulation make a triumphant return? Arch Neurol. 2005;62;533.
83. Doshi PK, Chhaya N, Bhatt MH. Depression leading to attempted suicide after bilateral subthalamic nucleus stimulation for Parkinson's disease. Mov Disord. 2002;17;1084.
84. Soulas T, Gurruchaga JM, Palfi S, et al. Attempted and completed suicides after subthalamic nucleus stimulation for Parkinson's disease. J Neurol Neurosurg Psychiatry. 2008;79;952.
85. Burkhard PR, Vingerhoets FJ, Berney A, et al. Suicide after successful deep brain stimulation for movement disorders. Neurology. 2004;63;2170.
86. Voon V, Krack P, Lang AE, et al. A multicentre study on suicide outcomes following subthalamic stimulation for Parkinson's disease. Brain. 2008;131;2720.
87. Myslobodsky M, Lalonde FM, Hicks L. Are patients with Parkinson's disease suicidal? J Geriatr Psychiatry Neurol. 2001;14;120.
88. Odekerken VJ, van Laar T, Staal MJ, et al. Subthalamic nucleus versus globus pallidus bilateral deep brain stimulation for advanced Parkinson's disease (NSTAPS study): a randomised controlled trial. Lancet Neurol. 2013;12;37.
89. Schuepbach WM, Rau J, Knudsen K, et al. Neurostimulation for Parkinson's disease with early motor complications. N Engl J Med. 2013;368;610.
90. Rodriguez-Oroz MC, Obeso JA, Lang AE, et al. Bilateral deep brain stimulation in Parkinson's disease: a multicentre study with 4 years follow-up. Brain 2005; 128:2240.
91. Weintraub D, Duda JE, Carlson K, et al. Suicide ideation and behaviours after STN and GPi DBS surgery for Parkinson's disease: results from a randomised, controlled trial. J Neurol Neurosurg Psychiatry 2013; 84:1113

Chapter 12

Memory and Cognitive Impairment in Parkinson's Disease

Sumit Singh, Roopam Singh

Parkinson's disease (PD) is a neurodegenerative condition which is characterized by asymmetrical, slowly progressive akinetic rigid syndrome with tremors. Strictly speaking, cognitive impairment or memory problems is not a feature of Idiopathic Parkinson's disease (IPD). In fact, dementia within the first few years of onset of PD is one of the exclusion criteria in diagnosis.[1] Over the last few decades, there have been several studies where patients with IPD have been demonstrated to have significant cognitive deficits. To further complicate the issue, it has been reported that dopaminergic drugs are associated with either induction or worsening of cognition in PD patients. The instruments used for the evaluation of cognitive functions in PD are not standardized, lack homogeneity or reproducibility for evaluating patients in different stages of PD and different studies have used different criteria for their evaluation of cognition in PD. This leads to a wide variation in the test results. The present communication attempts to evaluate the magnitude and the spectrum of memory and cognitive impairment in patients of PD in a manner that is evidence based.

Table 1: Types of cognitive dysfunctions seen in Parkinson's disease

S.No.	Type of Cognitive Dysfunction
1	Executive function impairment
2	Memory deficits
3	Bradyphrenia
4	Confusion
5	Minimal cognitive impairment
6	Dementia

DOI: 10.1201/9781003505488-12

The neuro-patho-chemical substrate for cognition in PD

Autopsy studies evaluating selective brain atrophy in patients with dementia with Lewy bodies (DLB) and IPD have exhibited distinctive patterns of cortical atrophy,[2] depending on the duration of the disease and clinical features. Mild cognitive impairment of PD patients might be because of simultaneous non-epinephrine and cholinergic deficiencies observed in the neuronal pathways of these patients. The anatomical substrate for cognitive impairment in PD might be because of deafferentation of frontotemporal cortices due to prolonged dopaminergic deficiency. The medial temporal lobe atrophy (MTA) on MRI seen in patients with PD is similar to Alzheimer's disease (AD). Frontal lobar atrophy has also been observed in patients with PD associated with dementia (PDD) and a more profound loss was observed in patients with DLB. The dementia seen in PD is similar to that seen in patients with DLB. The executive dysfunction in patients with PD is probably a result of nigrostriatal dysfunction, resulting in denervation of the mesocortical fibers from the frontal lobes. Several studies have demonstrated a correlation between dementia and Alzheimer type pathology of PD patients, as there is a marked increase in the distribution of Lewy bodies, resulting in cholinergic deficits.[3]

Spectrum of cognitive and memory impairment in PD

Epidemiology

Cognitive impairment is common in patients of PD on detailed clinical evaluation with reference to working memory and executive functions.[4,5,6] Various studies report that mild cognitive impairment (MCI) is present in upto 40% of patients with PD.[7] MCI is defined as a cognitive decline with a level of performance 1-2 SD below the mean for the study cohort in one or more cognitive domains. It has been observed that PD with MCI (PD MCI) has a greater risk of developing dementia as compared to PD with normal cognition (PD NC).[8] In a cohort follow up study, where the duration of the study was for eight years, it was found that up to 62% of patients in the PD MCI group progressed to dementia over a period of four years as compared to 20% in the PD patients with normal cognition.[9] Various observational studies have demonstrated executive dysfunction in up to 15% of patients with PD,[10] however, these are all clinic based studies. Management with dopaminergic drugs might have variable effect on the executive functions with reports of improvement, no change or even deterioration in these functions after initiation of treatment in dopamine naïve patients. Data is sparse in terms of population based studies from India to evaluate cognitive deficits in patients of PD. Case control studies from India which were hospital based have demonstrated that PDD is seen in 27% of PD patients and in 36% of PD MCI patients. On follow up, over a period of seven years, 49.9% of these patients eventually developed PDD.[10]

Mild cognitive impairment (MCI)

MCI implies a deterioration in cognition which is not normal or commensurate with the age of the patient, but not severe enough to be termed as dementia. MCI in PD patients itself is an independent predictor of poor quality of life. Many authorities feel that this is a stage of transition between normal cognition and dementia. There has been a lack of agreement on the proposed diagnostic criteria for mild cognitive impairment in PD. The movement disorders society task force has put forward two levels of diagnostic criteria for MCI in PD (Table 2).[11] It needs to be understood that these evaluation parameters are time consuming, require specialized personnel to perform them and lack homogeneity and reproducibility. A tool to evaluate MCI in PD, which less time consuming and can be done with ease by clinicians even in a busy out patients clinic is still needed. Two criteria have been used in patients to evaluate MCI in PD. In a prospective study, 43 patients were evaluated by this tool. PD MCI was found in 23.3% of the patients using the brief assessment (criteria Level 1) and 60.5% of the patients had PD MCI using the comprehensive assessment tools (criteria Level 2). On follow-up, 42.3% of patients with PD MCI developed dementia. Other studies have reported that 47% of patients develop cognitive impairment within two to six years of onset of PD, and all patients with MCI finally develop dementia within five years of onset of MCI. Severity of the disease, as reflected by Hoehn and Yahr's stage, and the education level of the patients were the two risk factors for developing dementia.[12]

The MCI in Parkinson's disease encompasses both non-amnestic and amnestic domains, and may involve single domain or multiple domains. Even non demented patients complain of difficulty in multitasking, planning, word finding deficits and in word processing speed. MCI in PD usually involves amnestic domain in single subtype with the executive functions being most commonly affected. The incidence of either type of PD MCI varies according to the population studied, the neuropsychological tests used and the normative data for the population. PD MCI of the non-amnestic single domain type usually have deficits in memory, attention, psychomotor speed, executive function and visuospatial skills. In the CamPaIGN study, of the 58% with single domain MCI, 34% had frontostriatal deficits, while deficits of the temporal lobe were seen in 24% patients. Multiple-domain non amnestic MCI was seen in 42%.[13] Muslimovic et al.[14] proposed that PD MCI occurred in about 24% patients having deficits in executive function, memory and psychomotor speed evaluation. Various studies have proved that amnestic MCI was seen less commonly in PD as compared to non-amnestic MCI. Caviness et al.[15] reported that single domain affection was observed in 67% of their study cohort. Executive and frontal functions were most affected. In their study, the distribution of MCI was non-amnestic single domain>non-amnestic multiple domain=amnestic single domain>amnestic multiple domain. Other studies have also reported that single domain non-amnestic MCI was most likely (62% in their study group) as compared to other subtypes of the PD MCI.[16] The study group was small and the neuropsychological evaluation was in a non-standardized manner. In some studies, male sex, associated depression, advanced PD, onset of PD at an older age and worse motor symptoms have been found associated with PD MCI, whereas other studies refute these observations.

The diagnosis of PD MCI is essentially clinical with patients exhibiting impairment of cognition in absence of dementia. Various neuropsychological tests have been used to evaluate patients to detect PD MCI. There are no standardized tests and the incidence of PD MCI changes depending upon the "cut-off" used in each of these tests. It is obvious that if 2 SD below the mean for the population is taken as cut-off, the incidence of PD MCI will be lower than if the cut off is taken at 1.5 SD below the normal. Most of the tests that have been proposed as screening tests for PD-MCI and/or PDD, like the PD Cognitive Rating Scale (PD CRS), Parkinson Neuropsychometric Dementia Assessment (PANDA), MoCA, SCOPA-COG and Mattis DRS among many others take about 10 to 30 minutes to administer.[17] Imaging of brain including PET scan have little role in routine evaluation of patients with PD MCI and are still under evaluation. Positron emission tomography (PET) studies using FDG glucose shows decreased FDG uptake in the prefrontal area and posterior parietal lobes – both in the single and multiple domain type PD MCI. Various biomarkers, especially genetics, CSF and neuroimaging are also currently under evaluation to help in diagnosis of PD MCI.

Clinical trials evaluating therapeutic options are limited for PD MCI. Acetyl Cholinesterase inhibitors (AChEI) like rivastigmine and memantine have not been evaluated in PD-MCI. Rasagline has been evaluated in 48 non-demented PD patients in a randomized, double blind, placebo controlled, multi centre, open label study on cognitive dysfunction. These patients had 1.5 SD impairment in their performance on executive functions and memory tests, but could not fulfil the DSM IV criteria for dementia. There was significant improvement in a few cognitive tasks which were relevant to executive function and attention in the intervention group, in contrast to the placebo group. Atomoxetine, a selective norepinephrine reuptake inhibitor used for attention deficit hyperactivity disorder (ADHD), was evaluated in 12 PD patients with executive dysfunction for a period of eight weeks. It improved scores in 75% of the patients on the Clinical Global Impression-Change Scale and on behavioural measures of executive dysfunction. None of these drugs are FDA approved for the management of PD MCI. Apathy, depression, anxiety and sleep dysfunction which are often associated with PD, need to be treated independently. Drugs, which impair cognition and are often prescribed in PD for motor symptoms or non-motor symptoms, need to be administered with caution in patients with MCI. These include anticholinergics like benzhexol and trihexyphenidyl, and alpha-blockers and antipsychotics.[18]

Executive function deficits and PD

Cognitive functions which cater to decision making capacity of a person in his daily life are known as executive functions. These include abstract thinking, planning, monitoring, sequencing of thoughts, and retrieval of recent memory. Dysfunction of executive functions have been reported in PD in various studies. Wisconsin Card Sorting Test (WCST), Reiten's Trail Making Test (RTMT), Tower of London test (TOL) task, Raven's progressive matiricies and Form Sequence Learning test are the various neuropsychological tests which have been utilized for assessment of executive functions in PD patients. Executive Dysfunction (ED) has been noted in the early stages of PD based on the performance in TOL task, trail making tests, and Stroop tests. As the disease progresses deterioration is observed in almost all the neuropsychological tests. A neuropsychological battery called as the Behavioural Assessment of the Dysexecutive Syndrome (BADS) has been reported to be the most sensitive for evaluation of impairment of executive function in PD patients.[19] Patients who have advanced disease tend to have a more severe ED. Lower intelligence, old age and bradykinetic PD patients are other clinical variables associated with ED as compared to their counterparts.[20] A metaanalysis (that included 33 studies) a naive, non-demented patients of PD, demonstrated unequically abnormalities in WCST, WF, Stroop test and trail making. Neuropsychological profiling of Indian patients in small samples have exhibited ED in PD patients similar to those in other similar studies.[20] There are various hypotheses about the pathogenesis of ED in PD patients. One of these is the "dopamine toxicity" theory. This concept proposes that in the initial stages of PD, the ventral striatum remains relatively "dopamine preserved", even if these patients are having motor deficits. Exposing these patients to dopaminergic drugs in therapeutic doses might alleviate the motor symptoms in PD, but at the cost of executive functions due to relative toxic effect known as "dopamine overdose" to ventral striatum fibres. These fibres have projection to the frontal lobes apart from other areas of the brain. Selective overstimulation of these frontal fibres, is proposed to result in executive dysfunction in PD patients. This theory, however, fails to explain the presence of ED in treatment of naive non-demented PD patients.

Apart from a detailed neuropsychological evaluation, there are no specific tests which can diagnose the presence of ED.

Memory dysfunction and dementia in PD

PD patients have more significant deterioration of short term memory as compared to the long term memory which remains intact. In various studies, visual memory was found to be more affected than the verbal short term memory, and this correlates with the UPDRS scores of patients with higher scores having a more profound defect. Visual, spatial and verbal memory have all been found to be impaired in advanced disease. Tests used for evaluation of the visual memory in PD include the Benton's visual retention test, Reiten's Progressive matrices, Form Sequence learning test, etc. Those used for the verbal memory are the Rey Auditory Verbal Learning Test (RAVLT), Digit Span Test and Pattern Recognition Tests. There have been studies which have evaluated the impact of Levodopa on the memory in "dopamine naive" PD patients with conflicting reports. In case control studies early PD patients have been reported to have deterioration in the visual memory with relatively preserved verbal memory, after treatment with levodopa is initiated.[22] It is difficult to find isolated memory impairment in PD patients with preservation of other domains of cognition. Almost always memory impairment is part of MCI or PDD in PD patients.

Unlike other cognitive dysfunctions in PD, dementia is fairly well studied and evaluated. In various studies, PDD has been reported to have prevalence of 20% in cross-sectional studies and 80% in longitudinal studies depending upon the definition used to describe dementia.[23] Development of dementia in PD results in deterioration in the patients' quality of life and causes them to become more dependent in activities of daily living. Coexistent depression, anxiety and behavioural dysfunction worsen dementia and the disability arising out of it. As a result of dementia, these patients need more detailed evaluation, closer follow up and intensive treatment. It has been found that the age of the patient is associated with a higher chance of having dementia in PD, regardless of motor disability, while other studies have shown PDD correlates with the severity of disease, male gender, low education level, visual hallucinations, axial bradykinesia, REM sleep behavioural disorder, autonomic dysfunction, excessive daytime sleepiness, apart from the age.[24] Older people with advanced disease have 12 times increased chances of developing dementia than younger patients with milder disease. Genetic factors have been proposed to be associated with PDD. Mutations in apolipoproteinE ε4 (ApoE ε4) allele are associated with a higher risk of PDD. However, ApoE ε2 gene has been found to be protective for PDD. Similarly, H1 haplotype of the microtubule associated protein tau (MAPT) gene, heterozygous mutations in the glucocerebrosidase gene and duplication of the α-synuclein gene have all been reported to increase the risk of dementia as compared to those without this genetic constitution.

The clinical features of PDD are those of a classical subcortical dementia. Most prominent defects are observed in the executive functions. There is deterioration of short term memory with relative preservation of language function. As compared to AD, patients with PDD have a more marked deterioration of attention and visuospatial deficits. Memory dysfunction in PPD, unlike AD is more in free recall of memory as compared to recognition of past events. Various behavioural problems are seen in PDD, and at least one of them might be reported by about 70-80 % of patients. These can vary from apathy, withdrawal, agitation to frank delusions and hallucinations. Hallucinations are usually formed and paranoid delusions are common.[25] Histopathology of the brain of PDD resembles that seen in AD and diffuse Lewy body disease. There is spreading of the Lewy bodies to areas beyond the brainstem – as is seen in PD without dementia – to the neocortex. The nucleus basalis of Maynert exhibits the Lewy bodies in abundance. There is presence of neurofibrillary plaques across the neocortex. Major neurotransmitter found to be deficient in PDD is acetylcholine, along with a minor depletion of serotoninergic and noradrenergic activity.[26]

There are no reliable neurochemical or radiological markers which confirm the diagnosis of PDD. MRI shows cortical atrophy which is similar to AD and involves temporoparietal lobes. PET scan shows hypo metabolism in similar areas, along with severe depletion of cholinergic activity in frontotemporoparietal region of brain using MP4A PET. None of the biomarkers or the imaging modalities are diagnostic of PDD, and therefore the diagnosis remains essentially clinical. Prior to the MDS Task Force criteria, the diagnosis of PDD was based on the DSM IV – R criteria. These criteria lacked the specificity for PDD.

The MDS Task force has outlined two sets of diagnostic criteria for PDD [Table 3]. One of these is a simple algorithm to diagnose PDD in a clinical setting. The other entails a more detailed and thorough evaluation of PDD patients and can be used for the clinical trials and for research purposes. The fundamental step in the diagnosis of PDD is the clinical diagnosis of PD, the clinical diagnosis of dementia and the development of dementia after the diagnosis of PD. Presence of associated behavioural features like apathy, delusions, hallucinations and depression along with neuropsychiatric manifestations involving attention, executive functions, visuospatial skills and memory in various combinations classifies a patient as having "probable" or "possible PDD". Presence of any other disorder which is known to cause dementia, confusion or altered sensorium is taken as an exclusion criterion for PDD.[27]

Apart from neurobehavioral therapy and psychosocial support, various drugs are available for the management of PDD. Cognitive training –involving training of individuals with PDD on specific cognitive tasks and cognitive stimulation which involves the patients in various tasks intending to stimulate their cognitive skills have both been evaluated in PD. Systematic reviews have reported positive impact of cognitive stimulation in PDD patients in their overall ADAS – cog scores. Physical exercise of all types has been associated with significant positive impact, i.e., slowing down the cognitive deterioration in AD as well as in patients with MCI [31]. There are no studies of such type in PD patients, but it is worth encouraging patients with PDD for an active exercise schedule.

Depression, mood changes and hallucinations are very common in PDD and need to be tackled accordingly. Most patients with depression respond well to selective serotonin reuptake inhibitors like escitalopram or paroxetine. Hallucinations respond well to rivastigmine. More aggressive hallucinations might need quetiapine or clozapine. Pharmacological interventions do not result in cure for PDD. Neuropsychological parameters have shown favourable response to available medications.[28] In a Cochrane database systematic review, six randomized trials were evaluated, where 1236 participants were randomized for evaluation of the role of acetylcholine esterase inhibitors (AChEI) in PDD. For global assessment of clinical condition of the PDD patients, three of these trials evaluated the difference in the Alzheimer's Disease Cooperative Study-Clinical Global Impression of Change (ADCS-CGIC) score of - 0.38, favouring the Acetyl Cholinesterase inhibitors as compared to placebo. About 20% patients had a clinically meaningful recovery with AChEI, in contrast to 14% patients in the placebo group. There were observed treatment benefits in patients treated with AChEI on a composite cognitive score, and the MMSE and the behavioural disturbance. The ADCS-ADL and the Unified Parkinson's Disease Rating Scale (UPDRS)-ADL rating sub scales showed better response with cholinesterase inhibitors as well. The side effect profile of Donepezil and Rivastigmine were more in the treatment arm as expected with more side effects observed in Rivastigmine group compared to the Donepezil group. Memantine has been evaluated in trials for PDD without any favourable response. Currently only Rivastigmine – capsules or transdermal patch are FDA approved for treatment of PDD.[29]

Effect of deep brain stimulation for PD or PDD

Deep brain stimulation is an established treatment for PD, and there is data which is evolving where DBS may be offered to patients with PD early on during their disease. While the motor symptoms of PD have shown a major improvement after DBS, the non-motor symptoms might not show the response. It has been seen that post DBS, the patients might not show improvement in their cognitive deficits, and there may be an appearance of new deficits post DBS. These include apathy, depression, anxiety, impulsivity and alienation of relationships. Patients develop increased goal directed activity, an elevated mood and racing of their thoughts.

Role of DBS on impulse control disorders (ICD) is debatable. While some studies indicate that ICD might improve after STN DBS, others have failed to reproduce the results, and report worsening of these disorders. Patients who have pre-existing cognitive impairment are unlikely to show improvement in their cognitive status, and in fact deterioration might be observed in some of the cognitive faculties. Frank dementia is a clear contraindication for DBS, and patient who are above the age of 70 are more likely to have post-operative cognitive impairment. Post-operative cognitive dysfunction is more likely in patients with axial disease, high UPDRS scores and a high dopaminergic requirement pre-operatively. Hallucinations and delusions improve in most patients after STN DBS, consequent to the reduction in the dopaminergic drugs. The targets used for DBS have been compared for their effect on cognition. A recent meta-analysis indicated that PD patients undergoing STN DBS exhibit minor deterioration in memory, psychomotor speed, executive functions, attention and overall cognition apart from deficits in phonemic and semantic fluency. On the other hand, GPi-DBS led to fewer neurocognitive declines as compared to STN-DBS. GPi-DBS and STN-DBS were comparable in terms of having lower levels of depression after surgery.[30]

Table 2: Diagnostic criteria for mild cognitive impairment in Parkinson's disease

I. Inclusion criteria **Diagnosis of Parkinson's as per UK PD Brain Bank Criteria** Gradual decline in patients with established PD, in cognitive ability reported by either the patient or informant/observed by the clinician Cognitive deficits on either formal neuropsychological testing or a scale of global cognitive abilities Cognitive deficits are not sufficient to interfere with functional independence, subtle difficulties on complex functional tasks may be present
II. Exclusion criteria **Diagnosis of PDD as per the MDS Task Force proposed criteria** Other causes for cognitive impairment (e.g., stroke, delirium, major depression, adverse effects of medication) Other PD-associated comorbid conditions (e.g., motor impairment or depression, severe anxiety, excessive daytime sleepiness, or psychosis) which might influence cognitive testing
III. Specific guidelines for PD-MCI level I and level II categories **A. Level I (abbreviated assessment)** Impairment on a scale of global cognitive abilities validated for use in PD on two or more tests, on a prescribed neuropsychological battery (battery should include less than two tests within each of the five cognitive domains, or less than five cognitive domains are assessed) **B. Level II (comprehensive assessment)** Neuropsychological testing that includes two tests within each of the five cognitive domains (i.e., attention and working memory, executive, language, memory, and visuospatial) Impairment on at least two neuropsychological tests, represented by either two impaired tests in one cognitive domain or one impaired test in two different cognitive domains Impairment on neuropsychological tests may be demonstrated by: o Performance approximately 1 to 2 SD below appropriate norms or o Significant decline demonstrated on testing over a period of time on follow up or o Significant decline from estimated premorbid levels
IV. Subtype classification for PD-MCI (requires two tests for each of the five cognitive domains assessed and is strongly suggested for research purposes) PD-MCI single-domain—abnormalities on two tests within a single cognitive domain (specify the domain), with other domains unimpaired or PD-MCI multiple-domain—abnormalities on at least one test in two or more cognitive domains (specify domains).

Table 3: MDS Task force criteria for diagnosis of PDD in a clinical setting

I. For a diagnosis of PDD, two of the following must be present:	(i) a diagnosis of PD according to UKPDS diagnostic criteria (ii) PD developed before the documentation of dementia DD to differentiate from DLBD: in PDD, motor symptoms develop at least one year before development of dementia, while in DLBD, the motor symptoms occur within the first year prior to the onset of dementia and frequently after the onset of dementia.
II. Dementia, being defined as:	(i) impairment in at least two cognitive domains and (ii) cognitive deficiency severe enough to impair daily life (social, occupational or personal care) that must be independent of impairment because of PD motor symptoms. The MDS Task Force recommended that the Mini-Mental State Examination be used for assessment of Dementia
I. Core features	Diagnosis of PD + Dementia syndrome
II. Associated clinical features	Impairment of at least two of four cognitive domains (May be supported by behavioural symptoms)
III. Presence of features which make diagnosis uncertain	Cognitive and behavioural symptoms presenting as a result of other conditions, for example: • Co-existence of any abnormality that could itself cause cognitive impairment, but not cause dementia • Unknown time interval between onset of motor and cognitive symptoms
IV. Presence of features which make diagnosis impossible	• Acute confusion due to systemic diseases/ abnormalities or drug intoxication • Major depression according to DSM IV • Features of probable vascular dementia according to NINDS-AIREN

PROBABLE PDD – I +II, POSSIBLE PDD – I+II+III, IMPOSSIBLE FOR PDD – I+II+IV

References

1. Rizzo G, Copetti M, Arcuti S, Martino D, Fontana A, Logroscino G. Accuracy of clinical diagnosis of Parkinson disease: A systematic review and meta-analysis. Neurology. 2016;86(6);566-76.
2. Double KL, Halliday GM, McRitchie DA, Reid WG, Hely MA, Morris JG. Regional brain atrophy in idiopathic Parkinson's disease and diffuse Lewy body disease. Dementia. 1996;7(6);304-13.
3. Seishin Shinkeigaku Zasshi, Tachibana H[1]. Cognitive impairment in Parkinson's disease. 2013;115(11);1142-9
4. SJG Lewis, A Dove, TW Robbins, RA Barker, AM Owen. Cognitive impairments in early Parkinson's disease are accompanied by reductions in activity in frontostriatal neural circuitry. The Journal of Neuroscience. 2003;23(15);6351–6356.
5. Monchi O, Petrides M, Mejia-Constain B, Strafella AP. Cortical activity in Parkinson's disease during executive processing depends on striatal involvement. Brain. 2007;130(1);233–244.
6. Owen AM, Aarsland D, Kurz MW. Cognitive dysfunction in Parkinson's disease: the role of frontostriatal circuitry. Neuroscientist. 2004;10(6);525–537.

7. Litvan I, Goldman JG, Troster AI, et al. Diagnostic criteria for mild cognitive impairment in Parkinson's disease: movement disorder society task force guidelines. Movement Disorders. 2012;27(3);349–356.
8. Aarsland D, Andersen K, Larsen JP, Lolk A, Kragh Sorensen P. Prevalence and characteristics of dementia in Parkinson disease: an 8-year prospective study. Archives of Neurology. 2003;60(3);387–392.
9. Williams-Gray CH, Evans JR, Goris A et al. The distinct cognitive syndromes of Parkinson's disease – 5 year follow-up of the CamPaIGN cohort. Brain 2009;132(11);2958–2969.
10. Sanyal J, Banerjee TK, Rao VR. Dementia and cognitive impairment in patients with Parkinson's disease from India: a 7- year prospective study. Am J Alzheimers Dis Other Demen. 2014;29(7);630-6.
11. Ding W, Ding LJ, Li FF, Han Y, Mu L. Neurodegeneration and cognition in Parkinson's disease: a review European Review for Medical and Pharmacological Sciences. 2015;19;2275-2281
12. Galtier I1, Nieto A1, Lorenzo JN2, Barroso J1. Mild cognitive impairment in Parkinson's disease: Diagnosis and progression to dementia. J Clin Exp Neuropsychol. 2016;38(1);40-50.
13. Foltynie T, Brayne CE, Robbins TW, Barker RA. The cognitive ability of an incident cohort of Parkinson's patients in the UK. The CamPaIGN study. Brain. 2004;127(3);550–60.
14. Muslimovic D, Post B, Speelman JD, Schmand B. Cognitive profile of patients with newly diagnosed Parkinson disease. Neurology. 2005;65(8);1239–45.
15. Caviness JN, Driver-Dunckley E, Connor DJ, Sabbagh MN, Hentz JG, Noble B, et al. Defining mild cognitive impairment in Parkinson's disease. Mov Disord. 2007;22(9);1272–7.
16. Aarsland D, Bronnick K, Janvin C, et al. Frequency and profile of mild cognitive impairment (MCI) in Parkinson's disease: a multicentre meta-analysis of neuropsychological data. Movement Disorders. 2009;24(1).
17. Litvan I, Goldman JG, Troster AI, Schmand BA, Weintraub D, Petersen RC, Mollenhauen B, Adler CH, Marder K, Williams-Gray CH, Aarsland D, KulisewskyJ, Rodriguez-Oroz MC, Burn DJ, Barker RA, Emre M. Diagnostic criteria for mild cognitive impairment in Parkinson's disease: Movement Disorder Society Task Force guidelines. Mov Disord. 2012;27;349-356.
18. Marsh L, Biglan K, Gerstenhaber M, Williams JR. Atomoxetine for the treatment of executive dysfunction in Parkinson's disease: a pilot open-label study. Mov Disord. 2009;24(2);277–82.
19. Bennett PC, Ong B, Ponsford J. Assessment of executive dysfunction following traumatic brain injury: comparison of the BADS with other clinical neuropsychological measures. J Int Neuropsychol Soc. 2005;11;606-613.
20. Pattern of cognitive impairment in patients with Parkinson's disease and psychosis: A critical review. Lenka A, Hegde S, Arumugham SS, Pal PK. Parkinsonism Relat Disord. 2016;S1820-31.
21. Georg Dirnberger, Marjan Jahanshahi. Executive dysfunction in Parkinson's disease: A review. Journal of Neuropsychology. 2013;7;193–224.
22. Verbal and visual memory in patients with early Parkinson's disease: effect of levodopa. Singh S, Behari M. Neurol India. 2006;54(1);33-7.

23. Hely MA, Reid WG, Adena MA, Halliday GM, Morris JG. The Sydney multicenter study of Parkinson's disease: the inevitability of dementia at 20 years. Mov. Disord. 2008;23;837-844.

24. Anang JB, Gagnon JF, Bertrand JA, Romenets SR, Latreille V, Panisset M, et al. Predictors of dementia in Parkinson disease: a prospective cohort study, Neurology. 2014;83(14);1253–1260.

25. Aarsland D, Brønnick K, Ehrt U, De Deyn PP, Tekin S, Emre M, et al. Neuropsychiatric symptoms in patients with Parkinson's disease and dementia: frequency, profile and associated care giver stress, J. Neurol. Neurosurg. Psychiatry. 2007;78(1);36–42.

26. Horvath J, Herrmann FR, Burkhard PR, Bouras C, Kövari E, Neuropathology of dementia in a large cohort of patients with Parkinson's disease, Parkinsonism Relat. Disord. 2013;19(10);864–868.

27. Dubois B, Burn D, Goetz C, Aarsland D, Brown RG, Broe GA, et al. Diagnostic procedures for Parkinson's disease dementia: recommendations from the movement disorder society task force, Mov. Disord. 2007;22(16);2314–2324.

28. Jennifer YY. Szeto and Simon J.G. Lewis Current Treatment Options for Alzheimer's Disease and Parkinson's Disease Dementia Current Neuropharmacology. 2016;14;326-338.

29. Wang HF, Yu JT, Tang SW, Jiang T, Tan CC, Meng XF, Wang C, Tan MS, Tan L. Efficacy and safety of cholinesterase inhibitors and memantine in cognitive impairment in Parkinson's disease, Parkinson's disease dementia, and dementia with Lewy bodies: systematic review with meta-analysis and trial sequential analysis. J Neurol Neurosurg Psychiatry. 2015;86(2);135-43.

30. Philip E. Mosley,Rodney Marsh, The Psychiatric and Neuropsychiatric Symptoms After Subthalamic Stimulation for Parkinson's Disease. J Neuropsychiatry Clin Neurosci. 2015;27;19–26.

Chapter 13

Autonomic Dysfunction in Parkinson's Disease

Chaturbhuj Rathore

Introduction

Parkinson's disease (PD) is usually described by the motor symptoms; however, non-motor symptoms, including symptoms of autonomic dysfunction are being increasingly identified in PD over the last two decades, causing major disability to the patient. Despite this, majority of patients and clinicians emphasize the motor symptoms of PD, leading to underreporting and under recognition of autonomic dysfunction in PD. Autonomic dysfunctions are quite common in PD and being reported in approximately 90% of PD patients during their life.[1] Autonomic dysfunctions are associated with significantly impaired quality of life (QOL) and increased mortality in PD.[1,2] Timely recognition and proper management of autonomic dysfunctions can improve the QOL in patients with PD. This chapter will discuss the frequency, pathophysiology, symptoms and management of various autonomic dysfunctions encountered in PD (Box 1).

Box 1: Highlights of Autonomic Dysfunctions in PD

- Autonomic dysfunctions occur in 20-70% of patients with PD.
- Autonomic symptoms usually occur in advanced PD.
- These symptoms result from the direct involvement of autonomic centers.
- Autonomic symptoms usually do not respond to dopaminergic therapy.
- Effective symptomatic treatment of autonomic symptoms in PD can significantly improve quality of life.

Magnitude of the problem

The prevalence of autonomic symptoms is not well documented in PD due to the lack of properly conducted, prospective, community based studies. Majority of studies have focused on the patient reported symptoms and have reported variable prevalence of autonomic dysfunctions in PD. This is largely related to the variability in study designs, use of different questionnaires and the inclusion of patients at variable disease stages. In a questionnaire based study involving 141 PD patients and

DOI: 10.1201/9781003505488-13

50 matched controls, the frequency of various autonomic symptoms such as orthostatic dizziness, dysphagia, constipation, erectile dysfunction, bladder problems and hyperhidrosis was significantly more in patients.[3] Frequency of various symptoms in patients with PD ranged from 20% (hyperhidrosis) to 70% (bladder dysfunction). Overall 50% of patients had significantly impaired QOL due to autonomic symptoms. In another study, Verbaan and colleagues evaluated the frequency of various autonomic symptoms in a group of 420 PD patients using SCOPA-AUT (SCales for Outcomes in PArkinson's disease-Autonomic) questionnaire.[4] Overall scores were significantly higher in PD patients as compared to age matched controls, indicating significant autonomic dysfunction in patients. Approximately 50-90% patients had urinary symptoms, 50-70% patients had gastrointestinal symptoms, 50% patients had orthostatic dizziness, 40% had symptoms of thermoregulatory dysfunction and 30-50% patients had sexual dysfunction. Greater severity of disease, older age and greater doses of dopaminergic drugs were related to more autonomic problems in this study. Same factors were also found to be associated with autonomic dysfunctions in a study involving 61 patients.[5] In another study involving 532 patients, symptoms of autonomic dysfunction were noted in 30-60% of patients.[6] Autonomic dysfunctions were noted even in patients with early disease and in younger patients. In this study, constipation was the most frequent symptom (59%) followed by orthostatic symptoms (30%), seborrhea (27%), sweating disorders (24%), urinary dysfunction (22%), sexual disorders (18%) and hypersalivation/dry mouth (14%). Older age, greater disease severity and the presence of arterial hypertension were associated with a higher chance of autonomic dysfunction. Though common in older PD patients, majority of the patients in this study had mild symptoms and only one-fourth of patients required intervention for autonomic dysfunction. In a community based postal questionnaire survey of 150 patients and 150 controls, patients had two-fold higher risk of urinary and sexual dysfunctions and four-fold higher risk of other autonomic symptoms including orthostatic dizziness.[7] All these studies indicate that autonomic dysfunctions are present in 40-80% of PD patients and are more common in older patients with more severe disease.

However, autonomic symptoms are also present in patients with early and drug naïve PD. Muller and colleagues evaluated the frequency of various sensory and autonomic symptoms in 207 newly diagnosed PD patients and 175 controls.[8] In this study, autonomic dysfunctions were observed to be more common in the patient group. The most common symptoms were decreased olfaction (59%), urinary symptoms (47%), drooling (42%) and constipation (39%). However, in more than 90% of patients, the symptoms were mild and did not significantly affect the QOL.

Pathophysiology of autonomic dysfunction in PD

Autonomic dysfunction in PD results from the direct affection of the autonomic regulatory centers rather than dopaminergic neuronal loss. This is further corroborated by the fact that most of the autonomic symptoms in PD do not respond to dopaminergic therapy. Lewy bodies and neuronal loss have been documented in various autonomic centers including hypothalamus, locus coeruleus, parabrachial nucleus, raphe nucleus and intermediate reticular zone of the medulla.[9,10] In addition, Lewy bodies have also been demonstrated in preganglionic, ganglionic and postganglionic parasympathetic and sympathetic neurons including sites such as dorsal vagal nuclei and intermediolateral cell column.[10] Investigators have found Lewy bodies in the submucosal and myenteric plexuses of the gastro-intestinal tract which can explain gastrointestinal symptoms in PD patients.[11] As per the Braak staging of PD, dorsal motor nuclei of vagus nerves and olfactory nuclei are affected in stage 1, while raphe nuclei, locus coeruleus and lateral tegmental nuclei are affected in stage 2.[12] These are followed by the appearance of Lewy bodies in dopaminegic neurons along with the appearance of motor symptoms of PD during Braak stages III and IV. Additionally, α-synuclein inclusions have also been found within the enteric plexus in patients with incidental Lewy body disease.[13] Due to early involvement of specific autonomic centers in PD pathology, few autonomic features such as constipation and impaired cardiovascular reactivity can predate the motor symptoms of PD.[11,13,14]

Rating scales for autonomic dysfunction in PD

While many popular rating scales are available for the motor symptoms of PD, there is relative paucity of validated and easy to use rating scales for autonomic dysfunction in PD. SCOPA program has developed different validated rating scales for various non-motor symptoms of PD including autonomic symptoms (SCOPA-AUT).[15] SCOPA-AUT is a 25 items self-administered questionnaire which assesses various autonomic dysfunctions: gastrointestinal (7 items), bladder (6 items), thermoregulatory (4 items), orthostatic (3 items), pupillomotor (1 item), and sexual (2 items each for men and women) dysfunctions. The scale has been found to have good test-retest reliability and external validity. Composite Autonomic Symptom Scale (COMPASS) is a 73 items questionnaire scale which assesses nine domains of autonomic dysfunction.[16] However, with 73 items, it is a complex and time consuming scale. There are other questionnaires and rating scales to assess non-motor functions in PD which also evaluate autonomic functions. Two of the widely used items are non-motor symptom scale (NMSS) and non-motor symptom questionnaire for Parkinson's disease (NMSQuest).[17,18]

Common autonomic dysfunctions in PD

Autonomic dysfunctions in PD can involve various organ systems, including cardio-vascular system, gastro-intestinal system, genito-urinary system and thermoregulatory system. Common symptoms of autonomic dysfunction in PD are summarized in Table 1.

Table 1: Common symptoms of autonomic dysfunction in PD

Cardiovascular dysfunction

- Orthostatic hypotension
- Postprandial hypotension
- Post-exercise hypotension
- Supine hypertension

Gastrointestinal dysfunction

- Sialorrhoea/drooling
- Dysphagia
- Gastroparesis: Bloating, early satiety, nausea
- Constipation
- Defecatory dysfunction: Straining, pain, incomplete evacuation

Genitourinary dysfunction

- Irritative symptoms: Frequency, urgency, urge incontinence
- Obstructive symptoms: Hesitancy, straining, poor stream

Sexual dysfunction

- Decreased libido
- Erectile dysfunction
- Hyper-sexuality (mostly drug induced)

Thermoregulatory dysfunction

- Cold/heat intolerance
- Hyper and hypohidrosis
- Hypothermia and hyperthermia

Cardiovascular dysfunction

Orthostatic hypotension can occur in 40-80% patients with PD, while syncope occurs in ~5% patients.[19] Orthostatic hypotension is defined as ≥ 20 mm Hg fall in systolic blood pressure or ≥ 10 mm Hg fall in diastolic blood pressure within three minutes of standing or on 60 degree head-up tilt. The symptoms of orthostatic hypotension result in the decreased perfusion of various organs, most notably the brain. These symptoms include dizziness, blurring of vision, fatigue, tiredness, neck and back pain, exercise intolerance, dyspnoea and syncope. These symptoms are exacerbated after meals or exercise. Some patients may experience only postprandial or post-exercise hypotension due to pooling of blood in the splanchnic circulation and to the muscles.[20,21] On the other hand, few patients with chronic orthostatic hypotension may be completely asymptomatic.

Patients with advanced PD are more likely to have symptomatic orthostatic hypotension, although some evidence of sympathetic dysfunction is present even in early PD.[14] Evidence of cardiac sympathetic denervation in PD has been shown by various investigations, including testing during Valsalva maneuver, 6-[^{18}F] fluorodopamine PET uptake, and myocardial scintigraphy of the ^{123}I-meta-iodobenzylguanadine (MIBG), which is a physiological analogue of norephinephrine.[22,23] This sympathetic denervation is present even in early PD and can be used to differentiate PD from multiple system atrophy where it is absent. In addition, direct involvement of preganglionic sympathetic neurons and sympathetic ganglia contribute to the sympathetic failure in patients with PD.

Orthostatic hypotension does not require any treatment if symptoms are mild. In symptomatic patients, nonpharmacological measures, including increase in salt intake and use of stockings should be tried initially. Pharmacological treatment becomes necessary when nonpharmacological measures are ineffective. Fludrocortisone in the dose of 0.1-0.2 mg per day is first line therapy and is quite effective in patients with mild to moderate symptoms. Pressor agents, including ephedrine (25-50 mg three times daily) or midodrine (2.5-10 mg three times daily) are second line drugs when fludrocortisone is ineffective.[24] The doses should not be given after mid-afternoon to avoid supine hypertension in night.

Supine hypertension is another autonomic dysfunction which can occur in patients with PD.[25] Some PD patients may have non-dipping of night blood pressure. Supine hypertension is usually asymptomatic, but can cause supine headaches in some patients. Careful titration of antihypertensive drugs is required to control supine hypertension without producing or aggravating orthostatic hypotension.

Gastro-intestinal dysfunction

Presence of gastro-intestinal dysfunction was even recognized by James Parkinson who described symptoms such as excessive drooling, constipation and dysphagia as common accompaniment of PD. In a survey of 98 PD patients, presence of nausea, excessive drooling, dysphagia, constipation, and difficulty in defecation were major gastrointestinal symptoms encountered.[26]

Sialorrhoea and drooling

Excess of saliva is encountered in 70% patients with PD which increases to 88% in advanced PD.[4] Excessive salivation is also present in 40% of patients with early PD.[8] Sialorrhoea results from inefficient swallowing and not due to increased saliva production as was thought previously. Nonpharmacological measures such as sucking a hard candy or chewing gum can improve conscious swallowing, and thus may help to reduce drooling. Anticholinergic drugs may help in reducing saliva production, but it can cause confusion, constipation and urinary retention. Use of Glycopyrrolate, another anticholinergic agent which does not cross blood brain barrier, can avoid central side effects but it still can cause peripheral side effects.[27] Topical atropine solutions can be used to decrease saliva production.[28] Intraparotid injections of botulinum toxin can be used to decrease saliva production in difficult cases.[29]

Dysphagia

Overall, 50% of patients with PD complain of dysphagia as a prominent symptom. However, on clinical assessment including barium swallow and video fluoroscopy, it is detected in 70-90% of patients with PD.[26,30,31] Though classically a symptom of advanced PD, dysphagia can occur in some patients with early PD. Dysfunctions of oral, pharyngeal and esophageal muscles cause dysphagia in PD. Impaired lingual function and motor control causes difficulty in forming and transferring bolus. Impaired coordination of pharyngeal muscles results in misdirected swallowing and presence of residual food particles in valleculae, thereby increasing the risk of aspiration. Various esophageal dysfunctions, including impaired peristalsis, spasms and dysfunction of esophageal sphincter also contribute to dysphagia in patients with PD.[32]

Modified barium swallow and video-flouroscopy are standard tests for the evaluation of dysphagia in PD. Treatment of dysphagia in PD poses important challenges. Typically, dopaminergic therapy has no effect on dysphagia.[33] Nonpharmacological approaches which teach safe swallowing such as chin down swallowing and expiratory muscle strength training (EMST) are the standard treatment options in dysphagia.[34] Patients with crico-pharyngeal dysphagia may benefit with myotomy or botulinum toxin injections in cricopharyngeus.[35,36]

Gastric dysfunction

Gastroparesis or slow gastric emptying is common in patients with PD and can result in various symptoms such as nausea, postprandial discomfort, early satiety and bloating. In one study of 28 patients with untreated PD, half gastric emptying time was 59 minutes in patients with PD, while it was 44 minutes in healthy controls.[37] Gastroparesis can also result in delayed levodopa absorption or even complete dose failure. For mild symptoms, nonpharmacological measures such as small frequent meals and lying down after eating can be employed. Patients should also avoid high fiber and fat diet. In more symptomatic cases, pharmacological treatment with prokinetic agents such as domperidone and mosapride is effective. These drugs can cause prolongation of QT interval and hence should be used cautiously in older patients.

Constipation

Approximately 50% of patients with PD have constipation.[26] It also occurs in early PD even several years earlier than the motor symptoms. A study involving 6,000 healthy men showed that the subjects with less than one bowel movement per day have 2.7 fold higher future risk of developing PD as compared to subjects who have one bowel movement per day. The risk is increased to four-fold when compared to men who have more than one bowel movement in a day.[38] Autopsy studies of patients with constipation but without PD have shown Lewy bodies in the enteric plexus and in the substantia nigra of these patients.[13] Constipation is due to slow colonic transit, which is two times prolonged in patients with PD.[39]

First line treatment of constipation in PD involves nonpharmacological measures such as increasing water and fiber intake, regular exercises and discontinuing medicines which can aggravate constipation. If required, both bulk and osmotic laxatives and prokinetic agents can be used when nonpharmacological measures are ineffective. Anecdotal reports suggest the usefulness of cholinomimetic drugs such as pyridostigmine and functional magnetic stimulation.[40]

Defecation dysfunction

Defecatory dysfunction is characterized by painful defecation, straining during defecation and incomplete evacuation. This can occur in 60-70% of patients with PD.[26] Lack of coordination in muscles of defecation and failure of proper relaxation of puborectalis muscle and internal and external anal sphincters during defecation produces a functional outlet obstruction, resulting in defecatory dysfunction in PD.[41] In addition, decreased spontaneous phasic contractions of colon also can contribute to defecatory dysfunction in PD. Treatment strategies aimed at enhancing colonic transit such as the use of laxatives are not useful for defecatory dysfunction. Dopaminergic therapy can improve defecatory functions in some patients with PD. In few patients with troublesome symptoms, botulinum injections in external anal sphincter and puborectalis muscle may be helpful in improving defecatory dysfunction.[42]

Genito-urinary dysfunction

Urinary symptoms in PD can be divided in two broad categories: (1) storage or irritative symptoms including frequency, urgency and urge incontinence; and (2) voiding symptoms which include hesitancy, poor stream, straining and incomplete voiding. Urinary symptoms can occur in 50% of patients with PD.[43] The severity of urinary symptoms increases with disease severity. In fact, presence of urinary symptoms within one year of disease onset is helpful in differentiating PD from multiple system atrophy.

Irritative symptoms typically occur due to detrusor over-activity or detrusor hyper-reflexia. It is believed that loss of nigral neurons leads to disinhibition of ponto-mesencephalic micturition center which results in decreased D1-mediated inhibition of micturition. Detrusor hyper-reflexia has been shown to occur in 50-100% of PD patients.[43] However, all the patients with detrusor hyper-reflexia are not symptomatic. Muscarinic anticholinergic drugs such as oxybutynin and tolterodine can be used to treat detrusor hyper-reflexia.[44] In recent years, more bladder selective M3 blocking drugs such as trospium, solifenacin and darifenacin are available which have low potential of systemic side effects. In resistant cases, cystoscopy guided intra-detrusor botulinum toxin injections can provide symptomatic relief for 3-6 months. [45]

Voiding or obstructive symptoms can occur in one third cases of advanced PD. These are thought to result from detrusor underactivity and delayed relaxation of urethral sphincter (sphincter bradykinesia). Typically in PD, post void residual urine volumes are less than 100 ml, a feature helpful in differentiating from multiple system atrophy.[46] Prostatic hyperplasia as a cause of obstructive symptoms should be ruled out in all PD patients. Treatment of obstructive symptoms in PD can be quite difficult. If possible, contributing drugs such as anticholinergic agents should be stopped. Alpha receptor blockers (doxazosin, terazosin, and tamsulosin) and 5-alpha reductase inhibitors (finasteride and dutasteride) which are helpful in prostatic hyperplasia can be tried. In resistant cases, self intermittent catheterization may be the only option.

Sexual dysfunction

Majority of patients with PD develop reduced sexual interest and activity which has been noted to occur in 40% of men and 70% of women with PD.[47-49] Similarly, erectile dysfunction occurs in 50-80% of men with PD. Sexual dysfunction in PD is often multi-factorial, and various factors such as motor disability, pain, depression and anxiety often contribute to sexual dysfunction.[47] As a first step in the management of sexual dysfunction in PD, these factors should be identified and appropriately treated. cGMP-specific phosphodiesterase type 5 inhibitors such as sildenafil, vardenafil, and tadalafil can be used for treating erectile dysfunction in men. The safety and efficacy of sildenafil has been specifically proven in PD.[50] These drugs can cause significant orthostatic hypotension, and patients should be warned about this. Patients with testosterone deficiency can benefit with testosterone replacement therapy.[51]

Thermoregulatory dysfunction

Symptoms of thermal dysregulation, including hypohidrosis, hyperhydrosis and heat and cold intolerance occur in 30-55% of patients with PD.[4,6] Thermoregulatory dysfunction results from direct involvement of hypothalamus and sympathetic ganglia in the disease process. Episodic hyperhidrosis can occur in PD patients often during wearing off periods or during the periods of dyskinesia.[52] This usually occurs over the face, trunk and proximal limbs. This can be treated by proper adjustment of dopaminergic therapy. However, episodic hyperhidrosis can also occur unrelated to motor symptoms and is difficult to treat. Subthalamic deep brain stimulation has been shown to be effective in reducing episodes of paroxysmal sweating.[53] Hypohidrosis usually occurs in distal limbs which is consistent with the postganglionic pathology in PD. PD patients have excessive cold intolerance and are prone to develop hypothermia during extreme cold conditions for which they should be warned.[24] Malignant hyperthermia can occur in PD patients often in relation to sudden cessation of dopaminergic therapy.

Effect of antiparkinsonian drugs on autonomic functions in PD

Dopamine regulates systemic blood pressure by vasodilatation and by decreasing catecholamine release. However, levodopa largely does not affect the systemic blood pressure in healthy individuals.[54,55] Few reports have suggested that levodopa lowers resting blood pressure and causes mild orthostatic hypotension in PD patients, while other reports have suggested no effects of levodopa therapy on cardiovascular functions.[56-58] Dopamine agonists can also decrease blood pressure and can potentially lead to orthostatic hypotension.[54,59] Similarly, selegiline also can cause orthostatic hypotension.[59] On the other hand, catechol-O-methyl-transferase inhibitors do not significantly affect cardiovascular reactivity in patients with PD.[60] Effects of levodopa on bowel functions are not certain, although dopamine can reduce gastric motility. Both dopamine agonists and anticholinergic drugs can produce nausea and constipation.

Subthalamic deep brain stimulation can improve autonomic dysfunctions, including gastric motility, paroxysmal sweating and bladder problems.[61,62] This may be related to direct effects of subthalamic stimulation as well as to the reduced doses of dopaminergic drugs.

Autonomic function tests in PD

Role of autonomic function tests is limited in the clinical management of PD.[63] These tests can be used to diagnose the nature and extent of specific autonomic dysfunctions and sometimes to differentiate PD from other parkinsonian disorders such as multiple system atrophy. Various validated tests such as head up tilt test to assess orthostatic dysfunction, modified barium swallow and video-flouroscopy for the assessment of dysphagia, urodynamic studies and uroflowmetry for assessing urinary symptoms, defecating proctography for evaluating defecatory dysfunction and quantitative sudomotor axon reflex test (QSART) and sympathetic skin response (SSR) for the assessment of the sudomotor functions can be used. Other tests, including urethral and anal sphincter electromyography, myocardial scintigraphy of MIBG, plasma norepinephrine levels in response to head-up tilt and clonidine-induced growth hormone challenge test can be used to differentiate PD from other causes of autonomic failure.

Management of autonomic dysfunction in PD

Management of autonomic systems has been described along with specific symptoms and is summarized in Table 2. However, certain principles need to be followed while dealing PD patients with autonomic dysfunction.[24] Majority of autonomic symptoms are seen in advanced stages of PD and do not respond to dopaminergic therapy. Management of autonomic symptoms is mainly aimed at providing symptomatic relief. Every effort should be made to rule out treatable causes such as Zecher's diverticulum in patients with dysphagia, prostatic hyperplasia in patients with urinary symptoms and depression and anxiety in patients with sexual dysfunction before ascribing the symptoms to autonomic dysfunction. Nonpharmacological measures should be employed as first line therapy for all the symptoms. Every effort should be made to identify the drugs which can cause a particular symptom, such as anticholinergic drugs causing voiding symptoms. Physiotherapy and exercise programs have potential to improve majority of symptoms and should be encouraged for all the patients. In resistant cases involving other specialists such as urologists and gastroenterologists is quite useful.

Table 2: Management of autonomic dysfunction in PD

Symptom	Nonpharmacological measures	Pharmacological treatment
Orthostatic hypotension	• Omit unnecessary hypotensive medications • Increase salt intake • Avoid precipitating situations such as prolonged standing, getting up suddenly, post meal and post-exercise • Elevate head end of bed • Waist high stockings	• Fludrocortisone • Midodrine • Ephedrine • Phenylpropanolamine • Ergotamine/caffeine
Drooling	• Encourage conscious swallowing	• Trihexyphenidyl • Benztropine • Glycopyrrolate • Topical atropine solution • Intraparotid botulinum toxin
Dysphagia	• Increase consistency of food • Safe swallowing techniques such as chin down swallowing • Percutaneous gastrostomy	• Botulinum toxin injections in cricopharyngeus • Crico-pharyngeal myotomy
Gastroparesis	• Small frequent meals • Avoid high fat and fiber diet • Lying down after eating	• Domperidone • Mosapride
Constipation	• Increase fluid intake • Regular exercises • High fiber diet	• Bulk laxatives • Osmotic laxatives • Mosapride • Functional magnetic stimulation
Defecatory dysfunction	• Increase fluid intake • Regular exercises • High fiber diet	• Botulinum toxin
Irritative bladder symptoms		• Tolterodine • Oxybutynin • Darifenacin, solifenacin • Trospium • Intra-detrusor botulinum toxin
Obstructive bladder symptoms	• Self catheterization	• Terazosin • Doxazosin • Tamsulosin • Bethanechol chloride • Dutasteride and finasteride
Sexual dysfunction	• Treat depression and anxiety	• Sildenafil
Hyperhidrosis		• Adjust dopaminergic therapy • Deep brain stimulation

Conclusions

Autonomic dysfunctions are common in PD patients, which can significantly impair the quality of life in these patients. Though majority of these symptoms occur at advanced stage of PD, few symptoms like constipation may predate the motor symptoms by several years. These autonomic symptoms occur due to the direct involvement of autonomic regulatory centers and are not responsive to dopaminergic therapy. Effective symptomatic treatment is available for majority of autonomic symptoms at present. Availability of newer drugs for urological and gastrointestinal symptoms, which have low potential of side effects has further improved the management of these symptoms in patients with PD. Every effort should be made to identify other treatable causes for autonomic symptoms, and nonpharmacological measures should be employed in the management of autonomic symptoms for all the patients. A timely recognition and effective management of these symptoms can greatly improve the quality of life in patients with PD.

References

1. Witjas T, Kaphan E, Azulay JP, et al. Nonmotor fluctuations in Parkinson's disease: frequent and disabling. Neurology. 2002;59;408–13.
2. Stubendorff K, Aarsland D, Minthon L, Londos E. The impact of autonomic dysfunction on survival in patients with dementia with Lewy bodies and Parkinson's disease with dementia. PLoS One. 2012;7;e45451.
3. Magerkurth C, Schnitzer R, Braune S. Symptoms of autonomic failure in Parkinson's disease: prevalence and impact on daily life. Clin Auton Res. 2005;15;76-82.
4. Verbaan D, Marinus J, Visser M, van Rooden SM, Stiggelbout AM, van Hilten JJ. Patient-reported autonomic symptoms in Parkinson disease. Neurology. 2007;69;333-41.
5. van Dijk JG, Haan J, Zwinderman K, Kremer B, van Hilten BJ, Roos RA. Autonomic nervous system dysfunction in Parkinson's disease: relationships with age, medication, duration, and severity. J Neurol Neurosurg Psychiatry 1993;56;1090-5.
6. Korchounov A, Kessler KR, Yakhno NN, Damulin IV, Schipper HI. Determinants of autonomic dysfunction in idiopathic Parkinson's disease. J Neurol. 2005;252;1530-6.
7. Hobson P, Islam W, Roberts S, Adhiyman V, Meara J. The risk of bladder and autonomic dysfunction in a community cohort of Parkinson's disease patients and normal controls. Parkinsonism Relat Disord. 2003;10;67-71.
8. Müller B, Larsen JP, Wentzel-Larsen T, Skeie GO, Tysnes OB; Parkwest Study Group. Autonomic and sensory symptoms and signs in incident, untreated Parkinson's disease: frequent but mild. Mov Disord. 2011;26;65-72.
9. Gibb WR, Lees AJ. The relevance of the Lewy body to the pathogenesis of idiopathic Parkinson's disease. J Neurol Neurosurg Psychiatry. 1988;51;745–52.
10. Wakabayashi K, Takahashi H. The intermediolateral nucleus and Clarke's column in Parkinson's disease. Acta Neuropathol. 1997;94;287–9.
11. Braak H, de Vos RA, Bohl J, et al. Gastric alpha-synuclein immunoreactive inclusions in Meissner's and Auerbach's plexuses in cases staged for Parkinson's disease-related brain pathology. Neurosci Lett. 2006;396;67–72.

12. Braak H, Del Tredici K, Rüb U, de Vos RAI, Jansen Steur ENH, Braak E. Staging of brain pathology related to sporadic Parkinson's disease. Neurobiol Aging. 2003;24;197–210.

13. Abbott RD, Ross GW, Petrovitch H, et al. Bowel movement frequency in late-life and incidental Lewy bodies. Mov Disord. 2007;22;1581-6.

14. Orimo S, Takahashi A, Uchihara T, et al. Degeneration of cardiac sympathetic nerve begins in the early disease process of Parkinson's disease. Brain Pathol. 2007;17;24–30.

15. Visser M, Marinus J, Stiggelbout AM, van Hilten JJ. Assessment of autonomic dysfunction in Parkinson's disease: the SCOPA-AUT. Mov Disord. 2004;19;1306–12.

16. Suarez GA, Opfer-Gehrking TL, Offord KP, Atkinson EJ, O'Brien PC, Low PA. The Autonomic Symptom Profile: a new instrument to assess autonomic symptoms. Neurology 1999;52;523–8.

17. Chaudhuri KR, Martinez-Martin P, Schapira AH, et al. International multicenter pilot study of the first comprehensive self-completed nonmotor symptoms questionnaire for Parkinson's disease: the NMSQuest study. Mov Disord. 2006;21;916–23.

18. Chaudhuri KR, Martinez-Martin P, Brown RG, et al. Results from an international pilot study. Mov Disord. 2007;22;1901–11.

19. Senard JM, Raï S, Lapeyre-Mestre M et al. Prevalence of orthostatic hypotension in Parkinson's disease. J Neurol Neurosurg Psychiatr. 1997;63;584–9.

20. Micieli G, Martignoni F, Cavalinlini A, Sandrini G, Nappi G. Postprandial and orthostatic hypotension in Parkinson's disease. Neurology. 1987;37;386–93.

21. Smith GD, Mathias CJ. Postural hypotension enhanced by exercise in patients with chronic autonomic failure. QJM. 1995;88;251–6.

22. Goldstein DS, Holmes C, Li S-T, Bruce S, Metman LV, Cannon RO III. Cardiac sympathetic denervation in Parkinson disease. Ann Intern Med. 2000;133;338–47.

23. Li S-T, Dendi R, Holmes C, Goldstein DS. Progressive loss of cardiac sympathetic innervation in Parkinson's disease. Ann Neurol. 2002;52;220–3.

24. Dewey RB Jr. Autonomic dysfunction in Parkinson's disease. Neurol Clin. 2004;22(3);S127-39.

25. Goldstein DS, Pechnik S, Holmes C, et al. Association between supine hypertension and orthostatic hypotension in autonomic failure. Hypertension. 2003;42;136–42.

26. Edwards LL, Pfeiffer RF, Quigley EM, Hofman R, Balluff M. Gastrointestinal symptoms in Parkinson's disease. Mov Disord. 1991;6;151–6.

27. Arbouw ME, Movig KL, Koopmann M et al. Glycopyrrolate for sialorrhea in Parkinson disease: a randomized, double-blind, crossover trial. Neurology. 2010;74;1203–7.

28. Hyson HC, Johnson AM, Jog MS. Sublingual atropine for sialorrhea secondary to parkinsonism: a pilot study. Mov Disord. 2002;17;1318–20.

29. Nóbrega AC, Rodrigues B, Melo A. Does botulinum toxin injection in parotid glands interfere with the swallowing dynamics of Parkinson's disease patients? Clin Neurol Neurosurg. 2009;111;430–2.

30. Cereda E, Cilia R, Klersy C, et al. Swallowing disturbances in Parkinson's disease: a multivariate analysis of contributing factors. Parkinsonism Relat Disord. 2014; 20;1382-7.

31. Bushmann M, Dobmeyer SM, Leeker L, Perlmutter JS. Swallowing abnormalities and their response to treatment in Parkinson's disease. Neurology. 1989;39;1309-14.

32. Sung HY, Kim JS, Lee KS, et al. The prevalence and patterns of pharyngoesophageal dysmotility in patients with early stage Parkinson's disease. Mov Disord. 2010;25;2361-2368.

33. Menezes C, Melo A. Does levodopa improve swallowing dysfunction in Parkinson's disease patients? J Clin Pharm Ther. 2009;34;673–6.

34. Troche MS, Okun MS, Rosenbek JC et al. Aspiration and swallowing in Parkinson disease and rehabilitation with EMST: a randomized trial. Neurology. 2010;75;1912–9.

35. Restivo DA, Palmeri A, Marchese-Ragona R. Botulinum toxin for cricopharyngeal dysfunction in Parkinson's disease. NEJM. 2002;346;1174–5.

36. Born LJ, Harned RH, Rikkers LF, Pfeiffer RF, Quigley EM. Cricopharyngeal dysfunction in Parkinson's disease: role in dysphagia and response to myotomy. Mov Disord. 1996;11;53–8.

37. Hardoff R, Sula M, Tamir A, et al. Gastric emptying time and gastric motility in patients with Parkinson's disease. Mov Disord. 2001;16;1041-7.

38. Abbott RD, Petrovitch H, White LR, et al. Frequency of bowel movements and the future risk of Parkinson's disease. Neurology. 2001;57;456-62.

39. Sakakibara R, Odaka T, Uchiyama T, et al. Colonic transit time and rectoanal videomanometry in Parkinson's disease. J Neurol Neurosurg Psychiatry. 2003;74;268-72.

40. Chiu CM, Wang CP, Sung WH, Huang SF, Chiang SC, Tsai PY. Functional magnetic stimulation in constipation associated with Parkinson's disease. J Rehabil Med. 2009;41;1085–9.

41. Kim JS, Sung HY. Gastrointestinal Autonomic Dysfunction in Patients with Parkinson's Disease. J Mov Disord. 2015;8;76-82.

42. Albanese A, Brisinda G, Bentivoglio AR,Maria G. Treatment of outlet obstruction constipation in Parkinson's disease with botulinum neurotoxin A. Am J Gastroenterol. 2003;98;1439–40.

43. Winge K, Fowler CJ. Bladder dysfunction in parkinsonism: mechanisms, prevalence, symptoms, and management. Mov Disord. 2006;21;737–45.

44. Appell RA. Pharmacotherapy for overactive bladder: an evidence-based approach to selecting an antimuscarinic agent. Drugs. 2006;66;1361–70.

45. Kulaksizoglu H, Parman Y. Use of botulinim toxin-A for the treatment of overactive bladder symptoms in patients with Parkinsons's disease. Parkinsonism Relat Disord. 2010;16;531–4.

46. Sakakibara R, Hattori T, Uchiyama T, Yamanishi T. Videourodynamic and sphincter motor unit potential analyses in Parkinson's disease and multiple system atrophy. J Neurol Neurosurg Psychiatr. 2001;71;600–6.

47. Bronner G. Sexual problems in Parkinson's disease: the multidimensional nature of the problem and of the intervention. J Neurol Sci. 2011;310;139–143.

48. Koller WC, Vetere-Overfield B, Williamson A, Busenbark K, Nash J, Parrish D. Sexual dysfunction in Parkinson's disease. Clin Neuropharmacol. 1990;13;461–3.

49. Sakakibara R, Shinotoh H, Uchiyama T et al. Questionnaire-based assessment of pelvic organ dysfunction in Parkinson's disease. Auton Neurosci. 2001;92;76–85.

50. Zesiewicz TA, Helal M, Hauser RA. Sildenafil citrate (Viagra) for the treatment of erectile dysfunction in men with Parkinson's disease. Mov Disord. 2000;15;305–8.

51. Okun MS, McDonald WM, DeLong MR. Refractory nonmotor symptoms in male patients with Parkinson disease due to testosterone deficiency: a common unrecognized comorbidity. Arch Neurol. 2002;59;807–11.

52. Swinn L, Schrag A, Viswanathan R, et al. Sweating dysfunction in Parkinson's disease. Mov Disord. 2003;18;1459–63.

53. Sanghera MK, Ward C, Stewart RM, Mewes K, Simpson RK, Lai EC. Alleviation of drenching sweats following subthalamic deep brain stimulation in a patient with Parkinson's disease – a case report. J Neurol Sci. 2009;285;246–9.

54. Kujawa K, Leurgans S, Raman R, et al. Acute orthostatic hypotension when starting dopamine agonists in Parkinson's disease. Arch Neurol. 2000;57;1461–3.

55. Iwasaki S, Hamaguchi K, Iwasaki A, et al. Hypotensive effect of long-term levodopa in patients with Parkinson's disease. Eur Neurol. 1990;30;194–9.

56. Senard JM, Verwaerde P, Rascol O, et al. Effects of acute levodopa administrationon blood pressure and heart variability in never treated parkinsonians. Hypertens Res. 1995;18(1);S175–7.

57. Sachs C, Berglund B, Kaijser L. Autonomic cardiovascular responses in parkinsonism: effect of levodopa with dopa-decarboxylase inhibition. Acta Neurol Scand. 1985;71;37–42.

58. Calne DB, Brennan J, Spiers AS, et al. Hypotension caused by L-dopa. BMJ. 1970;1;474–5.

59. Haapaniemi TH, Kallio MA, Korpelainen JT, et al. Levodopa, bromocriptine and selegiline modify cardiovascular responses in Parkinson's disease. J Neurol. 2000;247;868–74.

60. Lyytinen J, Sovijarvi A, Kaakkola S, et al. The effect of catechol-O-methyltransferase inhibition with entacapone on cardiovascular autonomic responses in L-Dopa-treated patients with Parkinson's disease. Clin Neuropharmacol. 2001;24;50–7.

61. Priori A, Cinnante C, Genitrini S, et al. Non-motor effects of deep brain stimulation of the subthalamic nucleus in Parkinson's disease: preliminary physiological results. Neurol Sci. 2001;22;85–6.

62. Wolz M, Hauschild J, Fauser M, et al. Immediate effects of deep brain stimulation of the subthalamic nucleus on nonmotor symptoms in Parkinson's disease. Parkinsonism Relat Disord. 2012;18;994-7.

63. Asahina M, Vichayanrat E, Low DA, Iodice V, Mathias CJ. Autonomic dysfunction in parkinsonian disorders: assessment and pathophysiology. J Neurol Neurosurg Psychiatry. 2013;84;674-80.

Chapter 14

Balance and Gait in Parkinson's Disease

Rajeev Aggarwal, Madhuri Behari

Parkinson's disease (PD) is a non-remitting disabling condition with profound effect on balance and gait. People with Parkinson's disease (PWP) have high rate of falls compared to peer group of healthy persons. PWP has two times more risk of falling and sustaining fracture as compared to healthy controls.[1] The prevalence of fall rate in PWP is between 21-87%.[1,2] Even PWP who never had a fall, had a 21% risk of fall in three month follow up period.[2] Quality of life in PWP is drastically compromised by gait and balance disorders. Almost all scales on quality of life for PD have included mobility component as an essential domain.[3]

Normal gait cycle and its parameters

Gait is the rhythmical alternating movement of legs, arms and trunk that results in forward displacement of the body.[4] A normal gait cycle consists of rhythmical cyclical movement of lower limbs and counter movement of upper limbs for displacement. Gait cycle consists of stance phase and swing phase.

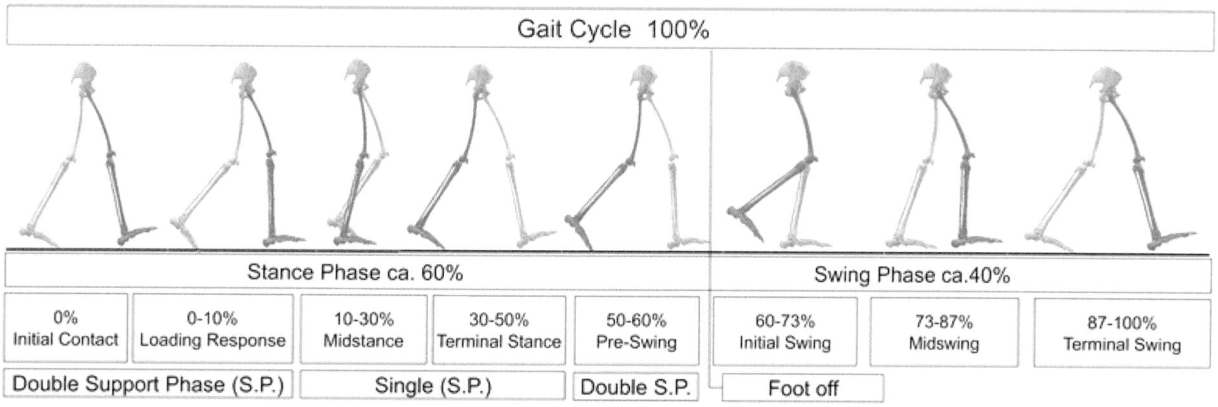

During stance phase, a foot remains in contact with floor, while in swing phase the foot remains in air. When both feet are in contact with floor, it is called as double support phase. In running, double support time is zero. The distance between two consecutive foot placements of the same foot is termed as stride length. One stride length is a sum total of successive right and left step length. The

DOI: 10.1201/9781003505488-14

distance between initial contact position of one foot to the initial contact position of other foot is termed as step length. Step length varies according to age, height and gender apart from pathological reasons. Normal step length in adults range from 55 cm to 75 cm. Cadence is the number of steps per minute and range from 90 to 110 steps per minute. Gait speed is the distance covered in unit time and average gait speed range 1-1.5 meters per second.

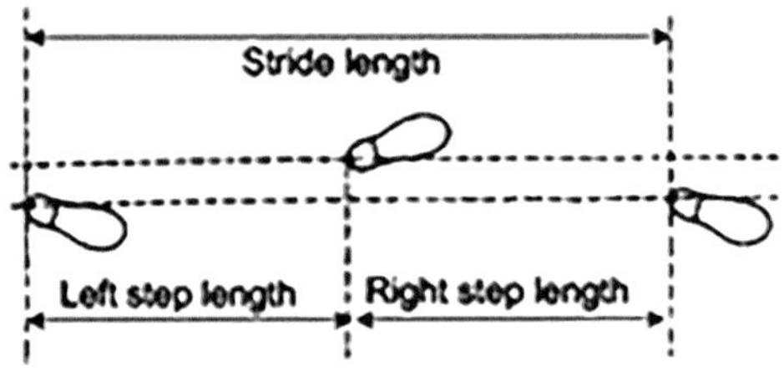

Neurogenic control of gait and balance

Good postural control and balance are the prerequisite for walking. Posture may be defined as any body position where body parts are aligned static in both antero-posterior and medio-lateral directions, whereas the ability to sustain the centre of gravity within base of support or limits of stability in both static and dynamic conditions is termed as balance.[4] In humans, balance and gait are result of complex and sophisticated interplay of voluntary and automatic control systems via large range of anatomical organs. Various organs involved in maintaining balance and gait are proprioceptors, muscle spindles, motor units, afferent and efferent neurons, spinal cord, brain stem, vestibular system, cerebellum and cerebrum. The control systems range from monosynaptic stretch reflex, spinal reflex (mass limb reflex - withdrawal response), primitive walking reflex (automaticity of gait) controlled by midbrain and sub-thalamic areas, postural tone control by brain stem and vestibular system and selective motor control by cerebral cortex. Of late, our understanding regarding balance, posture and gait has improved. Postural control is an outcome of interaction and integration of many sensory and motor processes. Postural control requires perception and integration of information from nervous system and musculoskeletal system. The two main goals of postural control are postural equilibrium and postural orientation. Postural orientation deals with body parts alignment and tone of muscles in relation to gravitational forces, supporting surface, base of support, visual surroundings and internal orientation. The integration of sensory information from visual, vestibular and somato-sensory systems provides spatial orientation. Postural equilibrium entails the stabilization of center of mass of our body during external or internal perturbation by coordinated efforts of various sensori-motor tactics to keep postural stability. Any damage or derangement to sensori-motor strategies for postural equilibrium or postural orientation results in various instabilities.[5]

Pathogenesis of postural instability, balance and gait in PWP

Instability of posture is a cardinal feature of PD affecting functional mobility and balance. The basal ganglia (main pathological seat of PD) controls agility of postural tone, modulates postural tone, selects postural strategies depending on environmental circumstances and automatizes postural responses and gait.[6] Initially, balance and gait impairment may be attributed to bradykinesia and rigidity that is responsive to dopamine, but as the disease progresses, other mechanisms like nonflexible postural set, deranged kinesthesia, loss of automaticity and impaired executive functions that are dopamine resistant dominates. Excessive non-modulating postural tone (rigidity) causes biomechanical alteration in posture (like stooped posture, reduced range of joint motion and protruded head) affecting automatic postural responses to external perturbation. Co-contraction of muscles due to rigidity impairs unilateral muscle activation and antagonistic relaxation, making person vulnerable to lose balance and fall. Automatic postural responses to unexpected external perturbation can lead to compensatory stepping reaction or reach to maintain postural control. Both of these strategies become smaller and slower due to bradykinesia. Orientation of body in space requires central sensory integration of visual, vestibular and somato-sensory inputs. In the absence of sensory inputs or conflicting sensory inputs, the nervous system ignores ambiguous information and relies on more useful information. In-spite of normal vestibular inputs, PWP have difficulty in maintaining balance on unstable surface with closed eyes, suggesting slowing of reweighting process of nervous system. PWP have impaired use of proprioception[7] as they are unable to perceive small changes in inclination (impaired kinesthesia). This impaired kinesthesia and slower reweighting of sensory inputs is compensated by over dependence on visual inputs for postural orientation.[8] Therefore, PWP are prone to loss of balance and falls in poorly lit area. Many features of impaired cognitive functions such as impaired cognitive focus, poor attention span, fear of fall and co-morbid dementia may disturb balance in PWP. The automaticity of postural control and gait during secondary cognitive or motor task is also compromised in PWP. The exact mechanism of this remains unclear.[8] The timings and extent of involvement of various mechanisms of postural control varies from person to person; therefore proper assessment of balance and gait are paramount for their management.

Gait abnormalities in PD

Typical gait of a PWP is often expressed as shuffling gait. Shuffling gait consists of abnormalities like reduced ground clearance, stride length, arm swing and gait speed, reversal of heel toe pattern to flat foot or toe heel pattern, asymmetrical gait, reduced base width and forward flexed trunk. Usually, PD gait is hypokinetic and is associated with akinesia with advancement of disease. Akinesia refers to inability of the patient to initiate the gait, i.e., freezing of gait. Freezing are the brief episodes when PWP are unable to generate effective stepping movement pattern.[4] PWP often experience freezing when they face sudden obstruction in the path or need to take a turn or travel through a crowded area. Sometimes, these patients exhibit involuntary speeding up of gait as if they are trying to catch their own centre of gravity, and then it is called as festinating gait. PWP on levodopa for long period usually in end stage disease walk with dyskinetic gait. Dyskinetic gait is manifested by abnormal dystonic or choreatic movements. Ankle plantar flexors, lateral flexors of trunk on one side and the wrist and hand flexors are commonly affected muscle groups. These abnormal movements may be persistent or episodic, leading a threat to maintain the balance and often results in tripping and falls. PWP often have large step-to-step variability compared to age matched healthy persons. Gait abnormalities worsen when patients' attention is diverted to a secondary task.[9] These secondary tasks may be a motor task like carrying a tray in hands or cognitive task like doing some mental work of calculation or planning. Patients often feel more difficulties in walking when others are observing them. It can be assumed that PWP loose automaticity in walking, but they can walk better if they are asked to walk consciously with correct gait pattern, i.e., heel toe pattern, reciprocal arm movement and increased step length. Patients can be explained about their gait; as they have to control their walking pattern in manual mode rather than automatic mode. Interestingly, these patients experience much less difficulties in stair climbing or cycling than ground level walking as these motor tasks include more cortical intervention or perceive stairs or cycle pads as external cues.[10]

Gait abnormalities in PD can be studied as deficits during straight walking on flat surface, initiation of gait, turning and ability to adjust to negotiate environmental threats like irregular surface, sudden jerks in crowded areas or responding to an imminent threat like crossing the road.[11]

1. Walking on flat surface

Walking on flat surface requires least attention. While walking on flat surface, PWP usually exhibit reduced velocity due to short step and stride length. Cadence is usually preserved or increased. Shorter steps are due to reduced torque development by plantar flexors (around 20% less compared to age matched controls) during push off phase of the gait. Power generation at ankle has been found to be strongly associated with stride length, step length and gait velocity in elderly population,[12,13] but in regulating step length PWP have to rely greatly on a complex mechanism including interplay of deficient central programming, rigidity and reduced muscle power. As PWP feel difficulty in maintaining their balance on one leg, double support time increases. Affected side upper limb moves less compared to unaffected side upper limb. There is generalized decreased range of motion of trunk in all planes. Pelvis on more affected side remains retracted, leading to reduced propulsive capacity of affected lower limb.[14]

2. Gait initiation

Gait initiation requires much more attention, i.e., more cortical control than walking straight on flat surface. Gait initiation involves anticipatory postural adjustment (APA), where a person shifts his postural weight forward as well as towards the stance leg, thereby unloading the stepping leg. After weight shifting, the stepping leg push offs to take first step. PWP exhibit reduced APAs while initiating a step. Lateral and forward shifting of weight are inadequate for unloading the stepping leg, resulting in delay in step execution. It leads to a reduction in push-off force along with reduced step length. The pattern of gait initiation remains unaffected in PWP inspite of hypometric gait initiation.

3. Turning

Turning demands higher level of cognitive control to alter pattern of gait to take turn and adjust dynamic postural control. In a healthy person, the eyes, head, trunk and feet move in a sequential manner while turning. This sequential movement of body is lost in PWP while turning. Turning in PWP is generally expressed as "enbloc" turn as they move their eyes, head, trunk and legs simultaneously rather than in sequential pattern. Turning speed in PWP is dependent on neck rigidity, often resulting in increased time to complete a turn. Axial rigidity and loss of inter-segmental flexibility are contributory to turning difficulty in PWP. Number of steps required to take a turn are generally more in PWP compared to healthy adults.

4. Ability to adjust to negotiate environmental threats

Negotiation of environmental threats requires highest level of cognitive control that includes anticipation, planning and execution of strategy to avoid or overcome the obstacle. PWP exhibits instability and more collisions while crossing obstacles. Inappropriate motor planning and break in execution leads to increased risk of losing balance and probability of fall.

Parkinson's disease and falls

Falls in PWP is the commonest preventable problem. Incidence and prevalence of fall increases with the progression of disease. There are many recognized risk factors for falls that increases susceptibility for falling in PWP. These factors are higher disease severity (HY stage 3 and above), lower functional capability, motor fluctuations (unexpected "offs"), dyskinesia, freezing, forward protruded head, flexed posture, postural instability, gait impairments, side effects of drugs, delayed reaction time, cognitive impairment, depression and fear of fall. Fear of fall increases with history of falls. Falls usually lead to evasion of activity, musculo-skeletal and cardio-respiratory deconditioning and increased hospital admission resulting in increased burden on family, society and state.

Assessment of posture, balance and gait in PD

Assessment of posture, balance and gait in PWP is a tedious task, as there may be fluctuation in clinical status many times within a day.[7] PWP apparently normal in "on" state may be severely disabled in "off" state. Intrinsic factors like cognitive impairment, fear of fall, depression, other non-motor symptoms and secondary task may further complicate assessment. Therefore, posture, balance and gait should be assessed in their best state and worst state of the day. For OPD settings validated clinical scales and questionnaire are in common use while more detailed objective evaluation can be done in laboratory settings.

Clinical scales and questionnaires

- Berg Balance Scale (BBS): BBS consists of 14-items, assessing mainly static balance and equilibrium. It assesses balance in various postures and conditions like sitting, standing, turning, single leg stand and eyes open and closed. Its score range from 0-56. Higher the score, better the balance. Score interpretation: 0-20 (high risk of fall); 21-40 (moderate risk of fall) and 41-56 (low risk of fall). It is widely used in balance assessment of PWP. Time required is 15-20 min. BBS shows ceiling effect. Cut off score of 47 for fallers among PWP have 72 % sensitivity and 75% specificity.[15]
- Six-minute walk test (SMWT): SMWT is a sub-maximal exercise test of endurance or aerobic capacity. It is the maximum distance covered by a person in six minutes while walking to and fro in a long corridor (preferably 100 feet long). Normal distance is 500-700 meters, depending on age, gender, height and body type. Higher the distance covered, better it is. No data is available to suggest as to what score can be considered normal in PWP.
- Balance Evaluation System Test (BESTest): It is 36-item balance test focusing on testing for biomechanical constraints (Centre of mass alignments and lower limb muscle strength), stability limits/ verticality (Functional reach tests), transitions/ anticipatory (sit to stand, modifying base of support and reach), reactive postural responses (propulsion and retropulsion tests), sensory orientation (Standing on unstable surface or ramp) and stability in gait (gait pattern with varied level of difficulties including walking in forward direction, with closed eyes, stepping over obstacles, varying gait speeds, with different head turns, and with a dual task). Score range is 0-108. Higher

the score, better is the balance. Time required 20-30 minutes. Cut-off score of 69% for fallers among PWP have 84% sensitivity and 76% specificity.[15]

- 360° turn test: 360° turn test is a simple tool to assess dynamic balance where the person is asked to turn a complete circle (360°). The number of steps taken and/or time to complete the 360° turn are recorded. In a study on older population (n= 511) who had a history of fall (n=139) took an average of 2.7 seconds while non faller group (n=372) took an average of 2.2 seconds to complete the test.[16] Number of steps taken to complete is ≤ 6.
- Ten meter timed walk test: The person is required to walk 10 meters without any assistance, but the time is recorded for the middle six meters to account for acceleration and deceleration. Gait velocity is calculated as distance divided by time.
- Tinetti Performance-Oriented Mobility Assessment (POMA): POMA test is used to assess risk of falls. POMA test has two subsets; Balance tests and Gait tests. Balance tests have nine items with score ranging from 0 to16, and Gait tests have seven items with score ranging from 0 to 12. Higher the score, better the outcome. POMA score ≥ 24, low risk of fall; 19-23, moderate risk and ≤ 18, high risk of fall.
- Freezing of gait Questionnaire (FOGQ): FOGQ consists of six items to assess freezing and its impact. Score range 0-24 with lower the score indicating better balance.[17]
- Activities-Specific Balance Confidence (ABC) Scale: ABC scale is a 16-item scale where patients rate their confidence level on 0-100 scale in performing day to day indoor as well as outdoor activities. Total score aggregates between 0-1600 is divided by 16 to get ABC scale score from 0-100. Score evaluation $\geq 80\%$ (suggests high level of functioning); 50-80% (moderate level of functioning) and $\leq 50\%$ (low level of functioning).
- Falls Efficacy Scale - International (FES-I): FES-I assesses the fear of persons for fall in performing day to day indoor as well as outdoor activities. It is a 16 items questionnaire with score range of 16-64. Lower the score, the better. A score of more than 23 has high risk of falls.[18]

Laboratory based quantitative assessment of gait

Gait and balance disorders in PWP are inconsistent depending on time (On-off fluctuation) and situation. PWP responses vary between laboratory setting and real time monitoring. Laboratory based motion capture systems mainly employ body markers, 3D-camera system and force plate to assess gait kinematics (study of joint range of motion and joint angles during ambulation). These gait analysis system are often clubbed with EMG system to assess gait kinetics (study of muscle recruitment and forces produced during ambulation). Instrumental walkways record force and timings of footsteps to provide data about spatiotemporal parameters like step length, stride length, gait speed, cadence, step time, stride time, inter-cycle variability and movement of centre of gravity. These tools are very precise and accurate in quantification of gait, but are expensive and confined for laboratory based assessment as well as have limited validity due to experimental set-up.

Technological advancement in gait assessment

Invention of Micro Electro Mechanical Systems (MEMS) has led to electronic human motion analysis in a new era of motion sensors. Various types of MEMS sensors (usual size 20 μm to 1 mm) such as accelerometers (measures linear acceleration and deceleration), gyroscopes (measures angular acceleration), magnetometers (measures the direction and strength of magnetic field), electronic compasses (combination of accelerometer and magnetometer) or pressure sensors (force sensing resistors) aggregates to constitute inertial measurement unit (IMU). This IMU is the basic unit of motion sensors. Each MEMS sensor provides some kinematic or movement information and their integration can estimate complex 3-dimensional real time movement. These motion sensors can be worn on body to measure body movements, including spatiotemporal gait parameters, freezing of gait, falls, sway, balance impairment, tremors and movements in day to day indoor as well as outdoor activities. Real time assessment of gait is much more informative and accurate for treatment planning and intervention.[19] Our team has developed a "PD shoe" that can do real time assessment of spatiotemporal parameters. PD shoe is developed on water shoe (that provides platform) with an insole that lodges three pressure sensing resistors at heel, balls of metatarsals and toes. A microprocessor receives the signals from resistors and sends these signals by wireless technology to a computer to record real time gait parameters.[20]

Management of posture, balance and gait in PD

Pharmacological management

Dopaminergic drugs target to increase availability of neurotransmitter dopamine are very effective for tremors, rigidity and bradykinesia. Balance and gait parameters related with bradykinesia improve with dopaminergic drugs.[7] Levodopa, the revolutionary drug in PD improves step length, trunk movements, arm swings and gait velocity, but temporal parameters like stride time, step time, double support time and cadence are not much improved with levodopa. Levodopa improves anticipatory postural adjustment (APA) in terms of increased push off force and larger first step for gait initiation, and thus helps in reducing gait akinesia.[21] In contrast, Curtze et al.[22] failed to demonstrate improvement in duration of APA and size of the first step. The number of steps and turning speed to complete the turn improves with dopaminergic drugs, but turning in place may not improve.[23] The ability to cross obstacles improves with dopaminergic medications, albeit with shorter and faster steps, putting them at risk of fall.[24] Dopaminergic medications reduce freezing duration and frequency,[25] but in some PWP levodopa induces freezing which is difficult to treat with medications.[26] Stride length and stride time variability, a hallmark of FoG do not improve with dopaminergic medications.[27] Dopamine agonists generally enhance the effect of Levodopa on gait parameters.[28] Tolcapone, COMT inhibitor also augments the favorable effects of levodopa drug on gait speed.[29] Overall, dopaminergic along with other drugs for PD are moderately effective in early PD (initial five years), but in advanced PD, these drugs are largely ineffective for balance and gait impairments.[30]

Deep Brain Stimulation (DBS) and balance and gait in PD

Deep brain stimulation mainly targets three major nuclei viz. thalamic ventro-intermediate nucleus (Vim), internal globus pallidus (GPi) and sub-thalamic nucleus (STN) by using high frequency stimulation. Six months post DBS (sub-thalamic nuclei), compensatory stepping and feet in place responses are weaker compared to pre-DBS irrespective of whether the stimulation is "on" or "off". DBS of GPi does not have any effect on postural responses, rather it may worsen in some cases.[31] DBS does not improve postural responses due to bradykinesia. The rule of thumb is that any symptom unresponsive to dopaminergic drugs does not improve following DBS. This is true for gait impairments as well. Overall there is still debate regarding safety, risks and benefits of DBS on gait and balance.

Physiotherapeutic management

Physiotherapeutic management of PD has evolved from anecdote to evidence. Initially, physiotherapeutic management was based on personal experiences and subjective reporting of PWP. Currently, there is growing evidence of physiotherapy in management of PD. Physiotherapy treatment focuses mainly on the following components:

1. Compensatory strategies for posture, balance and gait impairment
2. Optimization of joint motion, muscle length, strength and endurance
3. Neural recovery and prevention of degenerative injuries to brain

1. Compensatory strategies for posture, balance and gait impairment:

PD is a non-remitting neuro-degenerative disease without any definite cure, and therefore the main stay of treatment revolves around compensatory strategies. PWP retain their ability to move, but its activation is usually impaired. The automaticity of habitual movements like walking, speaking, writing etc. is impaired due to derangement in basal ganglia and associated neural circuits like thalamic-cortical-spinal loops, through brainstem pedunculo pontine nucleus along with reticulospinal tracts. Therefore, alternative neural circuits are activated to facilitate optimal movement. Sensory cues stimulate pre motor cortex rather than supplementary motor area or basal ganglia. Sensory cues training also facilitate learning motor skills by augmenting perceptual and sensory sensations. Rhythmical auditory cueing, attention strategies, proprioceptive cues (vibrations under sole of feet) and visual cueing are commonly adopted techniques for compensation. In a systematic review by Rocha et al[32] (2014) on effects of external cues on gait parameter, it was reported that on an average, 20 sessions each of approximately 30 minutes durations are required for adequate gait training. Sensory cues improve balance, freezing, risk of falls, gait speed, stride length, step length, cadence and psycho-motor performance of PWP.[32]

a. Visual cues: Visual cues are visual inputs through external visible cues at appropriate timing and place for proper activation of gait cycle. These cues stimulate visual motor pathways via cerebellar circuit.[33] Bright coloured tapes or white coloured cardboard strips can be placed on ground at appropriate distance as per patient's height and expected step length (approx. between 50-70 cm) along a long walkway, and PWP is asked to step over these tapes/stripes. These visual cues can be fixed on floor along common routes at home like bed to bathroom or living room to kitchen etc. Visual cues can also be provided in terms of written instructions on walls to focus on large steps or heel-toe pattern. Walkers and canes are also available with Laser to provide visual cues to improve gait and overcome freezing.

b. Auditory cues: Auditory cues consist of rhythmical auditory stimulation given by metronome or by rhythmical music. This rhythm compensates for internal rhythm deficit. Auditory cues are most commonly used cues for gait training in PWP. Initially, metronome rate should be set at patient's present cadence and asked to focus on increasing step length or heel toe pattern. Later on, metronome rate can be set at a rate of around 100 (near to normal cadence). PWP should practice to walk with metronome in different conditions like walking through narrow passage, turning, overcoming freezing episode, stopping and starting in different environmental constraints. In a recent study at our centre Harsha et al.[9] demonstrated effect of dual task on gait parameters while using metronome for auditory cues. Twenty-five PWP were tested for secondary motor task (coin transfer from one hand to another) or cognitive task (digit subtraction) while walking with auditory cues. Addition of secondary task reduced step length and gait velocity while cadence and number of steps were increased. Cognitive task was more detrimental compared to motor task.

c. Cognitive cues (Attention strategies): PWP are asked to concentrate on walking i.e., shifting walking from automatic to manual mode. Initially, instructions are given to pay attention on one component of gait (most commonly asked to take bigger/long steps). Alternatively, PWP may be asked to keep heel down first or count their steps to complete a given distance. If a patient is well versed with visual cues, he/she may be instructed to imagine visual cues while walking without cues. Cognitive impairment limits the efficacy of this technique.

d. Proprioceptive cues: The proprioceptive cues provide sensory stimulation like vibration at periphery (mostly on foot). Stimulation of proprioceptors activates pre-motor cortex through dorso-lateral tract, bypassing supplementary motor area's deficit to get rhythmic gait. Stimulation of mechanoreceptors through vibrators at ankle and foot via PD shoe are synchronized with every step to provide a feedback of walking. Initial results of this technique are promising,[34] as these cues can be provided in real time indoor as well as outdoor. In a clinical trial using PD shoe, we found it to be safe and effective technique. Two weeks training with PD shoe in PWP with FOG improved balance as reflected in improvement in Berg's balance scale score, gait as reflected in Timed up & go test and Falls Efficacy Score-International (FES-I) scores, and quality of life as shown in PDQ-39 self-report (Unpublished data).

2. Optimization of joint range of motion, muscle length, strength and endurance

a. Maintaining muscle length and postural correction

PWP gradually adopt forward bent posture (Camptocormia), flexion at hip, knee, shoulders and elbow with tightness of anterior group muscles. Many PWP adopt bending at one side that may be attributed to asymmetrical nature of disease. Stretching of anterior group muscles mainly neck flexors, pectorals, biceps, trunk side flexors, hip flexors and hamstrings helps in correction of forward bent posture. Usual dose of stretching is 30 seconds hold, 2-3 repetitions of each muscle, preferably on daily basis. Postural correction maneuvers, like standing against wall, standing in front of a mirror can help in perception of postural impairment and doing remedial exercises. Strengthening of postural muscles (Neck extensors, back extensors, hip extensors, quadriceps and gastrosoleus muscle) also contribute in correction of posture. Postural rehabilitation by proprioceptive and tactile stimulation of back using materials of different texture and kinesio taping showed significant improvement of posture, balance and gait parameters.[35]

b. Fitness training

King and Horak (2009)[36] conceptualized sensori-motor-agility exercise program by integrating movement principles of various fitness training program like tai chi, lunges, boxing, Pilates, Kayaking to improve mobility in PWP. Tai chi is a whole body exercise program that focuses on movement of centre of mass by increasing body kinesthesia sense, augmenting postural alignment and sequencing body movements. Kayaking includes large coordinated arm joints movements focusing on axial rotation and speed. Agility training emphasizes on quickly changing movement pattern; for example quick-turns, high knee exercises, walking sideways, backward and altering the speed of movements to overcome freezing. Boxing incorporates postural adjustments (anticipatory as well as reactive), backward walking, agility and speed to hit a moving bag. Lunges include movement of centre of mass in forward or sideways direction. When centre of mass falls outside the base of support, it forces the person to take a step to maintain the balance. It helps in improving the limits of multi-directional stability, postural correction by stepping and quick change of directions. Pilates exercises help PWP to improve trunk control, core-stability of spine, sequencing of movements, practice of transfer activities like rolling, transfers and sit to stand activities.

c. Resistance training:

Resistance training incorporates exercise of usually large group muscles against resistance to improve strength of muscle. It is an important component of fitness training in all age groups. In PWP, strength training produces similar results as in age match healthy controls. As mentioned earlier, strength of plantar flexors is main contributory in gait initiation and step length, Hass et al. (2012)[37] reported improvement in gait initiation by strength training of lower limb muscles. In a systematic review, Chung et al. (2016)[38] reported that resistance training in PD is effective in improving motor functions, muscle strength and balance, but without improving gait and quality of life parameters. The usual dose of resistance training for large group muscles of lower limb is three set of 10 repetitions each; two to three times every week for three to six months. Some researchers have also trained trunk muscles and upper limb muscles in PWP[38] with good results.

d. Treadmill training:

Treadmill training improves gait parameters, aerobic capacity and cardiac endurance for PWP. In a systematic review by Cochrane group in 2015[30] on use of treadmill training for PWP, it was concluded that treadmill training is effective for improving stride length and walking speed in PWP at HY stage 1 to 3. Although there is no information on duration of carryover effect of this training, but it is recommended in young and fit PWP for gait training. Usual dosimetry of treadmill training is a session of 20-40 minutes, 3-5 sessions per week and for 4-12 weeks. Treadmill training may be clubbed with body weight support for safety, visual and auditory cues to improve step length

3. Neural recovery and prevention of degenerative injuries to brain:

Both animal models and human study have confirmed that forced exercise and not just leisure activities is required to produce neural modulation. Forced exercises increase neurotrophic proteins, mainly glial-derived neurotrophic factor (GDNF) and brain-derived neurotrophic factors (BDNF) in substantia nigra and striatum.[39,40] Forced exercise may induce synaptogenesis, angiogenesis, reduce oxidative stress injuries and optimize mitochondrial activity.[40]In early PD, high intensity exercise can normalize excitability of cortico-motor area as estimated by duration of cortical silent period in response to single pulse trans-cranial magnetic stimulation[41] supporting dose dependent benefits of training. Exercise induced neural-plasticity helps in strengthening of synapses and improving neural transmission. Fisher et al.[42] reported increase in binding potential of D2 receptor in four PWP of early stage PD, using intensive treadmill training (three times per week for eight weeks) and demonstrated improvement in turning performance as a measure of postural control. Intensive exercises have been reported to be effective for improving dyskinesia and reducing levodopa dosage in advance stage PWP.[43]It is suggested that PWP should practice intensive aerobic training (static cycling, swimming, jogging, dancing, cross trainer) for continuous 30-40 minutes for five to seven days in a week on a regular basis. Before and after aerobic training, 10-15 minutes of warm up and cool down low level large group muscles range of motion exercises are essential. During aerobic training, a target heart rate of 60 to 80% of age predicted maximum heart rate (220-age) should be achieved.

Whole body vibration training

Whole body vibration (WBV) training is a recent modality where a person stands on a platform that provides vibrations, which helps in performing better on strength, balance and gait scales in PWP. There is moderate level of evidence that a single session of whole body vibration may improve muscle strength, but only weak level of evidence for improvement in health-related quality of life and proprioception in PWP.[44] On the other hand, there is minor evidence of long-term effects of WBV on improvement of proprioception, strength, balance and gait.[45]

Strategies for overcoming freezing

Freezing is the most troublesome and disabling symptom of PD. Freezing is associated with frequent falls. To overcome freezing, a person should anticipate probable cause of freezing and take remedial actions in advance, like keep stepping on turns or while crossing a door/narrow passage. Some patients start stamping in anticipation to continue stepping. Using compensatory strategies are very effective in freezing management. Individualized strategy need to be tested in each PWP before giving recommendation. In a pilot study, robot assisted gait training has been found to be effective for reducing freezing episodes.[46]Taking lager steps while turning is suggested to avoid freezing. In a cluttered space, patients should turn pivoting one foot while taking turn from other foot. Patients should avoid crossing of feet at any time while taking turns.

Falls prevention

Falls prevention remains the ultimate goal in management of PD. Individualized risk assessment and management is the key of success. Environmental constraint and hazards, like carpets, foot mats and loose wires on ground, polished flooring, wet bathrooms, uneven pedestal and small steps at home that go unnoticed should be identified and corrected as much as possible. Strength training, progressive balance training, postural correction, freezing management, avoidance of dual task, correction of vision impairment are important strategies for falls prevention.

Summary

Posture, balance and gait disorders are disabling features of PD. The pathogenesis of these disorders is far from our understanding of dopamine deficit model due to degeneration of nigral cells in substantia nigra. The wide variety of posture, balance and gait disorders in PWP, that too with intraday temporal variation, poses difficulties in assessment and treatment planning of these disorders. Clinical scales, laboratory based testing and body sensors are in use for assessment of these disorders. Both pharmacological and surgical management offer relief in early to moderate level balance and gait disorders. However, physiotherapy remains a main stay of management of these disorders in addition to pharmacological and surgical management. Compensatory strategies like use of visual, auditory, cognitive and proprioceptive cues help in overcoming deficit of internal rhythm for gait abnormalities. Optimization of muscle length, strength and endurance helps in maintaining good posture and overcoming balance disorders. Fitness training and treadmill training are effective strategies for balance and gait disorders in PD. Falls prevention and overcoming freezing episodes by patient tailored strategies are recommended. Forced exercise has been proposed for neural recovery and prevention of degenerative injuries of brain. Whole body and segmental vibration are new promising modalities to manage balance and gait disorders in PD.

References

1. Hely MA, Reid WGJ, Adena MA, Halliday GM, Morris JGL. The Sydney multicenter study of Parkinson's disease: the inevitability of dementia at 20 years. Mov Disord. 2008;23(6);837–44.
2. Pickering RM, Grimbergen YAM, Rigney U, Ashburn A, Mazibrada G, Wood B, et al. A meta-analysis of six prospective studies of falling in Parkinson's disease. Mov Disord. 2007;22(13);1892–900.
3. Aggarwal R. What Should Constitute a Health Related Quality of Life Scale for Parkinson's Disease? J Clin Diag Res. 2016;10(9);OC35-OC39.
4. Bloem BR, Marinus J, Almeida Q, Dibble L, Nieuwboer A, Post B, et al. Measurement instruments to assess posture, gait, and balance in Parkinson's disease: Critique and recommendations. Mov Disord. 2016;31(9);1342–55.
5. Horak FB. Postural orientation and equilibrium: what do we need to know about neural control of balance to prevent falls? Age and Ageing. 2006;35(2);ii7-ii11.
6. Takakusaki K, Habaguchi T, Ohtinata-Sugimoto J, Saitoh K, Sakamoto T. Basal ganglia efferents to the brainstem centers controlling postural muscle tone and locomotion: a new concept for understanding motor disorders in basal ganglia dysfunction. Neuroscience. 2003;119(1);293–308.
7. Nallegowda M, Singh U, Handa G, Khanna M, Wadhwa S, Yadav SL, et al. Role of sensory input and muscle strength in maintenance of balance, gait, and posture in Parkinson's disease: a pilot study. Am J Phys Med Rehabil. 2004;83(12);898–908.
8. Park J-H, Kang Y-J, Horak FB. What Is Wrong with Balance in Parkinson's Disease? J Mov Disord. 2015;8(3);109.
9. Chawla H, Walia S, Behari M, Noohu MM. Effect of type of secondary task on cued gait on people with idiopathic Parkinson's disease. J Neurosci Rural Pract. 2014;5(1);18–23.
10. Snijders AH, Toni I, Ružička E, Bloem BR. Bicycling breaks the ice for freezers of gait. Mov Disord. 2011;26(3);367–71.
11. Smulders K, Dale ML, Carlson-Kuhta P, Nutt JG, Horak FB. Pharmacological treatment in Parkinson's disease: Effects on gait. Parkinsonism & Related Disorders. 2016;31;3–13.

12. Judge JO, Davis RB, Ounpuu S. Step length reductions in advanced age: the role of ankle and hip kinetics. J Gerontol A Biol Sci Med Sci. 1996;51(6);M303-312.

13. Winter DA. Biomechanics and motor control of human gait.Waterloo, Ontario: Univ. of Waterloo Press; 1991.

14. Sofuwa O, Nieuwboer A, Desloovere K, Willems A-M, Chavret F, Jonkers I. Quantitative gait analysis in Parkinson's disease: comparison with a healthy control group. Arch Phys Med Rehabil. 2005;86(5);1007–13.

15. Leddy AL, Crowner BE, Earhart GM. Functional gait assessment and balance evaluation system test: reliability, validity, sensitivity, and specificity for identifying individuals with Parkinson disease who fall. Phys Ther. 2011;91(1);102–13.

16. Dai B, Ware WB, Giuliani CA. A structural equation model relating physical function, pain, impaired mobility (IM), and falls in older adults. Arch Gerontol Geriatr. 2012;55(3);645–52.

17. Giladi, Shabtai, Simon, Biran, Tal, Korczyn. Construction of freezing of gait questionnaire for patients with Parkinsonism. Parkinsonism Relat Disord. 2000;6(3);165–70.

18. Delbaere K, Close JCT, Mikolaizak AS, Sachdev PS, Brodaty H, Lord SR. The Falls Efficacy Scale International (FES-I). A comprehensive longitudinal validation study. Age and Ageing. 2010;39(2);210–6.

19. Hobert MA, Maetzler W, Aminian K, Chiari L. Technical and clinical view on ambulatory assessment in Parkinson's disease. Acta Neurol Scand. 2014;130(3);139–47.

20. Winfree KN, Pretzer-Aboff I, Hilgart D, Aggarwal R, Behari M, Agrawal S. An untethered shoe with vibratory feedback for improving gait of Parkinson's patients: the PDShoe. Conf Proc IEEE Eng Med Biol Soc. 2012;1202–5.

21. Burleigh-Jacobs A, Horak FB, Nutt JG, Obeso JA. Step initiation in Parkinson's disease: Influence of levodopa and external sensory triggers. Mov Disord. 1997;12(2);206–15.

22. Curtze C, Nutt JG, Carlson-Kuhta P, Mancini M, Horak FB. Levodopa is a Double-Edged Sword for Balance and Gait in People with Parkinson's Disease. Mov Disord. 2015;30(10);1361–70.

23. Franzén E, Paquette C, Gurfinkel VS, Cordo PJ, Nutt JG, Horak FB. Reduced Performance in Balance, Walking and Turning Tasks is Associated with Increased Neck Tone in Parkinson's Disease. Exp Neurol. 2009;219(2);430–8.

24. Doan JB, de Bruin N, Pellis SM, Suchowersky O, Whishaw IQ, Brown LA. Obstacle Avoidance amongst Parkinson Disease Patients Is Challenged in a Threatening Context. Journal of Neurodegenerative Diseases. 2013;2013;e787861.

25. Fietzek UM, Zwosta J, Schroeteler FE, Ziegler K, Ceballos-Baumann AO. Levodopa changes the severity of freezing in Parkinson's disease. Parkinsonism & Related Disorders. 2013;19(10);894–6.

26. Espay AJ, Fasano A, van Nuenen BFL, Payne MM, Snijders AH, Bloem BR. "On" state freezing of gait in Parkinson disease: a paradoxical levodopa-induced complication. Neurology. 2012;78(7);454–7.

27. Plotnik M, Giladi N, Balash Y, Peretz C, Hausdorff JM. Is freezing of gait in Parkinson's disease related to asymmetric motor function? Ann Neurol. 2005;57(5);656–63.

28. Brodsky MA, Park BS, Nutt JG. Effects of a Dopamine Agonist on the Pharmacodynamics of Levodopa in Parkinson Disease. Arch Neurol. 2010;67(1);27–32.

29. Napolitano A, Del Dotto P, Petrozzi L, Dell'Agnello G, Bellini G, Gambaccini G, et al. Pharmacokinetics and pharmacodynamics of L-Dopa after acute and 6-week tolcapone administration in patients with Parkinson's disease. Clin Neuropharmacol. 1999;22(1);24–9.

30. Mehrholz J, Kugler J, Storch A, Pohl M, Hirsch K, Elsner B. Treadmill training for patients with Parkinson's disease. In: The Cochrane Collaboration, editor. Cochrane Database of Systematic Reviews [Internet]. Chichester, UK: John Wiley & Sons, Ltd; 2015 [cited 2016 Oct 7]. Available from: http://doi.wiley.com/10.1002/14651858.CD007830.pub4

31. St George RJ, Carlson-Kuhta P, Nutt JG, Hogarth P, Burchiel KJ, Horak FB. The effect of deep brain stimulation randomized by site on balance in Parkinson's disease. Mov Disord. 2014;29(7);949–53.

32. Rocha PA, Porfírio GM, Ferraz HB, Trevisani VFM. Effects of external cues on gait parameters of Parkinson's disease patients: A systematic review. Clinical Neurology and Neurosurgery. 2014;124;127–34.

33. Azulay JP, Mesure S, Amblard B, Blin O, Sangla I, Pouget J. Visual control of locomotion in Parkinson's disease. Brain. 1999;122(1);111–20.

34. Winfree KN, Pretzer-Aboff I, Hilgart D, Aggarwal R, Behari M, Agrawal SK. The Effect of Step-Synchronized Vibration on Patients With Parkinson's Disease: Case Studies on Subjects With Freezing of Gait or an Implanted Deep Brain Stimulator. IEEE Trans Neural Syst Rehabil Eng. 2013.

35. Capecci M, Serpicelli C, Fiorentini L, Censi G, Ferretti M, Orni C, et al. Postural rehabilitation and Kinesio taping for axial postural disorders in Parkinson's disease. Arch Phys Med Rehabil. 2014;95(6);1067–75.

36. King LA, Horak FB. Delaying mobility disability in people with Parkinson disease using a sensorimotor agility exercise program. Phys Ther. 2009;89(4);384–93.

37. Hass CJ, Buckley TA, Pitsikoulis C, Barthelemy EJ. Progressive resistance training improves gait initiation in individuals with Parkinson's disease. Gait Posture. 2012;35(4);669–73.

38. Chung CLH, Thilarajah S, Tan D. Effectiveness of resistance training on muscle strength and physical function in people with Parkinson's disease: a systematic review and meta-analysis. Clin Rehabil. 2016;30(1);11–23.

39. Knaepen K, Goekint M, Heyman EM, Meeusen R. Neuroplasticity - exercise-induced response of peripheral brain-derived neurotrophic factor: a systematic review of experimental studies in human subjects. Sports Med. 2010;40(9);765–801.

40. Zigmond MJ, Cameron JL, Leak RK, Mirnics K, Russell VA, Smeyne RJ, et al. Triggering endogenous neuroprotective processes through exercise in models of dopamine deficiency. Parkinsonism Relat Disord. 2009;15 Suppl 3;S42-45.

41. Fisher BE, Wu AD, Salem GJ, Song J, Lin C-HJ, Yip J, et al. The effect of exercise training in improving motor performance and corticomotor excitability in people with early Parkinson's disease. Arch Phys Med Rehabil. 2008;89(7);1221–9.

42. Fisher BE, Li Q, Nacca A, Salem GJ, Song J, Yip J, et al. Treadmill exercise elevates striatal dopamine D2 receptor binding potential in patients with early Parkinson's disease. Neuroreport. 2013;24(10);509–14.

43. Frazzitta G, Morelli M, Bertotti G, Felicetti G, Pezzoli G, Maestri R. Intensive rehabilitation treatment in parkinsonian patients with dyskinesias: a preliminary study with 6-month followup. Parkinsons Dis. 2012;2012;910454.

44. Pozo-Cruz B del, Adsuar JC, Parraca JA, del Pozo-Cruz J, Olivares PR, Gusi N. Using whole-body vibration training in patients affected with common neurological diseases: a systematic literature review. J Altern Complement Med. 2012;18(1);29–41.

45. Pozo-Cruz B del, Adsuar JC, Parraca JA, Olivares PR, Herrera E, Gusi N. Whole-body vibration effects in patients affected with Parkinson's disease: a systematic literature review. Rev Andal Med Deporte. 2011;4(2);63–70.

46. Lo AC, Chang VC, Gianfrancesco MA, Friedman JH, Patterson TS, Benedicto DF. Reduction of freezing of gait in Parkinson's disease by repetitive robot-assisted treadmill training: a pilot study. Journal of Neuro Engineering and Rehabilitation. 2010;7(1);51.

47. Dibble LE, Christensen J, Ballard DJ, Foreman KB. Diagnosis of fall risk in Parkinson disease: an analysis of individual and collective clinical balance test interpretation. Phys Ther. 2008;88(3);323–32.

48. Nonnekes J, Goselink R, Weerdesteyn V, Bloem BR. The retropulsion test: a good evaluation of postural instability in Parkinson's disease? J Parkinsons Dis. 2015;5(1);43–7.

Appendix

- Functional reach test: Person standing parallel and near a wall with shoulder flexed at 90 degree, elbow extended and closed fist reaches forward as far as possible without taking a step along a measure on the wall. Third metacarpal head is taken as reference. Person having a score of more than 10 inch has low risk of fall; score of six to ten inch has moderate risk and score of less than six inch has high risk of falls.
- Lateral reach test: Person standing with feet 10 cm apart with shoulder abducted at 90 degree, elbow extended in front of a wall, but not touching to it reach sideways as far as possible without taking a step along a measure on the wall kept at the level of acromian. It is also known as modified functional reach test.
- Mini-BESTest: It is miniature version of BESTest. It has 14 items with a score range 0-28. Higher score represents better balance.
- Timed up and go (TUG) test: The person is required to get up from a standard chair, walk at a comfortable and safe speed to a line three meters away, then turn at the line and walk back to the chair to sit down. Use of assistive devices is allowed for this test.In healthy persons, time for TUG test is less than or equal to 10 sec. A person with TUG test time $\leq$ 20 seconds should be mobile and can go out alone. Time of more than 20 sec is associated with more difficulties in walking alone outside home and may require assistive device. A score of $\geq$ 14 sec is associated with high risk of falls.

- Dynamic Gait Index (DGI): It is an eight item scale to assess gait pattern with varied level of difficulties while walking. Score ranges from 0-24 with higher scores indicating better balance. Cut-off score of 19 for fallers among PWP have 64% sensitivity and 85% specificity.[47]
- Functional Gait Assessment (FGA): Functional gait assessment scale has been adapted from DGI. Its 10 items scale to assess gait pattern with varied level of difficulties including walking forward, backward, with eyes closed, stepping over obstacles, changing gait speeds, with different head turns, and with a narrow base of support. Seven items are same as of DGI while three new items are added. Score ranges from 0-30 with higher the score the better. Cut off score of 15 for fallers among PWP have 72% sensitivity and 78% specificity.[15]
- Retropulsion test: The retropulsion test is widely regarded as the gold standard to evaluate postural instability, and is therefore a key component of the neurological examination in PD. A normal retropulsion test differentiates Hoehn and Yahr stage 2 patients from stage 3 patients. It has three variants viz.,the pull test (as per MDS-UPDRS scale),unexpected shoulder pull without prior warning and the push-and-release test.[48]

Chapter 15

Speech in Parkinson's Disease

Sachin Sureshbabu

Speech abnormalities are one of the commonest but often overlooked manifestations of Parkinson's disease (PD) with nearly 90% of patients being affected.[1-4] The manifold disruptions of communication are explored in this chapter in an attempt to understand its pathological mechanisms, clinical characteristics, diagnostic methods and treatment strategies.

The spectrum of speech abnormalities in PD

The prototypical description of hypophonic and monotonous speech was reiterated by most early investigators of speech in PD. Patients have a low baseline volume of speech, as well a reduced ability to raise the voice, thus fitting the term "hypophonic". Another related phenomenon is the inability to sustain phonation for sufficient period of time. Darley, Aronson and Brown elaborated the symptomatology to include a harsh breathy voice quality, variability of voice pitch, reduced stress, imprecise consonant articulation and short rushes of speech which may be interceded by inappropriate pauses.[2,3]

Although, conventionally considered as a hypokinetic phenomenon resulting from stiffness, dyssynergy and immobility of the articulatory apparatus akin to the motor phenomenology, the spectrum of speech dysfunction can be best addressed by classification into hypokinetic, hyperkinetic and repetitive (iterative) abnormalities.[4-7]

Hypokinetic dysarthria

As the term indicates, the speech becomes progressively slow as the disease marches forward. The volume of speech is reduced (hypophonia) with concomitant decline in the melodic intonation (aprosodic, monotonous speech) and sustainability of phonation (words and sounds die in intensity with time). Unwarranted interruptions and pauses for respiratory effort further slows down the speech and deprives it of flow and luster along with the inevitable result of poor intelligibility and effectiveness to convey meaning as well as emotion.[4-7] It is important to remember that the production of voice

DOI: 10.1201/9781003505488-15

from the larynx is affected much before the dysfunction of the orofacial articulatory apparatus, since the assessment of the voice parameters both clinically and quantitatively using specific softwares can aid in early detection and appropriate intervention.[8,9] The quality of voice can be harsh, rough, breathy or tremulous.[4,5] Although some studies have brought to light tendency for Parkinsonian subjects to speak faster than anticipated, especially while reading short texts, most of the literature supports a progressive decline in rate of speech, which is defined as the number of syllables a person can produce in a unit of time, including pauses.[6,11] However, speech acceleration, especially of the last sentence was found to be higher in PD subjects on a 170 syllable reading task by Skodda et al. Oral festination and rushes of speech are reported in 22-45% cases in various studies.[12,13]

- Skodda has sub-classified hypokinetic dysarthria into four basic elements, namely:
- Hypophonia, characterized by low voice volume with trailing off of vocal production;
- Dysphonia, referring to a breathy, hoarse or harsh voice quality;
- Hypokinetic articulation, denoting imprecise consonants and vowels due to reduced range of articulatory movements; and
- Dysprosodia, which includes reduced voice pitch inflections or monotone and rushed, dysfluent, hesitant or stuttered-like speech.

Hyperkinetic dysarthria

There is debate as to what deviant speech characteristics are to be classified as hyperkinetic or hypokinetic, as the mechanisms involved could have an interplay of both components. The contribution of involuntary movements like tremor and dyskinesias to hyperkinetic speech in PD is also not clearly delineated. Harshness of voice, excessive variation in loudness, imprecise articulation and poor respiratory synergy with speech are some of the after-effects of this proposed pathological substrate.[1-3,7]

Repetitive speech abnormalities

Also known as iterative speech phenomena, these are characterized by repetition of syllables or words or phrases that follow each other words or phrases. Benke et al. observed that this tachyphrenia phenomenon is not uncommon, being found in 28% of those with advanced PD. They were able to delineate two subtypes of this least studied and least understood speech motor phenomenology in PD: Type 1 wherein the speech was apparently hyperfluent with increased rate, low and decreasing in volume as well as poorly articulated. This fitted closely with the early descriptions of palilalia in post-encephalitic Parkinsonism. In the second category (Type 2), the speech was non-fluent, but articulation was near normal and pronounced with constant loudness and rate. The repetitions were staccato-like and could also be classified as "stuttering" as described by other authors in PD.[7] More than half of the patients with stuttering have one or more affected family members.[14]

Objective measures[12,15,16]

Rate of speech: It is defined as the number of sounds a person can produce in a unit of time, say a second or minute, including pauses. This can be measured during normal conversation or reading of a standard paragraph.

Assessment of articulation

Indian Speech and Hearing Association (ISHA) articulation assessment

This employs sets of three Hindi words which are arranged in such a way that the alphabet of interest appears in the first, middle and last part of the word in each set. The alphabet thus studied belong to one of the categories: labials, lingual or palatals.

Acoustic analysis of speech

Objective speech analysis using specialized softwares like VAGHMI and PRAAT enable one to study the acoustic characteristics of speech which includes:

Fundamental frequency (F0): It refers to the frequency of vibration of vocal folds.

Loudness of conversational voice measured in decibels (Db).

Mean Phonation Duration (MPD), defined as the maximum time a subject can sustain tone on a single breath.

S/Z ratio: A comparison of the time for which a patient is able to sustain S sound and Z sound, the latter being more affected by laryngeal dysfunction.

Dysdiadochokinetic rate (DDK): In this test, agility of tongue is tested while quickly repeating certain syllables.

Maximum phonation capacity (MPC): It is a term applied to the range of pitch and loudness. This can be assessed by asking the subject to produce the vowel "a" as high pitched and as low pitched as possible, and as loud and as soft as possible. The mean difference between the maximum and minimum values after three trials is taken as the MPC. These calculations are made possible by recording the voice in a microphone, and then converting it to a digital format which allows storage of data and later analysis using specific speech software.

Natural history of speech dysfunction in PD

Despite being nearly as common as the motor symptoms, only a small percentage of patients report it as the first symptom, except in advanced stages. It is seldom considered disabling by patients and family. But ironically, it has been found that dysfunction of the vocal cords as evidenced by occurrence of changes in the fundamental frequency variability of vocal cords, predates the clinical diagnosis by as many as five years.[17] Speech production declines as the disease progresses; this involves all parameters of speech, including articulatory imprecision, etc. As the disease progresses, flexibility of the peripheral speech apparatus declines leading to worsening of performance in almost all acoustic and voice parameters. It is noted that involvement of speech starts from the larynx and moves up and anteriorly to palate, tongue and lips. Dysarthria can be considered as an axial symptom and enjoys a good deal of association with gait abnormalities. Both these abnormalities also share non-responsiveness to deep brain stimulation. Abnormalities of rate and prosody correlate with freezing of gait, and both in turn show correlation with severity of disease. Oral and gait festination also cluster together.[18] Dysprosody on the other hand does not show significant correlation with the progression of motor disability, thus raising the possibility of independent patho-mechanisms underlying the two.[19]

Pathomechanisms

Speech dysfunction in PD is a combined outcome of derailment of three systems, namely the respiratory, phonatory and articulatory systems that orchestrate the finesse of speech. All these systems use muscles, and hence are basically affected by all the motor system impairments of PD like rigidity, bradykinesia and even tremor. While the respiratory system contributes mainly to regulation of vocal intensity, the phonatory system regulates melody and tone, and the articulatory system precision and pauses.

The lack of coordinated contraction of agonist and antagonist muscles leads to poor flexibility of vocal apparatus, thus preventing smooth articulation. The contraction and relaxation of vocal cord is a vital part of speech, and apart from granting the right intonation, it also indirectly dictates pauses and micro-pauses and the rate of speech. The inability to adequately contract as well relax the vocal cord muscles leads to decreased duration and increased number of pauses and micro pauses. Compensatory mechanism employed by parkinsonian subjects to circumvent this dysfunction is an incomplete closure of vocal cords, which makes speech between less silence a rarity in their conversations.[6] Electromyography (EMG) of thyroarytenoid and cricothyroid muscles performed at rest has revealed abnormal firing of motor units which corroborates the clinical observations.[20] Bradykinesia/hypokinesia of the jaw and oral muscles leads to lack of precision and emphasis on syllables and words. When analyzed separately, the articulation defect is more pronounced for labial as well as lingual and palatal syllables.[15] Caligiri et al. observed that lip movements tend to become slower with faster rates of speech in a seemingly compensatory act to avoid errors of articulation.[21]

Respiratory mechanics of speech is disturbed in manifold ways. The chest wall movements are limited and variable; abdominal contribution is more, and the actual time of a breath spent in speaking is significantly lower.[22]

Failure of sensory processing mechanisms have been recently identified as another contributory factor in Parkinsonian dysarthria. Every treating neurologist would have experienced on several occasions that PD patients are kind of reluctant to talk loudly and do not agree with the opinion of others that their speech was hypohonic or unintelligible. Any form of motor function, including speech is not programmed and delivered from executive centers without the integration and processing of afferent sensory/proprioceptive information arriving from the peripheral organs like the vocal cord and orofacial musculature. In PD, this arc of sensory processing is abnormal in more ways than one. Sensory perception of the peripheral organs is reduced, the processing of afferent information is erratic and the responses to inputs are frequently abnormal.[22,23] This aspect of speech pathophysiology is extremely relevant in the development of therapeutic strategies to improve speech.

Like a mentor who urges you to perform your best, internal cues like a "voice from within" can correct and improve your day to day performance of motor tasks.[23] These vital cues are probably missing in PD subjects as reflected by their tendency to perform better with external cues when it comes to handwriting, gait and speech. Finally, the failure to pay constant attention to one's own speech a process called "vocal vigilance" that involuntarily happens in a normal person also accounts for the reduced volume and mono-pitch.[24]

Emotional processing of speech is significantly impaired in PD patients. They fail to make necessary modulations of loudness and pitch conducive to different emotions and their perception of the same in another man's speech may also be similarly affected.[25] However, the latter association is not consistent and may be related to age of the patient.[26]

An interesting proposition to explain the curious relationships between speech abnormalities and gait (as explained in the natural history) is that these are extrapyramidal "dysrhythmias" due to failing basal ganglionic-cortico-thalamo-cortical circuitry which can no longer process time and execute motor activities smoothly at set intervals. Treatment with levodopa contributes to these derangements significantly.[18]

Revelations from functional imaging

The authors have used f-MRI to evaluate the neural substrates activated during tasks such as sustained phonation and phoneme. Researchers have compared the findings with normal controls and progressive supranuclear palsy (PSP) patients and PD patients. An increased activation of widespread neural networks was observed in PD patients, compared to controls while this apparently compensatory phenomenon was not a feature of PSP.[27] Others have found marked increase in Blood Oxygen Level Dependent (BOLD) signals in primary sensorimotor cortex representing orofacial area in PD subjects during reading as compared to controls.[28] Functional MRI studies have also revealed lack of integrative capacity of brain networks in the performance of combined tasks. Cluster of regional networks involved in speech were not activated when speech task is combined with a hand movement task which the investigators postulate could be due to the greater preference given by the brain to the motor activity.[29] Even the beneficial effects of voice therapy, levodopa and DBS also have a central basis as demonstrated in PET and f-MRI studies.

Psychosocial impact

Communication is a fundamental need and function of every individual in the society, which is especially true in our generation due to growing social pressure. PD disrupts verbal, emotional and gestural attributes of communication. Miller et al. in their attempt to understand how the affected individuals perceive and cope with this handicap unravels some salient points. Their subjects reported that interactions and conversations were definitely hampered; they found it difficult to make others understand, and they struggled with their low voice. On the coping front, there was a tendency to avoid and withdraw socially, but there were also those who wanted to strive and overcome through concentration, perseverance and other efforts.[30]

On the other side, speech abnormalities in PD can be viewed from the perspective of a listener. In the analysis of Jaywant et al. they sounded less interested, less involved, less happy and less friendly as compared to healthy controls.[31] In other words, a person suffering from PD, albeit unintentionally appears "uninvolved, uninterested, unhappy and uninspiring" to his family and social circle. It seems person is not interacting, is not interested and is not happy. As the disease progresses, the caregivers start experiencing difficulty in grasping their conversation. These observations undermine the social handicap associated with speech dysfunction.

Treatment

Speech and voice therapy is the most recommended form for treatment of dysarthria in PD. Nevertheless, a recent review showed that the improvement occurred mainly in loudness and to a minor extent in pitch.[32,33] Having said that, there is evidence to suggest that when administered properly and intensely, patients are more likely to show improvement in all areas of speech including loudness, pitch, articulation, sustained phonation and fatigue. The most widely approved paradigm for speech therapy in PD is Lee Silverman voice treatment at Lee Silverman center for PD in Scottsdale, Arizona, USA. This method functions by scientific attempt to correct the defective circuitry of speech in PD with an emphasis on intensive, regular and consolidated long-term plans for best outcome.[33] Speech and voice therapy can improve or even restore normal speech in mild to moderated dysarthria. However, in advanced stages, behavioural adjustments aimed at improving intelligibility of speech remains the cornerstone of speech therapy. For best execution of speech and voice therapy, a conducive atmosphere is the first requirement, as it has been shown that PD patients poorly emulate the lessons learned in a physician's clinic. Regular therapy at home without the supervision of therapist, out of clinic sessions and simulated environments by using multi-talker devices are useful in this regard. Active participation of family members or caregivers should also be encouraged. Speech intervention should be targeted separately and specifically at each of the deranged components like loudness, imprecise articulation, prosody and the like. Speaking slowly along with a conscious effort to increase loudness can improve intelligibility.[1]

Response to dopaminergic treatment may modestly improve the different dimensions of speech in PD but may not be as dramatic as cited in this early description: "The voiceless, aphonic speech, blurred by pallilalia and unclear articulation, became forceful and clear as in a normal person" (Birkmayer and Hornykiewicz, 1961). Skodda et al. investigated short and long-term response to levodopa administration in 23 PD subjects. They observed absence of significant impact on any of the speech parameters such as phonation, intonation, articulation or speech velocity by levodopa. The only improvement noted was in articulation of vowel and that too only in individual patients.[34]

Deep brain stimulation has been found to benefit only small subset of patient with dysarthria. Certain abnormalities like stuttering can in fact worsen after DBS, but not invariably. STN-DBS may have a beneficial effect on voice tremor and in speech volume, but worsens overall intelligibility and speech quality.[35,36] Speech patterns after DBS has been classified as stuttering, breathy voice, strained voice and spastic-dysarthric subtypes, which were observed in 25, 24, 16, 18 and 17 percent of the 76 patients studied.[36] Inadvertent activation of cortico-bulbar pathways are implicated in the genesis of a strained or spastic speech after STN-DBS. Wertheimer et al. reported in their analysis of 751 subjects with PD of whom 257 underwent DBS that there was significantly more speech impairment, especially "slurring" following DBS which interfered with the social and emotional life of the patient.[37] About 20-35% of patients undergoing DBS develops speech problems and the risk is higher for left sided and bilateral procedures.[38]

Low frequency stimulation (<100 Hz) as opposed to the usually employed high frequency stimulation in subthalamic nucleus-deep brain stimulation (STN-DBS) has been shown to improve or stabilize axial symptoms like freezing of gait (FOG), postural instability and speech impairment. Stimulation of globus pallidus internus also achieves better results in this regard when compared with standard protocol STN-DBS.[39,40]

Another potentially useful treatment for dysarthria is repetitive transcranial magnetic stimulation (r TMS) of the orofacial sensorimotor area (SM1). Its utility is not yet proven, and more studies are needed.[41]

Conclusion

Speech impairment is an integral part of clinical constellation of PD and should receive emphasis during clinical examination and should be emphasized while treating patients. There are several easily available methods and tools to detect and quantify various abnormalities of speech which help in tailoring and individualizing the management protocol which primarily involves administration of speech and voice therapy. The pathophysiological exploration of dysarthria also helps the treating physicians as well as researchers to understand the deviant dimensions of speech distinctly and explicitly.

References

1. Tjaden K. Speech and Swallowing in Parkinson's Disease.Top Geriatr Rehabil. 2008;24(2);115-126.
2. Darley FL, Aronson AE, and Brown JR.: Differential diagnostic patterns of dysarthria. J Speech Hear Res. 1969;12;246-269.
3. Darley F., Aronson AE, and Brown JR. Clusters of deviant speech dimensions in the dysarthrias. J Speech Hear Res. 1969;12;462-496.
4. Logemann JA, Fisher HB, Boshes B, and Blonsky ER. Frequency and cooccurrence of vocal tract dysfunctions in the speech of a large sample of Parkinsonian patients. J Speech Hear Dis. 1978;4;47-57.
5. Critchley M. The clinical significance of dysprosody. Psychol. Med. 1981;1;9-10.
6. Skodda S. Aspects of speech rate and regularity in Parkinson's disease. J Neurol Sci. 2011;310(1-2);231-6.
7. Benke T, Hohenstein C, Poewe W, and Butterworth B. Repetitive speech phenomena in Parkinson's disease. J. Neurol. Neurosurg. Psychiatr. 2000;69;319-32.
8. Zwirner P, and Barnes GJ. Vocal tract steadiness: a measure of phonatory and upper airway motor control during phonation in dysarthria. J Speech Hear Res. 1992;35:;761-768.
9. Darley FL, Aronson AE, and Brown JR. Differential diagnostic patterns of dysarthria. J Speech Hear Res. 1969;12;246-269.
10. Ho AK, Iansek R, Marigliani C, Bradshaw JL, Gates S. Speech impairment in a large sample of patients with Parkinson's disease. Behav Neurol. 1998;11(3);131-13.
11. Martínez-Sánchez F, Meilán JJ, Carro J, Gómez Íñiguez C, Millian-Morell L, Pujante Valverde IM, López-Alburquerque T, López DE. Speech rate in Parkinson's disease: A controlled study. Neurologia. 2016;31(7);466-72.

12. Sachin S, Shukla G, Goyal V, Singh S, Aggarwal V, Gureshkumar., Behari M. Clinical speech impairment in Parkinson's disease, progressive supranuclear palsy, and multiple system atrophy. Neurol India. 2008;56(2);122-6.

13. C Moreau, C Ozsancak, JL, Blatt, P Derambure, A. Destee, L Defebvre, Oralfestination in Parkinson's disease: biomechanical analysis and correlation with festination and freezing of gait, Mov. Disord. 2007;1503e1506.

14. Picillo M, Vincos GB, Sammartino F, Lozano AM, Fasano A. Exploring risk factors for stuttering development in Parkinson disease after deep brain stimulation. Parkinsonism RelatDisord.2017;pii: S1353-8020(17)30065-2.

15. Lazarus JP, Vibha D, Handa KK, Singh S, Goyal V, Srivastava T, Aggarwal V, Behari M. A study of voice profiles and acoustic signs in patients with Parkinson's disease in North India.J Clin Neurosci. 2012;19(8);1125-9.

16. Rusz J, Cmejla R, Ruzickova H, Ruzicka E. Quantitative acoustic measurements for characterization of speech and voice disorders in early untreated Parkinson's disease. J Acoust Soc Am. 2011;129(1);350-67.

17. Harel B, Cannizzaro M and Snyder PJ. Variability in fundamental frequency during speech in prodromal and incipient Parkinson's disease: a longitudinal case study. Brain Cogn. 2004;56;24-29.

18. Ricciardi L, Ebreo M, Graziosi A, Barbuto M, Sorbera C, Morgante L, Morgante F Speech and gait in Parkinson's disease: When rhythm matters.Parkinsonism Relat Disord. 2016;32;42-47.

19. Skodda S, Rinsche H, Schlegel U.Progression of dysprosody in Parkinson's disease over time--a longitudinal study.Mov Disord. 200915;24(5);716-22.

20. Zarzur AP, Duprat AC, Shinzato G, Eckley CA. Laryngeal electromyography in adults with Parkinson's disease and voice complaints. Laryngoscope 2007;117;831-4.

21. Caliguiri MP. The influence of speaking rate on articulatory hypokinesia in Parkinsonian dysarthria. Brain Lang.1989;36;493-502.

22. Hammer, MJ, Barlow, SM. Laryngeal somatosensory deficits in Parkinson's disease: implications for speech respiratory and phonatory control. Experimental Brain Research.2010; 201(3);401–409.

23. A Ho, R Iansek, J Bradshaw Regulation of parkinsonian speech volume: the effect of interlocuter distance.J Neurol Neurosurg Psychiatry. 1999; 67(2);199–202.

24. Rowe J, Friston K, Frackowiak R, Passingham R. Attention to action: specific modulation of corticocortical interactions in humans. Neuroimage. 2002;17(2);988-98.

25. Möbes J, Joppich G, Stiebritz F, Dengler R, SchröderC.Emotional speech in Parkinson's disease. Mov Disord. 2008;23(6);824-9.

26. Martens H, Van Nuffelen G, Wouters K, De Bodt M.Reception of Communicative Functions of Prosody in Hypokinetic Dysarthria due to Parkinson's Disease. J Parkinsons Dis. 2016;6(1);219-29.

27. Sachin S, Senthil Kumaran S, Singh S, Goyal V, Shukla G, Mahajan H, Behari M. Functional mapping in PD and PSP for sustained phonation and phoneme tasks. J Neurol Sci. 2008;273(1-2);51-6.

28. Rektorova I1, Barrett J, Mikl M, Rektor I, Paus T.Functional abnormalities in the primary orofacial sensorimotor cortex during speech in Parkinson's disease. Mov Disord. 2007;22(14);2043-51.

29. Pinto S, Mancini L, Jahanshahi M, Thornton JS, Tripoliti E, Yousry TA, Limousin P. Functional magnetic resonance imaging exploration of combined hand and speech movements in Parkinson's disease. Mov Disord. 2011;26(12);2212-9.

30. Miller N, Noble E, Jones D, Burn D.Life with communication changes in Parkinson's disease.Age Ageing. 2006;35(3);235-9.

31. Jaywant A, Pell MD. Listener impressions of speakers with Parkinson's disease. J Int Neuropsychol Soc. 2010;16(1):49-57

32. Atkinson-Clement C, Sadat J, Pinto S. Behavioral treatments for speech in Parkinson's disease: meta-analyses and review of the literature. Neurodegener Dis Manag. 2015;5(3);233-48.

33. Sapir S, Ramig L, Hoyt P, O'Brien C, Hoehn M. Phonatory respiratory effort (LSVT) vs. respiratory effort treatment for hypokinetic dysarthria: comparing speech loudness and quality before and 12 months after treatment. Folia Phoniatr. 2002;54;296-303.

34. Skodda S1, Visser W, Schlegel U.Short- and long-term dopaminergic effects on dysarthria in early Parkinson's disease. J Neural Transm (Vienna). 2010;117(2);197-205.

35. Klostermann F, Ehlen F, Vesper J, Nubel K, Gross M, Marzinzik F, Curio G, SappokT. Effects of subthalamic deep brain stimulation on dysarthrophonia in Parkinson's disease.J Neurol Neurosurg Psychiatry. 2008;79(5);522-9.

36. Tsuboi T, Watanabe H, Tanaka Y, Ohdake R, Yoneyama N, Hara K, et al.Distinct phenotypes of speech and voice disorders in Parkinson's disease after subthalamic nucleus deep brain stimulation. J Neurol Neurosurg Psychiatry. 2015;86(8);856-64.

37. Wertheimer J, Gottuso AY2, Nuno M2, Walton C3, Duboille A3, Tuchman M3, Ramig L The impact of STN deep brain stimulation on speech in individuals with Parkinson's disease: the patient's perspective. Parkinsonism Relat Disord. 2014;20(10);1065-70.

38. Alomar S, King NK Tam J, Bari AA, Hamani C, Lozano AM. Speech and language adverse effects after thalamotomy and deep brain stimulation in patients with movement disorders: A meta-analysis. Mov Disord. 2017;32(1);53-63. doi: 10.1002/mds.26924.

39. Sidiropoulos C1, Walsh R, Meaney C, Poon YY, Fallis M, Moro E. Low-frequency subthalamic nucleus deep brain stimulation for axial symptoms in advanced Parkinson's disease. J Neurol. 2013;260(9);2306-11

40. St George RJ, Carlson-Kuhta P, Nutt JG, Hogarth P, Burchiel KJ, Horak FB. The effect of deep brain stimulation randomized by site on balance in Parkinson's disease. Mov Disord. 2014;29(7);949-53.

41. Eliasova I1, Mekyska J, Kostalova M, Marecek R, Smekal Z, RektorovaI.Acoustic evaluation of short-term effects of repetitive transcranial magnetic stimulation on motor aspects of speech in Parkinson's disease. J Neural Transm (Vienna). 2013;120(4);597-605.

Chapter 16

Olfaction as a Biomarker in Parkinson's Disease

Abhishek Pathak, Surya Prakash Pandey

Introduction

Parkinson's disease (PD) was first described by James Parkinson in 1817 in his thesis work, which he first noticed in the people of London. Today, the disease affects 1-2% of the population, who is more than 65 years of age.[1] PD is now recognized as a spectrum of alpha synuclein and tau related disorder.[2] Apart from major cardinal motor manifestations of bradykinesia, rigidity, rest tremor and postural instability, it has many non motor features such as altered smell, taste, vision, sleep, bladder and bowel dysfunction, cardiovascular abnormality, salivation, mood and cognition.[2,3] Olfactory dysfunction occurs in more than 90% of PD patients, and it often precedes the motor manifestation for years, marking its importance and it easily exceeds the cardinal motor manifestation of tremors (75%).[4] Though the loss of smell is not complete in 80% of patients, most subjects fail to recognize olfactory loss until tested.[4] Until 1980s not much investigatory tools were available for detecting olfactory dysfunction, and hence it hardly drew any attention. After University of Pennsylvania Smell Identification Test (UPSIT) was developed, several authors have shown olfactory dysfunction in PD patients.[5]

Olfactory anatomy and physiology

Roughly 6,000,000, bipolar olfactory receptor cells, which develop from olfactory placode are present within the human neuroepithelium[6]

Olfactory epithelium consists of four major types of cells; they are: bipolar olfactory cells, basal cells, supporting cells and microvillus cell. Of these, bipolar olfactory cells are the receptor cells.[7] Once the odourant binds to their receptors, they activate Guanosine Triphosphate(GTP) binding proteins, which in turn triggers cAMP, opening ligand gated channel causing influx of calcium and sodium ions, and thus depolarizing the receptors.[8]

DOI: 10.1201/9781003505488-16

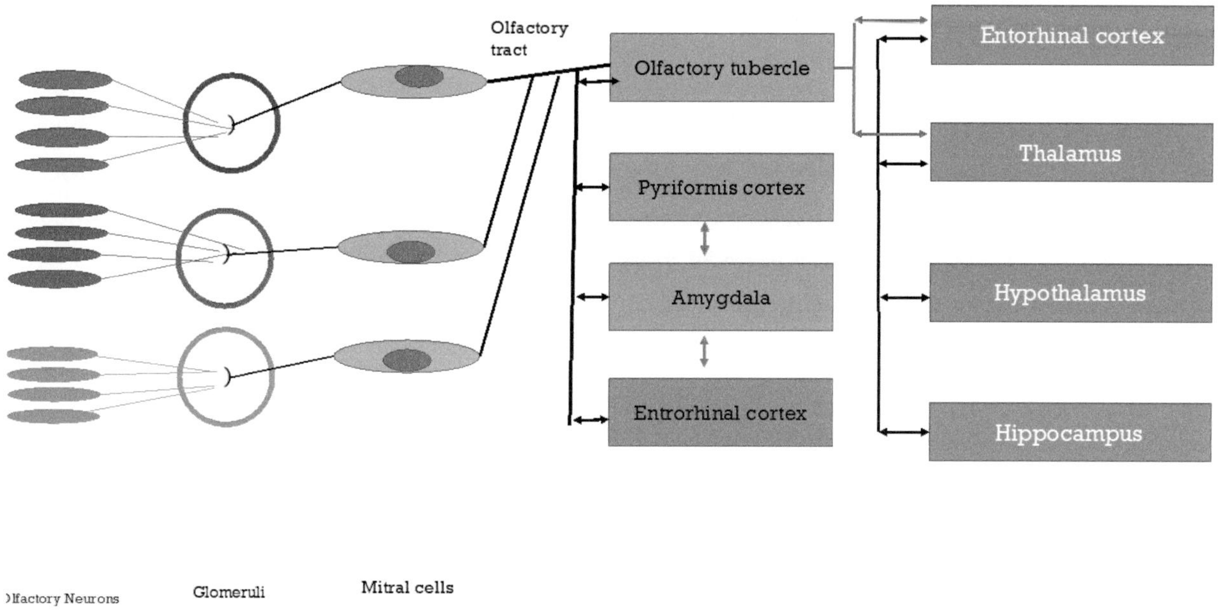

Pathology of olfactory system in PD

Dopamine deficiency in striatum causes the motor symptoms in patients of PD.[9] The mesostriatal dopamine pathway plays a vital role in motor abnormality, and thus when a big amount of dopaminergic neurons are lost in the substantia nigra, symptoms of PD arise.[10]

According to Braak's staging,[11] the pathological process progresses in PD in six stages. In stage 1, the pathology (inclusion body) initially occurs in the olfactory bulb and/or the dorsal motor nucleus of the glossopharyngeal and vagal nerves. Pathology in medulla oblongata and pontine tegmentum defines stage 2. In stage 3, amygdala and substantia nigra are also involved by the pathology, and it also marks the onset of clinical symptoms. Stage 4 marks the progression of the disease to temporal cortex. In the final stages 5 and 6, the pathological process appears in the neocortex, resulting in cognitive problems associated with advanced PD. The staging theory signifies the caudal to rostral progression of the disease which is consistent with the natural history of the disease, signifying the appearance of non-motor symptoms of PD much earlier than the motor symptoms.[12]

Hyposmia as a biomarker

Hyposmia precedes the motor symptoms for years, making it a suitable biomarker for PD. Bohnen et al. in one of the first studies of its kinds compared the accuracy of various clinical motor and non-motor tests for the diagnosis of PD, and concluded that odour identification is the most accurate predictor of PD.[13] Subsequently, Deeb et al. compared the diagnostic accuracy of UPSIT (odour identification) and DaT Scan, and concluded that UPSIT is a cheap and sensitive tool to diagnose PD.[14] The sensitivity of UPSIT and DaT Scan was roughly 90%. Ponsen et al. studied olfaction in around 361 relatives of patients with PD. Hyposmics (n=40) and normosmics (n=39) were followed up for two years with Single Photon Emission Tomography (SPECT) to look for nigro-striatal degeneration. In the two year follow up, four patients who had SPECT abnormalities at baseline developed clinical PD as compared to other relatives who were normosmics.[15]

In another landmark, prospective cohort study, Honolulu-Asia aging study, olfactory assessment was done in men of age 71 to 95 years without any evidence of PD or dementia. They were followed up for eight years, and 35 men developed PD. The odds ratio for PD in the lower quartile of odour identification was 5.2 compared with top two quartiles, proving olfaction may be a useful screening tool to detect risk assessment for PD.[16] In another landmark study, Baba et al. in a three year follow up study showed that a severe olfactory dysfunction is a risk factor for development of dementia in patients of PD.[17] In another follow up of population of 2267 patients, it was found that in over eight years of follow up, 36 patients developed PD, all of whom had had olfactory dysfunction.[18] Eventually, an observational study that evaluated associations between olfactory deficit and the risk of development of Parkinson disease Dementia with an observation period of three years concluded that the changes in olfactory area have a close relationship with the decline in the cognitive domains.[19] The study used functional imaging to demonstrate the same.[20,21] Finally, the study concluded that more severe hyposmia represents an important factor as a predictor for the development of dementia. An ongoing study, Parkinson at Risk Study (PARS) is using UPSIT to screen a cohort of thousands of first degree relatives of PD patients. The first degree relatives who had hyposmia constituted a population who were at risk of developing PD and were an asymptomatic population.[22] This subset of patients then underwent further testing with the more expensive and more specific SPECT-DAT neuroimaging procedure to identify subjects with hyposmia and evidence of nigrostriatal degeneration.[22] The patients who showed reduced tracer uptake were followed up for the risk of developing PD. Preliminary analysis of this cohort has demonstrated that hyposmic first-degree relatives shared several risk factors for the development of PD.[23] Goldstein and colleagues evaluated various clinical and physiologic markers of autonomic dysfunction in patients of anosmic PD, and compared them with the patients with subjects with only mild to moderate hyposmia.[24] The study concluded that the markers were significantly worse in the former. A potential relationship is emerging between olfactory dysfunction and cognitive problems, which are common and disabling in advanced PD. Bohnen and colleagues showed that cholinergic denervation, represented by acetylcholine PET correlated well with the olfactory dysfunction, and also showed decreased acetyl cholinesterase activity in hippocampus and other cortical areas.[25] This also translated to lower performance on several cognitive measures in patients with the worst olfactory function. Though these cross-sectional studies have a suggestion of a correlation between hyposmia and various other non-motor symptoms, further predictions can be assessed only by prospective studies.

Olfaction as a marker of disease progression

Olfactory dysfunction does not improve with any symptomatic drugs given for the PD. No correlation has been observed between olfactory dysfunction and the duration of the disease. Olfactory dysfunction proceeds the motor dysfunction years before the motor dysfunction occurs, and possibly the present tests are incapable of predicting the progression if any; or, by the time motor symptoms appear, most, if not all of the olfactory receptors have degenerated and not many receptors are left to degenerate after motor symptoms appear - a ceiling effect. These might be the possible explanations for the absence of any obvious correlation between olfactory dysfunction and the disease severity. Though, few studies have looked into the radiological and pathological examination of olfactory system as a possible biomarker of the disease. Hummel et al. used volumetric MRI to measure the olfactory bulb volume in a few patients of PD, non-PD hyposmic individuals and normal controls.[26] The study did not reveal any difference between the groups. Similarly, a study of seven patients by Witt M et al. did not find any evidence of Lewy bodies, though the small number may have limited the disease.[27]

Olfactory dysfunction as differential diagnosis of Parkinsonism

Olfactory dysfunction may help in differentiating different form of Parkinsonian and other related disorders. Indeed, Hummer et al., while conducting finer and more sensitive tests of olfaction like olfactory threshold, identification and discrimination in patients of sporadic ataxia and MSA, observed that olfactory thresholds were abnormally high in 16% of the patients. Odour discrimination and odour identification tests were impaired in 44% and 74% of the patients, respectively.[28] Olfactory tests therefore, provide a simple non-invasive tool in differentiating between PD from Parkinsonism plus syndromes (PPS) and as a biomarker of PD. Wenning and colleagues looked at olfactory function among different Parkinsonian syndromes, and concluded that Lewy body PD was associated with more severe hyposmia/anosmia than multiple system atrophy (MSA), compared with relatively normal scores in corticobasal degeneration (CBD) or patients with progressive supranuclear palsy (PSP).[29] While Doty et al. described nearly normal olfaction in PSP patients,[4] Muller demonstrated normal olfactory scores in CBD and PSP patients[30] Few genetic forms of PD (*SNCA*, *LRRK2* or *PINK1* mutations) are associated with olfactory dysfunction, whereas patients with *Parkin* or *DJ-1* mutations have normal olfactory functions.[31,32]

Olfaction assessment methods

Olfaction can be tested by several methods. Psychophysical tests include sensitivity, identification, discrimination and memory. UPSIT is one of the widely accepted instruments to assess odour function[5] as it was the first one to be developed. The UPSIT consists of a 40-item panel of odours (e.g., cherry menthol, motor oil, etc.). These can be tested by "scratch and sniff" cards, followed by the selection of the correct odour in a four-item forced-choice format.[23] Shorter versions based on the same principles (the Brief Smell Identification Test [BSIT]/Cross-Cultural Smell Identification Test [CC-SIT]) have been developed and validated across cultures and are commercially available from Sensonics, Inc. (NJ, USA)[33] (web page Sensonics Inc.). It has two available versions, A and B, designed specifically for Alzheimer's disease (AD), and PD respectively. Doty has also developed an odour-detection threshold assay (the Smell Threshold Test). He has developed a series of 17 bottles with phenyl-ethyl alcohol in gradually increasing concentrations, though high cost limits its routine use in clinical practices. Sniffin stics uses an easy to use pen-like device to identify odour discrimination and identification. Kobayashi et al. devised the Odour Stic Identification Test for the Japanese (OSIT-J), and concluded that the average olfactory score was significantly lower in Americans as compared to the Japanese, reflecting the cultural differences in odour testing.[34] The odour event-related potential (OERP), which is an electrical response to the olfactory stimuli can be recorded and can show the sequential activation of the olfactory pathway. PD patients show prolonged latencies, which may also be confounded by sniff air flow rate and volume. Though useful, clinical use of OERP is restricted due to its limited availability and complicated equipment. An Indian smell kit (INSIT) developed by Behari et al. found it to be non inferior to the Sniffin stics in Indian PD patients, with significant agreement.[35] Different methods of olfactory assessments with various advantages and disadvantages have been shown in table 1.

Table 1

Tests	Modality of odour tested	Disadvantages	Number of odour
University of Pennsylvania Smell Identification Test	Identification	Long time for administration	40
Brief Smell Identification Test	Identification	Malingers cannot be differentiated	12
Smell Threshold Test	Threshold	Expensive and longer time to administer	1 with multiple dilutions
Sniffin' Stics®	Identification, threshold and discrimination	Longer time for administration	16
INSIT	Identification, and discrimination	Short time to administer	10

Conclusion and summary

Olfactory dysfunction pre-dates the motor symptoms of PD by several years. Increased prevalence, low cost and ease of measurement makes it a potential biomarker for the disease (PD). Though the future advances in investigating tool is required to expand its current application, its validity and better sensitivity Alternatively, different methods studying pathology in the olfactory system, including tests which are non-invasive like imaging methods might be used as a marker of neurodegeneration. While biopsy of the olfactory bulb may provide more robust information, it may not be acceptable to many patients and neurologists alike.

References

1. Alves G, Forsaa EB, Pedersen KF, Dreetz Gjerstad M, Larsen JP. Epidemiology of Parkinson's disease. Journal of neurology. 2008;255 Suppl 5;18-32.
2. Barker RA, Rowe DB. Non-dopamine lesions in Parkinson's disease. Oxford University Press; 2010.
3. Lim SY, Lang AE. The nonmotor symptoms of Parkinson's disease--an overview. Movement disorders : official journal of the Movement Disorder Society. 2010;25(1);S123-130.
4. Doty RL, Deems DA, Stellar S. Olfactory dysfunction in parkinsonism: a general deficit unrelated to neurologic signs, disease stage, or disease duration. Neurology. 1988;38;1237-1244.
5. Doty RL, Shaman P, Dann M. Development of the University of Pennsylvania Smell Identification Test: a standardized microencapsulated test of olfactory function. Physiology & behavior. 1984;32;489-502.
6. Xiao Q, Chen S, Le W. Hyposmia: a possible biomarker of Parkinson's disease. Neuroscience bulletin. 2014;30;134-140.
7. Moran DT, Rowley JC, 3rd, Jafek BW, Lovell MA. The fine structure of the olfactory mucosa in man. Journal of neurocytology. 1982;11;721-746.

8. Breer H. Odor recognition and second messenger signaling in olfactory receptor neurons. Seminars in cell biology. 1994;5;25-32.
9. Kish SJ, Shannak K, Hornykiewicz O. Uneven pattern of dopamine loss in the striatum of patients with idiopathic Parkinson's disease. Pathophysiologic and clinical implications. The New England journal of medicine. 1988;318;876-880.
10. Ondo WG. Restless Legs Syndrome: Diagnosis and Treatment: CRC Press;2006.
11. Braak H, Bohl JR, Muller CM, Rub U, de Vos RA, Del Tredici K. Stanley Fahn Lecture 2005: The staging procedure for the inclusion body pathology associated with sporadic Parkinson's disease reconsidered. Movement disorders : official journal of the Movement Disorder Society. 2006;21;2042-2051.
12. Chaudhuri KR, Naidu Y. Early Parkinson's disease and non-motor issues. Journal of neurology. 2008; 255(5);33-38.
13. Bohnen NI, Studenski SA, Constantine GM, Moore RY. Diagnostic performance of clinical motor and non-motor tests of Parkinson disease: a matched case-control study. European journal of neurology. 2008;15;685-691.
14. Deeb J, Shah M, Muhammed N, Gunasekera R, Gannon K, Findley LJ, Hawkes CH. A basic smell test is as sensitive as a dopamine transporter scan: comparison of olfaction, taste and DaTSCAN in the diagnosis of Parkinson's disease. QJM : monthly journal of the Association of Physicians. 2010;103;941-952.
15. Ponsen MM, Stoffers D, Booij J, van Eck-Smit BL, Wolters E, Berendse HW. Idiopathic hyposmia as a preclinical sign of Parkinson's disease. Annals of neurology. 2004;56;173-181.
16. Ross GW, Petrovitch H, Abbott RD, Tanner CM, Popper J, Masaki K, Launer L, White LR. Association of olfactory dysfunction with risk for future Parkinson's disease. Annals of neurology. 2008;63;167-173.
17. Baba T, Takeda A, Kikuchi A, Nishio Y, Hosokai Y, Hirayama K, Hasegawa T, Sugeno N, Suzuki K, Mori E, Takahashi S, Fukuda H, Itoyama Y. Association of olfactory dysfunction and brain. Metabolism in Parkinson's disease. Movement disorders : official journal of the Movement Disorder Society. 2011;26;621-628.
18. Haehner A, Hummel T, Reichmann H. Olfactory loss in Parkinson's disease. Parkinson's disease. 2011;450939.
19. Westervelt HJ, Stern RA, Tremont G. Odor identification deficits in diffuse lewy body disease. Cognitive and behavioral neurology : official journal of the Society for Behavioral and Cognitive Neurology. 2003;16;93-99.
20. Berendse HW, Booij J, Francot CM, Bergmans PL, Hijman R, Stoof JC, Wolters EC. Subclinical dopaminergic dysfunction in asymptomatic Parkinson's disease patients' relatives with a decreased sense of smell. Annals of neurology. 2001;50;34-41.
21. Siderowf A, Newberg A, Chou KL, Lloyd M, Colcher A, Hurtig HI, Stern MB, Doty RL, Mozley PD, Wintering N, Duda JE, Weintraub D, Moberg PJ. [99mTc]TRODAT-1 SPECT imaging correlates with odor identification in early Parkinson disease. Neurology. 2005;64;1716-1720.
22. Stern MB, Siderowf A. Parkinson's at risk syndrome: can Parkinson's disease be predicted? Movement disorders : official journal of the Movement Disorder Society. 2010;25(1);S89-93.

23. Siderowf A, Jennings D, Connolly J, Doty RL, Marek K, Stern MB. Risk factors for Parkinson's disease and impaired olfaction in relatives of patients with Parkinson's disease. Movement disorders : official journal of the Movement Disorder Society. 2007;22;2249-2255.

24. Goldstein DS, Sewell L, Holmes C. Association of anosmia with autonomic failure in Parkinson disease. Neurology. 2010;74;245-251.

25. Bohnen NI, Muller ML, Kotagal V, Koeppe RA, Kilbourn MA, Albin RL, Frey KA. Olfactory dysfunction, central cholinergic integrity and cognitive impairment in Parkinson's disease. Brain : a journal of neurology. 2010;133;1747-1754.

26. Hummel T, Witt M, Reichmann H, Welge-Luessen A, Haehner A. Immunohistochemical, volumetric, and functional neuroimaging studies in patients with idiopathic Parkinson's disease. Journal of the neurological sciences. 201;289;119-122.

27. Witt M, Bormann K, Gudziol V, Pehlke K, Barth K, Minovi A, Hahner A, Reichmann H,Hummel T. Biopsies of olfactory epithelium in patients with Parkinson's disease. Movement disorders : official journal of the Movement Disorder Society. 2009;24;906-914.

28. Abele M, Riet A, Hummel T, Klockgether T, Wullner U. Olfactory dysfunction in cerebellar ataxia and multiple system atrophy. Journal of neurology. 2003;250;1453-1455.

29. Wenning GK, Shephard B, Hawkes C, Petruckevitch A, Lees A, Quinn N. Olfactory function in atypical parkinsonian syndromes. Acta neurologica Scandinavica. 1995;91;247-250.

30. Muller A, Mungersdorf M, Reichmann H, Strehle G, Hummel T. Olfactory function in Parkinsonian syndromes. Journal of clinical neuroscience:official journal of the Neurosurgical Society of Australasia. 2002;9;521-524.

31. Khan NL JS, Lynch JM et al. Mutations in the gene LRRK2 encoding dardarin (PARK8) cause familial Parkinson's disease: clinical, pathological, olfactory and functional imaging and genetic data. Brain : a journal of neurology. 2005;128;2786-2796.

32. Ferraris A, Ialongo T, Passali GC, Pellecchia MT, Brusa L, Laruffa M, Guidubaldi A, Paludetti G, Albanese A, Barone P, Dallapiccola B, Valente EM, Bentivoglio AR. Olfactory dysfunction in Parkinsonism caused by PINK1 mutations. Movement disorders : official journal of the Movement Disorder Society. 2009;24;2350-2357.

33. Doty RL MA, Lee WW. Development of the 12-item Cross-Cultural Smell Identification Test (CC-SIT). Laryngoscope. 1996;106;353–356.

34. Kobayashi M, Saito S, Kobayakawa T, Deguchi Y, Costanzo RM. Cross-cultural comparison of data using the odor stick identification test for Japanese (OSIT-J). Chemical senses. 2006;31;335-342.

35. George J, Jose T, Behari M. Use of Indian smell identification test for evaluating olfaction in idiopathic Parkinson's disease patients in India. Neurology India. 2013;61;365-370.

Chapter 17

Motor Complications of Parkinson's Disease

Ashwani Uttam

Introduction

Parkinson's disease (PD), caused by dopamine deficiency due to degeneration of striato-nigral fibres and accumulation of Lewy bodies in dopaminergic striatal neurons, is generally treated by dopamine replacement therapy. In the initial stage the response to levodopa/carbidopa therapy is smooth, patient being unaware when the response comes and when it dwindles. The response is good initially if the dose of levodopa/carbidopa is high enough and hence, the patient feels he/she is free from any PD symptoms throughout the day. However, over several years the response becomes less optimum and patients experience that they are not the same throughout the day. They experience the response to start several minutes (about 10-15 minutes) after the ingestion of dose and peak in few minutes (about 30 minutes or so), remain good for 3-5 hours and then slowly wear off. Even when the effect is present, it is not 100% as it used to be initially. They remain "off" i.e. when Parkinsons' symptoms appear between two doses of levodopa/carbidopa. Over years the effect becomes further less and the therapeutic window also becomes smaller, effect starting about an hour after ingestion of dose, lasting about 2-3 hours and when present it is only 40-50% of normal. Initially they are not perturbed about this behaviour of treatment effect, but when "off" period become longer and the therapeutic effect is sub optimal, they seek modification of treatment. Thus, we come to the concept of "wearing off" effect and "on-off" periods. These motor fluctuations form integral part of treatment of PD with levodopa/carbidopa. Further, some of the doses may not be able to kick off effect at all ("no on") or may kick off after a prolonged period ("delayed on"). After some years, "on" period may come on suddenly and "off" period may also come suddenly. This is likened to a switch turning "on and off". "Wearing off" effect has been described earlier when the shift from "on" to "off" is gradual. Over certain years some patients develop abnormal choreic movements. If they appear when the effect of levodopa/carbidopa dose is at the peak, it is known as "Peak dose dyskinesia". At times, it starts as the effect of a dose kicks in and again as it wears off - "Biphasic dyskinesia". Some patients during off period get dystonic postures especially of toes and feet. These are "off period dystonias". These may be painful or painless.

DOI: 10.1201/9781003505488-17

Advanced PD is characterized by motor complications. About half of PD patients with levodopa therapy for many years of stable response experience motor fluctuations and dyskinesia.[1] Similar incidences of motor complications are described in Indian patients in one study.[2] Study revealed that 50% of PD patients tend to develop motor fluctuations after about five years of disease onset. Motor fluctuations were seen more with early introduction and longer use of levodopa. Dyskinesia was more commonly seen in patients with earlier age of onset. Recently a relatively rare form of motor complication has been recognized called as "Punding". It is a stereotypic motor behaviour characterized by intense gratification with handling objects.

Table 1: *Types of motor complications*

Motor fluctuations(Off)
"Wearing off" phenomena
Unpredictable "off" periods
Freezing of gait (FOG)
Failure of the "on" response
Acute Akinesia
Dyskinesia
Peak dose dyskinesia
Diphasic dyskinesia
Off period dystonia

Motor complications in PD are classified as below:

- Wearing off/motor fluctuation
- Dyskinesia
- Dystonias
- Motor Blocks
- Freezing of gait
- Speech arrests
- Punding

Wearing off/motor fluctuations

Pathophysiology: With disease progression, further degeneration of nigro-striatal dopaminergic happen leading to reduced ability of neurons to store and release dopamine physiologically in continuous manner. Exogenous dopamine supply in the form of levodopa tablets causes stimulation of postsynaptic dopamine receptors in pulsatile manner because of its short half-life i.e. 90 minutes. Other important factors responsible for fluctuation are erratic intestinal absorption of levodopa related to delayed gastric emptying, slow intestinal motility and high dietary proteins intake (protein-related offs).

Following types of motor fluctuations are seen commonly:

1. **"Wearing off" phenomenon:** It is the predictable return of parkinsonian symptoms in advance of the next scheduled antiparkinsonian dose.
2. **"Unpredictable off":** Sometimes patients become "off" suddenly irrespective of time at which he or she had taken last levodopa dose.

3. **Failure of the "on" response:** Patients with motor fluctuations sometimes fail to turn "on" following a dose of levodopa. It may either be due to late effect of previous dose which did not kick off due to poor absorption severe off or decreased absorption of levodopa because of poor gastric motility.
4. **Acute akinesia:** Sudden and severe exacerbation of PD symptoms also includes an akinetic state which may last for many days and there is no response to antiparkinson drugs. This is different from "wearing off" effects and may sometimes occur in patients not previously treated with levodopa.[3]

Two more types of "off" phenomena though not motor phenomenon is worthy to mention here, because of common pathophysiology. One is "sensory offs" and other is "behavioural offs". Sensory off presents as pain and paraesthesias and behavioural off is present as depression, panic and anxiety.

Dyskinesias

Levodopa-related dyskinesia comprise of involuntary movements, which include chorea, dystonia, ballism and myoclonus.[4] Dyskinesias are common in "on" state and are usually choreiform or ballistic. Dyskinesia when found in the "off" state, is more commonly dystonic. "On" period dyskinesias usually last for few minutes to hours after a dose of levodopa. It is some time reported as tremor by patient and caregiver.

Pathophysiology of dyskinesia

With ongoing degeneration of nigro-striatal dopaminergic neurons, activation of post synaptic receptors depends on exogenous dopamine supply in the form of levodopa tablets. Although the exact neural pathway causing the development of dyskinesia is not known, non-physiological, pulsatile stimulation leading to super sensitive response of striatal glutamate receptor is believed to induce onset of dyskinesia due to dysregulation of function of genes and proteins in downstream basal ganglia-induced neuronal activity, and abnormalities and imbalances in several non-dopamine transmitter systems. These combined changes lead to abnormal firing patterns of neurons between basal ganglion and cerebral cortex.

Following types of dyskinesias are seen commonly:

1. **Peak-dose dyskinesia:** It usually starts 30 to 90 minutes after a dose of levodopa or at peak of therapeutic response. It is usually chorieform in nature. It presents with restlessness and continuous jerky, involuntary movements of the limbs, head, face, trunk, or respiratory muscles. Patients want to remain "on" rather than "off" and therefore they do not usually complain about dyskinesia until severe enough to produce significant problem in daily activities.
2. **Diphasic dyskinesia:** It has two peaks after each levodopa dose, the first when patients turn "on" and the second when they begin to turn "off". It consists of usually large-amplitude dyskinetic movements of the lower body.

Other drugs used in treatment of PD are much less likely to produce these motor abnormalities, but some, such as the dopamine agonists may synergistically exacerbate dyskinesia once they have already occurred following treatment with levodopa. Dyskinesia occurs in 30 - 40 percent of patients with levodopa treatment during the first five years of use and 60 percent or more by ten years.[5] Dyskinesia is more often seen in patients with PD of young-onset (<50 years of age) than in older patients.[6]

Previously, levodopa was used in relatively higher doses when it was first introduced as therapy for PD in the late 1960s and early 1970s and dyskinesia was seen more often. Later on, carbidopa was introduced and combined with levodopa, leading to use of smaller dose of levodopa and consequently less incidence of dyskinesia.

Dystonia

Dystonia is characterized by muscle contractions, sustained or intermittent, causing abnormal, often repetitive, movements, postures, or both. The movements are patterned, twisting and tremulous. It usually occurs in the limbs, but can be seen in face, neck or trunk.

Dystonia is usually seen as withdrawal or wearing off dystonia. It typically occurs in the early morning after long interval without medication at night and it causes painful flexion and inversion of the feet and toes (usually on the side of greater parkinsonian involvement); this usually resolves shortly after taking the first dose of levodopa. Less commonly dystonia can be seen in untreated early PD and rarely of peak levodopa effect; therefore a good history is required for the differentiation of "off" and peak dose dystonia.

Motor blocks

Motor blocks are usually seen in patients with prolonged illness.[2]

1. *Freezing of gait (FOG):* During an episode of FOG, patients become immobilized for seconds to minutes, usually when initiating walking, in a confined space such as a doorway or a closet, or when getting on or off an elevator. An episode of FOG can lead to falls, which is usually forward as the individual tries to move forward and his feet remain glued behind.

2. *Speech arrest:* It is less well studied phenomenon, but often seen in clinical practice. It can present in many ways like palilalia (repetition of first syllable of word), tachyphrenia (hastened speech) and vocal arrests. Patients have word finding difficulties, may have manifestation of cognitive freezing or extreme bradyphrenia.[6]

Punding

"Punding" is relatively rare motor complication of dopaminergic therapy in which patient has intense fascination of repetitive handling of house hold objects. Patient will examine, sort and re-arrange books, watches, pen/pencil, radios, cell phone etc. It is different from obsessive compulsive disorder as punding is not distressing to patients. It occurs with high dose of levodopa (>800-1000mg/day). It is not specific to levodopa and has been previously described with use of central stimulant drugs such as amphetamine and cocaine.[7] Punding has also been described with use of quetiapine.[8] One of Indian study found that punding was present in 12.4% of all PD patients, who had impulse control behavior disorder.[9] Independent predictive factors of punding are high impulsivity, poor disease-related quality of life, young age of onset and high daily dose of levodopa.[10]

Management

The basic principles of management of motor complications include-

1. Adjustment of the levodopa doses and dosing schedule
2. Addition of long acting dopa-mimetic agents

3. Addition of another antiparkinsonian medication
4. Taking levodopa on empty stomach, if contribution of dietary protein is under suspicion for poor drug absorption
5. Considering deep brain stimulation if medical management fails.

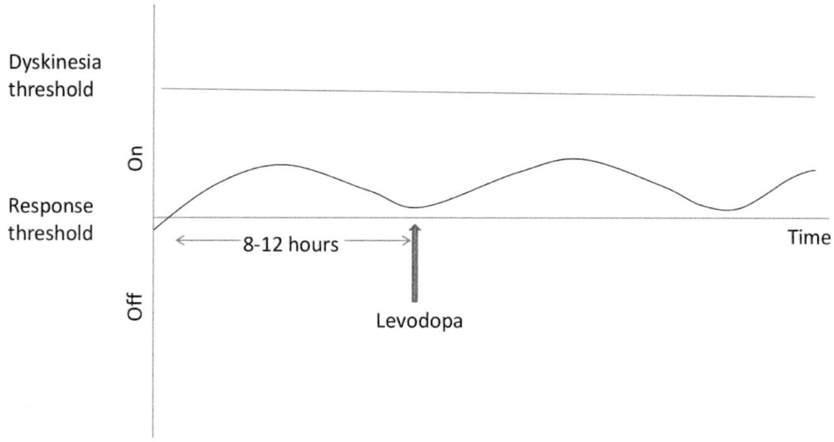

Figure 1: *Early Parkinson's drug disease response curve*

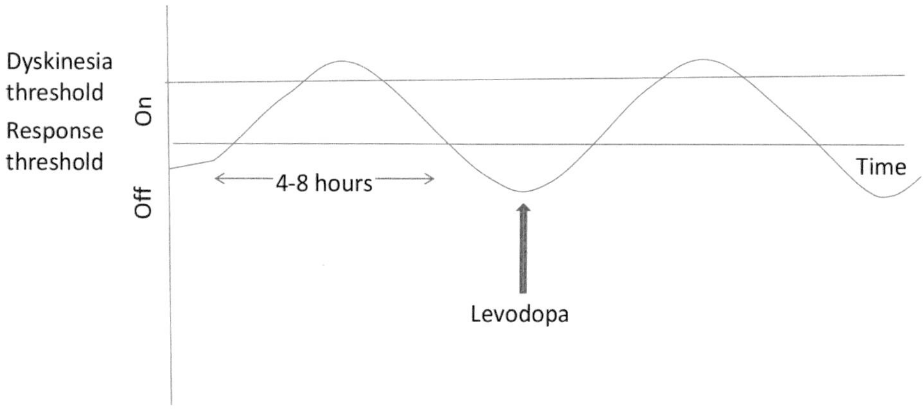

Figure 2: *Advancing Parkinson's drug disease response curve*

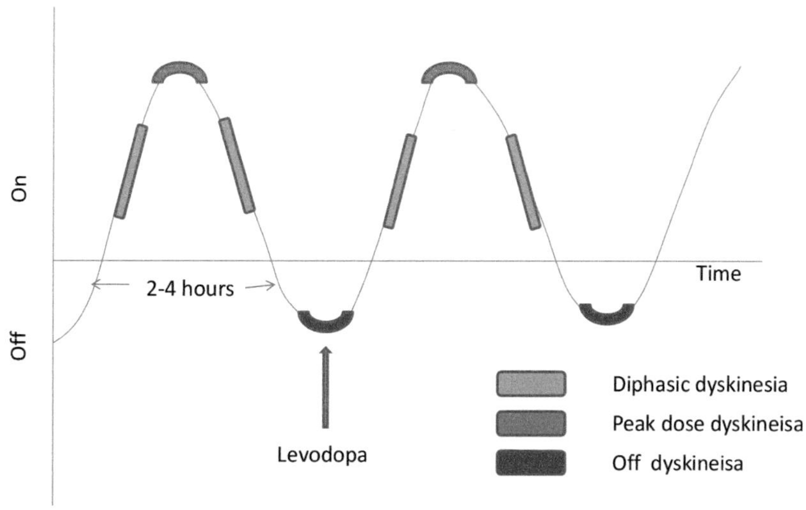

Figure 3: *Advanced Parkinson's drug disease response curve*

Management of "off"/Motor fluctuations

"Wearing off" phenomenon

Patients with PD often start appreciating end-of-dose effect less than four to five hours after a dose of levodopa three to five years after starting levodopa. Strategies that may be helpful include the following:

- Patients should be enquired about their diet and advise them to avoid high protein meals around the scheduled time of levodopa as it competes with absorption of levodopa and blocks the effect of a dose of levodopa if taken soon after or before a meal. They should also be advised to take levodopa empty stomach or at least half an hour before or one to two hours after a meal. Crushing levodopa tablet and mixing it with carbonated beverage also hastens absorption.[11]

- In majority of patient, shortening inter-dose interval fixes "wearing off" problem. In some patients increasing levodopa dose works especially if they are on relatively low dose and not having dyskinesia.[12-14] In the presence of pre-existing dyskinesia, at times it is difficult to titrate the dose and some patients show an "all or none" response. Lower doses produce no clinical response and slightly higher doses trigger a high response with dyskinesia. The precise mechanism for this "all or none" response is uncertain. One hypothesis is that in advanced PD, the pharmacologic response threshold becomes higher as compared to earlier PD. However, this explanation was challenged by findings from a longitudinal study that monitored the response to levodopa infusions in 20 subjects with PD, which found that the onset and offset of antiparkinsonian response and dyskinesia were similar when dyskinesia first emerged.[12] This result suggests that the thresholds for dyskinesia and for the antiparkinsonian response are similar from early in long-term therapy and that a different therapeutic window antiparkinsonian response and dyskinesia threshold no longer exists once dyskinesia emerges.

- Some patients may benefit from alternative levodopa formulations, such as controlled-release levodopa-carbidopa and may reduce "wearing off" phenomenon and could increase "on" period by about one and half hour per day.[13] However, the overall evidence is inconsistent and two systematic reviews concluded that the previous tablet formulation of controlled-release levodopa-carbidopa did not reliably reduce "off" time compared with immediate-release formulations.[14,15] In addition, the tablet formulation of controlled-release carbidopa-levodopa is less well absorbed than immediate-release carbidopa-levodopa; so, a dose has to be increased by one third of routine dose to achieve the same clinical response and that may increase the risk of late day dyskinesia.

- Levodopa-carbidopa in gel form can be delivered as a continuous infusion through a percutaneous gastrojejunostomy tube by battery powered pump. It may be used in place of oral levodopa-carbidopa to decrease off time. In a 12-week double-blind randomized controlled trial of 71 patients with advanced PD,[16] comparing with intermittent dosing of immediate-release oral carbidopa-levodopa showed continuous gel infusion of carbidopa-levodopa led to a significantly greater reduction in mean off-time (4.0 versus 2.1 hours) and an improvement in mean on-time without troublesome dyskinesia (4.1 versus 2.2 hours). Disadvantage of this method includes the need for surgical placement of the percutaneous tube and adverse events related to the device, such as skin infection, peritonitis, gastric reflux, and aspiration.[17] Sudden cessation of levodopa effect may also occur, if the duodenal tube backs up into the stomach, where some of the levodopa gel becomes inactivated by gastric acidity.

- Recently, capsule formulation of extended-release levodopa-carbidopa has been approved and marketed in US (Rytary). It is proposed to be effective than standard levodopa-carbidopa for reducing motor fluctuations. Extended-release (ER) carbidopa-levodopa (CD/LD) is an oral preparation with carbidopa to levodopa ratio 1:4. It rapidly achieves plasma concentration of levodopa at therapeutic level and maintains the level longer than immediate release (IR) of CD/LD.[18] A capsule contains a combination of immediate released (IR) and extended release CD/LD microbeads containing levodopa in strengths of 95, 145, 195 and 245mg. Tartaric acid has been added to formulation to increase intestinal absorption of levodopa distal to jejunum.[19] US FDA approved it for the treatment of PD in 2015. In early PD, it is used in fixed TID dosage.[20] In advanced PD, regimen is based on dosage ranges of previous levodopa treatment.[21] Unlike the previous controlled-release preparation of CD/LD, newer extended-release capsules can be opened up and the beads can be taken with apple sauce to ease administration for patients who have difficulty swallowing pills. However, the extended-release capsule formulation is more expensive than standard formulations of CD/LD.

- Adding an oral dopamine agonist (pramipexole, ropinrole or rotigontine) is other option, but one has to watch for excessive dopaminergic effects (e.g. visual hallucinations, confusion, somnolence, hypotension, impulse control disorders, or an increase in dyskinesia) and levodopa dose should be lowered accordingly.

- Subcutaneous apomorphine is newer addition in drugs for management of "wearing off" especially severe one. This is basically a rescue therapy. Patient must be premedicated with prophylactic antiemetics because of apomorphine's tendency to cause nausea and vomiting. It should be started with lowest dose and escalated for appropriate individual dose. It is expensive and not widely available drug.

- Adding catechol-O-methyl transferase (COMT) inhibitors entacapone or tolcapone prolongs and potentiate levodopa effect. Entacapone is preferred over tolcapone because of the small risk of hepatotoxicity with tolcapone. Entacapone 200mg is given with each dose of levodopa, up to a maximum of eight doses in a day. Dose of tolcapone is 100mg three times a day; the clinical effect is immediate, but trial and error titrations are required. Common side effects of COMT inhibitors are dyskinesia, visual hallucinations, nausea, vomiting, diarrhea and orthostatic hypotension. Lowering the dose of levodopa either before or after the addition of tolcapone or entacapone is usually sufficient for management of side effects. Both drugs cause brown-orange discoloration of urine.

- Adding the monoamine oxidase-B (MAO-B) inhibitors rasagiline, selegiline and sefinamide which prolong the half-life of dopamine in the brain are also beneficial. Dopamine is catabolized by MAO- B enzyme and these inhibitors cause complete and irreversible inactivation of this enzyme leading to enhanced dopamine transmission. Rasagiline is effective for motor complications of PD.[22] In LARGO trial Rasagiline increased "on" period by one hour without increasing troublesome dyskinesia.[22] Rasagiline was found beneficial irrespective of age of patient and simultaneous use of dopamine agonist.

Safinamide was also effective in randomized controlled trials for motor complications of PD when used as adjunctive treatment with levodopa.[23] Compared with placebo in these trials, safinamide increased mean daily "on" time without troublesome dyskinesia and improved motor function. As an example, a 24-week trial of 549 patients with PD taking oral levodopa who had "off" time ≥1.5 hours/day at study entry found that treatment with safinamide increased the mean "on" time without troublesome dyskinesia by approximately 1.4 hours, compared with 0.6 hours for placebo.[24]

Selegiline is another selective MAO-B inhibitor. It also augments levodopa effect in milder form.[25]

Other strategies

Anticholinergic drugs and amantadine have very little effect if any, in improving "wearing off" effect. However, amantadine is effective as an adjunctive medication in reducing dyskinesia.

In some studies, it is suggested that eradication of Helicobacter colonization improves levodopa absorption and reduces motor fluctuations in patients with PD.[26] However, a Cochrane systemic review in 2011 concluded that confirmation in larger clinical trials is needed before routine testing for *H. pylori* and antibiotic eradication can be recommended.[27]

Unpredictable "off" periods

Transitions from being "on" to being "off" can be sudden and unpredictable in some patients and usually bear no obvious relationship with the timing of levodopa dosing. This phenomenon is typically seen in patients with advanced PD, who also have other motor fluctuations and severe dyskinesias. The mechanism for these "off" episodes is not clear, but erratic absorption of levodopa from the gut and pharmacodynamic changes in the brain at the level of the synapse or dopamine receptors are likely culprits. Levodopa pharmacokinetics remains same throughout the course of the disease. Strategies that may be helpful include the following:

- Document the "off" periods, which are in fact unpredictable and long lasting. Although this may seem to be the case based upon the history provided by the patient or caregiver, it often turns out that the "off" periods actually are occurring as an end-of-dose phenomenon that has not been appreciated by the patient or caregiver. True "off" periods usually affect the whole body and should be differentiated from episodes of sudden transient freezing of gait, which can be random and not related to levodopa fluctuations.
- Avoid taking levodopa with high protein meals
- Consider adjusting levodopa dosing schedule. Plasma levodopa levels may be falling below the therapeutic threshold, especially if erratic absorption from the gut is suspected. However, controlled-release preparations are usually not helpful and occasionally exacerbate the situation.
- Alternatively, consider lowering the levodopa dose or adding and titrating a dopamine agonist or COMT inhibitor. In rare cases, sudden "off" episodes may be due to excessive levodopa effects.

Failure of "on" response

Patients with motor fluctuations sometimes fail to turn "on" following a dose of levodopa. Strategies that may be useful in managing failure of "on" response are:

- Avoid taking levodopa with high protein meals and to take it on empty stomach if levodopa induced nausea-vomiting is not an issue.
- Examine gastrointestinal absorption using isotope testing of gastrointestinal transit time.
- Prevent "wearing off" episodes; failure or delay of the "on" responses often occurs after prolonged "wearing off" periods.
- Failure of "on" response commonly occurs in mid-afternoon. Therefore, it may be possible to adjust the dosing schedule to prevent such episodes, such as by adding a dose of levodopa one hour before the usual failure of "on" time, if it is predictable.

The prokinetic drug can be used to improve gastric motility and hence levodopa absorption. Metoclopramide should not be used because it is a dopamine receptor blocker, crosses blood brain barrier and can worsen the symptoms of PD. Domperidone may be useful as a prokinetic agent in patients with PD.[28] Though it is a dopamine receptor blocker, it has selective peripheral activity in upper gastro-intestinal tract and does not cross blood brain barrier. Thus, it does not aggravate parkinsonian symptoms.

Acute akinesia

Acute akinesia is an exacerbation of PD characterized by severe motor retardation lasting for several days and it does not respond adequately to antiparkinsonian medications. This condition is precipitated by systemic infections like urinary tract infections and pneumonia, surgical stress, fractures or medication error (e.g. substituting levodopa-carbidopa 100/25 instead of 250/25, or inadvertently using a dopamine receptor blocking agent such as metoclopramide or an antipsychotic drug with dopamine blocking properties).[3] These episodes may have serious consequences requiring hospitalization to know and treat the precipitating factors.

Akathisia

Akathisia or motor restlessness is a form of levodopa withdrawal. It resembles restless leg syndrome, occurs at night and several hours after levodopa dose. The treatment is slow release levodopa or a dopamine agonist before retiring. Of note, some patients with PD have what seems to be clear restless leg syndrome and not akathisia, even though they are very similar.

Management of dyskinesias

All dyskinesias do not require treatment.[29] Approaches to manage troublesome dyskinesia often begin with adjusting the levodopa regimen or the use of adjunctive medications.

Peak-dose dyskinesia

Lowering individual levodopa dose and administration of lower dosage more frequently if associated with "wearing off", is usually sufficient for management of peak dose dyskinesia in early course of PD. Other measures which are helpful include shifting to a controlled-release preparation of levodopa, or reducing adjunctive dopaminergic drugs such as dopamine agonists, MAO-B inhibitors, or anticholinergic drugs. However, in patients with advanced disease when response is brittle, reducing the dose of levodopa below therapeutic threshold may result in complete failure to become "on". In these patients, levodopa dose should be reduced and the dose of dopamine agonist should be increased to compensate for the reduced levodopa dose. Dyskinesia is less associated with dopamine agonists than levodopa.

Diphasic dyskinesia

Diphasic pattern of dyskinesia is notoriously difficult to manage and usually requires more frequent doses of levodopa to prevent "wearing off" and the re-emergence of "off" dyskinesia prior to each dose. However, this strategy may lead to progressively increasing peak-dose dyskinesia as the day goes on. Addition of a dopamine agonist and significant reduction in the levodopa dose is helpful in such patients or in some cases, if tolerated; frequent doses of levodopa can be kept high, above the dystonia threshold. The main problem with this unusual form of motor fluctuation is management of occasionally prolonged dyskinesia at the end of the day when the patient goes to bed. Adjusted enteric levodopa infusion or deep brain stimulation should be considered if other strategies for management are unsuccessful.

Controlled-release levodopa should not be used in diphasic dyskinesia because its absorption is often delayed, and dyskinesia tends to increase progressively in later part of day.

Amantadine is useful for treating dyskinesia in PD.[30] It's short and long-term benefit have been demonstrated in many studies.[31,32] It's dosage ranges from 100mg to 400mg per day. Side effects include pedal edema, psychosis, livedo reticularis (mottled skin), and hallucinations, all reversible when the drug is stopped.

Clozapine an atypical antipsychotic in low doses (12.5 to 75mg) reduces levodopa induced dyskinesia.[33] The usefulness of clozapine is limited by its uncommon but potentially serious side effect i.e. granulocytopenia. Clozapine should be started when base line absolute neutrophil count (ANC) is above 1,500/mm^3. ANC should be monitored weekly for the first six months of continuous treatment, followed by biweekly monitoring thereafter.[34]

Management of dystonia

Withdrawal or "off" dystonia most commonly occurs in the early morning leading to painful flexion and inversion of feet and toes and this usually responds to first dose of levodopa. These "off" period dystonia can be prevented either by taking long acting controlled-release levodopa before going to bed or by taking levodopa or a dopamine agonist in midnight or on awakening. "Off" period dystonia during the day is managed by shortening the levodopa dose intervals or adding a dopamine agonist.

Peak-dose dystonia is managed similarly as peak-dose dyskinesia.

Management of motor blocks

Freezing of gait

Freezing of gait (FOG) can occur as a transient "off" phenomenon, treatable by the strategies used for wearing off or randomly at variable frequency in patients with advanced PD. Random freezing is poorly responsive to any of the available treatment modalities, including antiparkinson medications and deep brain stimulation. Management needs evaluation to find if freezing is troublesome or not. Attempts to treat it with medication adjustments may not be necessary if it is brief and occurs only occasionally. Physical and occupational therapy to determine if there are specific precipitants such as turning, pivoting or getting on an elevator may lead to suggestions on how to avoid FOG. Movement strategies or cues to that may help the patient "get into gear" include shifting weight, making wider turns, concentrating on taking larger steps forward or using visual imagery as if to step over an obstacle. A cane or special laser cane may also be helpful for FOG. [35]

Management of more severe and prolonged FOG includes medication adjustments, non-medication strategies and assessing comorbidities. In such cases, it is crucial to determine when and how FOG relates to the timing of levodopa doses. Dopa-responsive FOG occurs during end of dose "wearing off" periods and usually responds to shortening levodopa dose intervals or increasing the levodopa dose so as to avoid "wearing off" episodes. Adding or increasing dopamine agonists is usually not as effective and may even aggravate FOG. There are conflicting data from two small randomized controlled trials regarding the utility of methylphenidate for gait freezing and hypokinesia.[36,37] Subthalamic nucleus deep brain stimulation (DBS) is another strong option for levodopa-responsive FOG.[27]

Levodopa-related FOG is less common, accounting for only about five percent of cases of FOG, but much more difficult to treat. Fortunately, it is usually far briefer in duration than dopa-responsive FOG. It occurs at peak levodopa effect and may respond to reducing the dopamine agonist first, followed by reducing the levodopa dose (if necessary) although this step may compromise the levodopa response of other motor symptoms. The value of pedunculo-pontine nucleus DBS is being considered recently, but unproven in the management of levodopa-related FOG, as so far results have been inconsistent.[38,39]

Speech arrest

Similar to FOG, it is also often unresponsive to all pharmacotherapy. Some improvement can be achieved with rehabilitation interventions.[40]

Management of "punding"

Management of punding includes reduction of dopaminergic medications.[41] Entacapone can be added in case of worsened motor condition due to dopaminergic medications reduction.[42] Selegiline enhances levodopa function, but it generates amphetamine-like metabolites, which may exacerbate punding and hence should not be used in PD patients with punding.[43] Amantadine reduces punding by blocking N-methyl-D-aspartic acid receptors.[44]

Failure of medical management

Nigrostriatal degeneration and dopamine deficiency cause excessive subthalamic nucleus (STN) excitation of internal globus pallidus interna (GPi). This leads to excessive inhibition of thalamus, in turn causing reduced thalamic stimulation to cortex. This effect clinically translates to bradykinesia and rigidity.

Pallidotomy a neurosurgical procedure produces lesion in GPi and causes reduction in pallidal inhibitory effect which leads to improve Parkinsonism. Later on it was discovered that chronic electrical stimulation of GPi had similar and significant positive effect on bradykinesia, rigidity and tremors. Deep brain stimulation (DBS) was first used by Benabid and Pollack in 1987 for the treatment of medically refractory tremor in patients with PD and due to its reversible nature it has replaced lesion surgeries like pallidotomy and thalamotomy. In DBS, chronic and usually high frequency electrical stimulation is given to subthalamic or internal globus pallidus through electrodes implanted deep in subcortical regions of the brain. The electrodes are connected with the stimulating device usually implanted in the anterior chest wall through subcutaneous wires. Through stimulating battery, changes in current's frequency, pulse width, amplitude, and depth of stimulation modifications are done according to motor symptoms and sign.

Main advantage of DBS is that it does not make any significant or permanent lesion. Disadvantage includes requirement of intensive adjustments in the postoperative period, lifelong maintenance with need for battery replacement, risk of infection and hardware complications due to mechanical disruption. DBS is expensive too.

This has been proven in randomized control trial that deep brain stimulation is better than best medical therapy in improving motor fluctuations and dyskinesia associated with advanced PD.[45] Therefore, selected patients with advanced PD who have severe motor fluctuations or dyskinesia that are unresponsive to the medical interventions outlined in the sections above can be referred for treatment with deep brain stimulation.

Conclusion

- Patients with PD, who take levodopa for many years, develop motor fluctuations and dyskinesia as the disease progresses. Younger patients on levodopa have higher risk of complications. Dyskinesia can occur in any patient, if the dose is high enough. Motor fluctuations are alterations between periods of being "on", during which the patient enjoys a good response to a dose of levodopa and being "off" during which the patient experiences worsening parkinsonian symptoms. Dyskinesia consists of various types of abnormal involuntary movements mainly choreiform, which are due to the effects of levodopa.
- Basic principle for management of motor complications include adjusting the levodopa regimen (including changes to the amount of levodopa and inter-dose intervals or the levodopa formulation), and adding adjunct medication such as dopamine agonists, catechol-O-methyl transferase (COMT) inhibitors or monoamine oxidase (MAO) B inhibitors. Adjusting dietary protein is appropriate for those in whom a careful history reveals that protein interference is a problem.
- "Wearing off" can be managed initially by increasing the dose of levodopa. However, this may induce or exacerbate dyskinesia, confusion, hallucinations, insomnia or other side effects. Shortening inter-dose interval, while administering lower doses (i.e. smaller but more frequent doses) is also a consideration. Extended-release carbidopa-levodopa may be more effective than standard carbidopa-levodopa for reducing motor fluctuations but increases expense. Adding a second drug is suggested if adjustments to the levodopa regimen are not successful for reducing "wearing off". Options for adjunct treatment include dopamine agonist, COMT inhibitor or MAO-B inhibitor drugs.
- Acute akinesia should prompt a search for systemic infection or other intercurrent medical problems that are capable of causing a sudden worsening of parkinsonism.
- Not all dyskinesia require treatment, as milder dyskinetic movements are remarkably well tolerated by most patients with PD. Early in the course of PD, peak-dose dyskinesia can be managed by lowering individual levodopa dose, switching to a controlled-release preparation of levodopa or

reducing adjunctive drugs. For more advanced patients with PD, the dose of dopamine agonist should be titrated up and the levodopa dose correspondingly reduced.

- "Off" period dystonia that occurs early in the morning is managed either by taking controlled-release levodopa just before going to bed or by taking levodopa or a dopamine agonist during the night or on awakening in morning. Peak dose dystonia is managed similarly to peak dyskinesia.
- Motor Blocks like FOG and speech arrests are usually pharmaco-unresponsive.
- Medically refractory patients can be evaluated and referred for deep brain stimulation surgery.

References

1. Olanow CW, Watts RL, Koller WC. An algorithm (decision tree) for the management of Parkinson's disease: treatment guidelines. Neurology 2001;56;S1
2. Denny AP, Behari M. Motor fluctuations in Parkinson's disease. J Neurol Sci. 1999;165(1);18-23
3. Onofrj M, Thomas A. Acute akinesia in Parkinson disease. Neurology 2005;64;1162.
4. Quinn N. Drug treatment of Parkinson's disease. BMJ 1995;310;575.
5. Ahlskog JE, Muenter MD. Frequency of levodopa-related dyskinesias and motor fluctuations as estimated from the cumulative literature. Mov Disord. 2001;16;448.
6. Van Gerpen JA, Kumar N, Bower JH, et al. Levodopa-associated dyskinesia risk among Parkinson disease patients in Olmsted County, Minnesota, 1976-1990. Arch Neurol. 2006;63;205.
7. Schiorring E. Psychopathology induced by "speed drugs". Pharmacol Biochem Behav 1981;14:109-22.
8. Miwa H, Morita S, Nakanishi I, Kondo T. Stereotyped behaviors or punding after quetiapine administration in Parkinson's disease. Parkinsonism Relat Disord 2004;10;177-80.
9. Sharma A, Goyal V, Behari M, Srivastva A, Shukla G, Vibha D. Impulse control disorders and related behaviours (ICD-RBs) in Parkinson's disease patients: Assessment using "Questionnaire for impulsive-compulsive disorders in Parkinson's disease" (QUIP). Ann Indian Acad Neurol. 2015; 18(1);49–59.
10. Lawrence AJ, Blackwell AD, Barker RA, Spagnolo F, Clark L, Aitken M R, et al. (2007). Predictors of punding in Parkinson's disease: results from a questionnaire survey. *Mov. Disord.* 22, 2339–2345.
11. Daroff RB, Jankovic J, Mazziota JC, Pomeroy SL. Bradley's Neurology in Clinical Practice: Elsevier; 2016.
12. Nyholm D, Stepien V. Levodopa fractionation in Parkinson's disease. J Parkinsons Dis 2014;4;89.
13. Hutton JT, Morris JL, Román GC, et al. Treatment of chronic Parkinson's disease with controlled-release carbidopa/levodopa. Arch Neurol 1988;45;861.
14. Gershanik OS. Improving L-dopa therapy: the development of enzyme inhibitors. Mov Disord. 2015;30;103–113.
15. Nausieda PA, Hsu A, Elmer L, et al. Conversion to IPX066 from standard levodopa formulations in advanced Parkinson's disease: experience in clinical trials. J Parkinsons Dis. 2015;5;837–845.
16. Hauser RA, Hsu A, Kell S, et al. Extended-release carbidopa-levodopa (IPX066) compared with immediate-release carbidopa-levodopa in patients with Parkinson's disease and motor fluctuations: a phase 3 randomised, double-blind trial. Lancet Neurol 2013;12;346.

17. Stocchi F, Hsu A, Khanna S, et al. Comparison of IPX066 with carbidopa-levodopa plus entacapone in advanced PD patients. Parkinsonism Relat Disord. 2014;20;1335
18. Hauser RA, Ellenbogen AL, Metman LV, et al. Crossover comparison of IPX066 and a standard levodopa formulation in advanced Parkinson's disease. Mov Disord. 2011;26;2246–2252.
19. Hayashi M, Tomita M. Mechanistic analysis for drug permeation through intestinal membrane. Drug Metab Pharmacokinet. 2007;22:67–77.
20. Pahwa R, Lyons KE, Hauser RA, et al. Randomized trial of IPX066, carbidopa/levodopa extended release, in early Parkinson's disease. Parkinsonism Relat Disord. 2014;20;142–148.
21. Hauser RA, Hsu A, Kell S, et al. Extended-release carbidopa-levodopa (IPX066) compared with immediate-release carbidopa-levodopa in patients with Parkinson's disease and motor fluctuations: a phase 3 randomised, double-blind trial. Lancet Neurol. 2013;12;346–356.
22. Rascol O, Brooks DJ, Melamed E, et al. Rasagiline as an adjunct to levodopa in patients with Parkinson's disease and motor fluctuations (LARGO, Lasting effect in Adjunct therapy with Rasagiline Given Once daily, study): a randomised, double-blind, parallel-group trial. Lancet 2005; 365:947.
23. Schapira AH, Fox SH, Hauser RA, et al. Assessment of Safety and Efficacy of Safinamide as a Levodopa Adjunct in Patients With Parkinson Disease and Motor Fluctuations: A Randomized Clinical Trial. JAMA Neurol. 2017;74;216.
24. Borgohain R, Szasz J, Stanzione P, et al. Two-year, randomized, controlled study of safinamide as add-on to levodopa in mid to late Parkinson's disease. Mov Disord. 2014;29;1273.
25. Waters CH, Sethi KD, Hauser RA, et al. Zydis selegiline reduces off time in Parkinson's disease patients with motor fluctuations: a 3-month, randomized, placebo-controlled study. Mov Disord 2004;19;426.
26. Pierantozzi M, Pietroiusti A, Brusa L, et al. Helicobacter pylori eradication and l-dopa absorption in patients with PD and motor fluctuations. Neurology. 2006;66;1824.
27. Rees K, Stowe R, Patel S, et al. Helicobacter pylori eradication for Parkinson's disease. Cochrane Database Syst Rev. 2011;CD008453.
28. Jost WH. Gastrointestinal motility problems in patients with Parkinson's disease. Effects of antiparkinsonian treatment and guidelines for management. Drugs Aging. 1997;10;249.
29. Hung SW, Adeli GM, Arenovich T, et al. Patient perception of dyskinesia in Parkinson's disease. J Neurol Neurosurg Psychiatry 2010;81;1112.
30. Odin P, Ray Chaudhuri K, Slevin JT, et al. Collective physician perspectives on non-oral medication approaches for the management of clinically relevant unresolved issues in Parkinson's disease: Consensus from an international survey and discussion program. Parkinsonism Relat Disord 2015;21;1133.
31. Thomas A, Iacono D, Luciano AL, et al. Duration of amantadine benefit on dyskinesia of severe Parkinson's disease. J Neurol Neurosurg Psychiatry. 2004;75;141.
32. Wolf E, Seppi K, Katzenschlager R, et al. Long-term antidyskinetic efficacy of amantadine in Parkinson's disease. Mov Disord. 2010;25;1357.

33. Durif F, Debilly B, Galitzky M, et al. Clozapine improves dyskinesias in Parkinson disease: a double-blind, placebo-controlled study. Neurology. 2004;62;381.
34. Bennett JP Jr, Landow ER, Schuh LA. Suppression of dyskinesias in advanced Parkinson's disease. II. Increasing daily clozapine doses suppress dyskinesias and improve parkinsonism symptoms. Neurology. 1993;43;1551.
35. Nonnekes J, Snijders AH, Nutt JG, et al. Freezing of gait: a practical approach to management. Lancet Neurol. 2015;14;768.
36. Espay AJ, Dwivedi AK, Payne M, et al. Methylphenidate for gait impairment in Parkinson disease: a randomized clinical trial. Neurology 2011;76;1256.
37. Moreau C, Delval A, Defebvre L, et al. Methylphenidate for gait hypokinesia and freezing in patients with Parkinson's disease undergoing subthalamic stimulation: a multicentre, parallel, randomised, placebo-controlled trial. Lancet Neurol. 2012;11;589.
38. Ferraye MU, Debû B, Fraix V, et al. Effects of pedunculopontine nucleus area stimulation on gait disorders in Parkinson's disease. Brain. 2010;133;205.
39. Thevathasan W, Cole MH, Graepel CL, et al. A spatiotemporal analysis of gait freezing and the impact of pedunculopontine nucleus stimulation. Brain 2012;135;1446.
40. Park HK, Yoo JY, Kwon M, et al. Gait freezing and speech disturbance in Parkinson's disease. Neurol. Sci. 2014 35;357–363.
41. Avila A, Cardona X, Martin-Baranera M, Bello J, and Sastre F. Impulsive and compulsive behaviors in Parkinson's disease: a one-year follow-up study. J. Neurol. Sci. 2011. 310;197–201.
42. Evans AH, Strafella AP, Weintraub D, and Stacy M. Impulsive and compulsive behaviors in Parkinson's disease. Mov. Disord. 2009. 24;1561–1570.
43. Shin HS. Metabolism of selegiline in humans. Identification, excretion, and stereochemistry of urine metabolites. Drug Metab. Dispos. 1997. 25;657–662.
44. Fasano A, Ricciardi L, Pettorruso M, and Bentivoglio AR. Management of punding in Parkinson's disease: an open-label prospective study. J. Neurol. 2011. 258;656–660.
45. Fasano A, Daniele A, Albanese A. Treatment of motor and non-motor features of Parkinson's disease with deep brain stimulation. Lancet Neurol 2012; 11;429.

Chapter 18

Medical Management of Parkinson's Disease in the Early, Mid and Advanced Stages

Geetika Bajpai

Introduction

In 1817, 200 years ago James Parkinson's published an essay on "shaking palsy" - the first scientific description of this disease.[1] Today, Parkinson's disease (PD) is the second most common neurodegenerative disorder with a natural history of gradual but relentless progression. The basic pathophysiology of the disease is of dopamine depletion, which explains the major motor manifestations. So a simple solution would be to replenish dopamine. Levodopa, a synthetic precursor of dopamine, was discovered 140 years later.[2] It got converted to dopamine in the brain and helped in dramatic improvement in symptoms of PD. But, not all apparently simple looking problems have simple solutions. Some have deep rooted enigmas which unfold with time, and it so happened with PD. We discovered gradually that Levodopa had its own set of complications, and following Braak's staging, it was revealed that PD actually begins in non-dopaminergic cells opening other treatment targets.

Treatment principles

The goal of medical management is to control the signs and symptoms while minimizing the adverse effect. The basic principles are:

1. Neuroprotective agents should be given a priority.
2. Dopaminergic, as well as non-dopaminergic drugs, should be considered for motor as well as non-motor symptoms.
3. Therapy should be individualized.
4. Integrated therapy, including medicine, surgery and most importantly physical therapy, should be offered in combination for the best results.

Neuro-protective therapies are still in their early stage of development. Many drugs were studied but none have proved to be effective in modifying the course of disease till date. New drugs are being

DOI: 10.1201/9781003505488-18

explored and more evidence is being collected for the promising agents giving hope for the future treatment of PD.[3]

Table 1: Neuro-protective agents tested so far and their status

Agent	Mechanism of action	End point of study	Result
Vitamin E	Antioxidant	Time needed to start levodopa	Not useful
Riluzole	Glutamate anta-gonist	As above	Not useful
TCH 346	Preventing apoptosis	As above	Not useful
CEP1347	As above	As above	Not useful
Minocycline	Anti -inflammatory	Change in total UPDRS score	Not useful
CoQ10	Mitochondria respiratory enzyme	As above	Not useful
Creatine	Respiratory function facilitator	Long term follow-up	Not useful
Neurturin	Trophic factor	Change in total UPDRS score	Not useful
GDNF	Trophic factor	As above	Not useful
Rasagiline	Preventing apoptosis	Delayed start	Not useful
Rasagiline	As above	Delayed start	Useful
Pramipexole	As above	As above	Not useful
Pramipexole	As above	β-CIT SPECT	Not useful
Selegiline	As above	Change in total UPDRS score (wash out)	Useful
Selegiline	As above	Time to start levodopa	Useful
Ropinirole	As above	β-CIT SPECT	Useful

Therapeutic choices available

There are two categories of drugs used mainly for both motor as well as non-motor symptoms – dopaminergic and non-dopaminergic drugs.

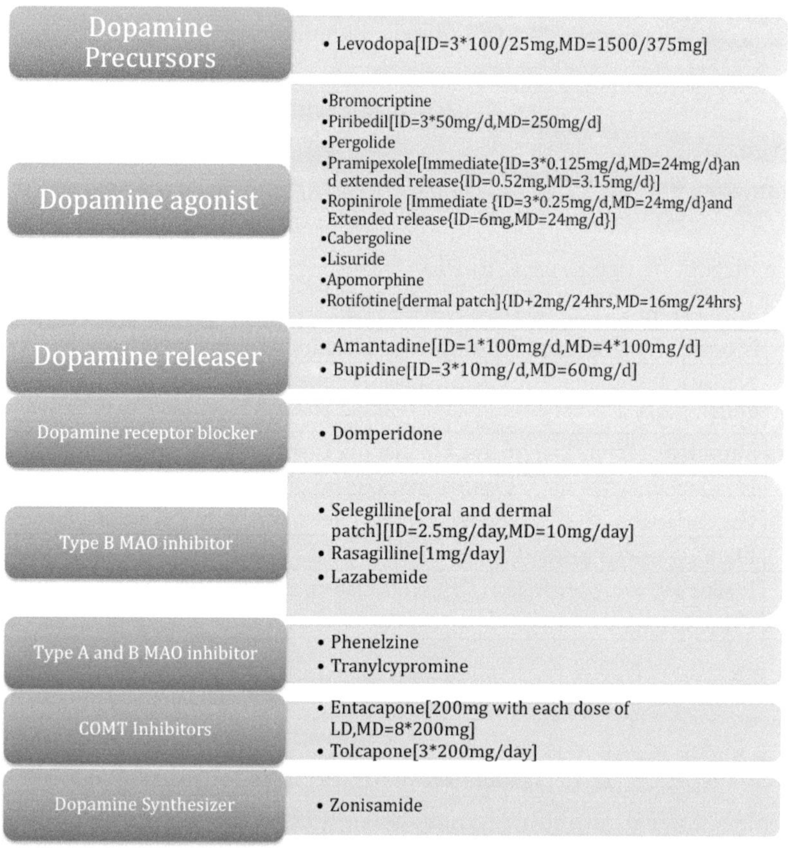

Figure 1: Dopaminergic agents used for motor manifestations

Table 2: Non-dopaminergic agents for motor symptoms[4]

Symptoms	Mechanism	Example
Tremor	Antimuscarinic	Trihexphendyl, Benztropine, Ethopropazine
Motor fluctuations and parkinsonism	Adenosine A2A receptor antagonists	Istradefylline (Approved in Japan) Tozadenant is currently in phase III
	Mixed activity that includes inhibition of sodium/calcium channels and monoamine oxidase type B (MAOB) activity	Safinamide is approved in Europe. Zonisamide is approved for use in Japan
Gait disorder, falls and freezing of gait	Cholinesterase inhibitors	Donepezil and rivastigmine
	Noradrenaline reuptake inhibitor	Methylphenidate-Higher doses needed upto 80mg /day
Stiffness	Muscle relaxant	Cyclobenzaprine, Baclofen, Diazepam

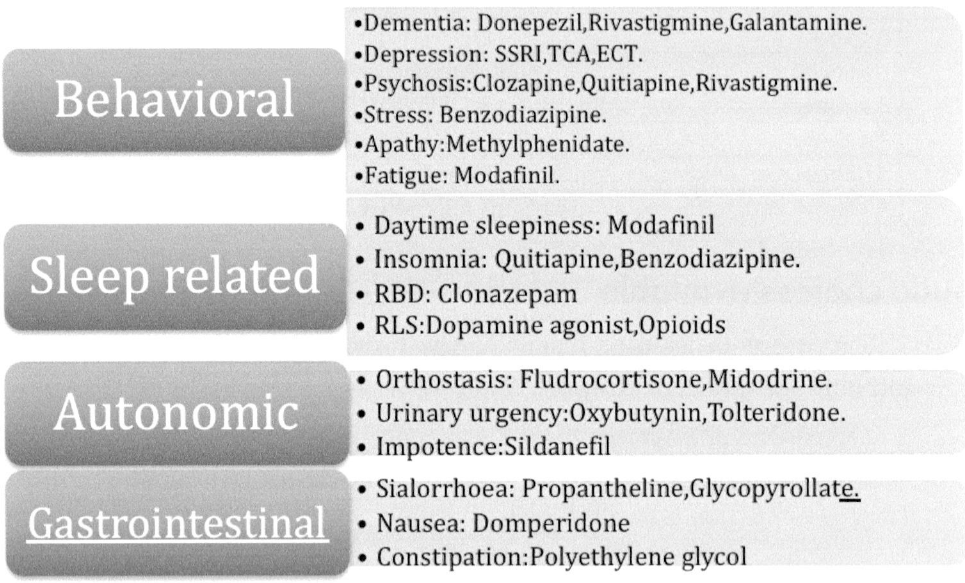

Figure 2: Non-Dopaminergic drugs for non-motor symptoms[5]

Table 3: Major side effects of drugs used in PD

Drug	Side effects
Levodopa	Nausea, Orthostatic Hypotension, Dyskinesia, Hallucination
Dopamine Agonist	Nausea, Hallucination, Orthostatic Hypostension, Impulse control disorders, Edema, Sleep attacks
Selegiline	Stimulant effect, Dizziness, Headache, Confusion, Exacerbation of LD adverse effects
Rasagiline	Headache, Arthralgia, Dysphagia, Depression, Flu like symptoms, Constipation, Exacerbation of LD adverse effects
COMT inhibitors	Dark coloured urine, Exacerbation of LD adverse effects, Hepatotoxicity (Tolcapone)
Amantadine	Hallucination, Confusion, Blurred vision, Ankle edema, Livedo reticularis, Dry mouth, Constipation

Few side effects need special mention because of their unique association with the management of PD:

1. Impulse control disorder: It is associated with dopamine agonist use. It implies behaviour comprising of pathological gambling, hypersexuality, binge eating and compulsive shopping. This requires gradual replacement of Dopamine Agonst by Levodopa.

2. Punding is described as repetitive, often purposeless stereotyped behaviour such as sorting or disassembling. It is also associated with dopamine over activity. Amantidine and Quetiapine were beneficial in some open labelled studies.
3. Dopamine Dysregulation Syndrome: It is also an addictive behaviour which impairs physical and social functioning. It is also attributed to dopaminergic drug over use. Hence, it requires gradual dose reduction of these drugs.

All dose adjustments should be gradual to avoid precipitating withdrawal syndrome in case of Dopamine agonist, and Neuroleptic malignant syndrome in case of Levodopa.

Stages of PD[4]

PD can be divided into five stages based on symptomology. The first two stages are categorized as early and the last two form the severe or advanced stage. This classification helps in setting realistic goals for both the patient and the physician, better prognostication and systematic management of the disease.

Table 4: Stages of PD[7]

Early PD		**Moderate PD**	**Severe PD**	
Stage 1	**Stage2**	**Stage 3**	**Stage 4**	**Stage 5**
Unilateral involvement	Bilateral involvement	Loss of balance and slowness of movement	Severe symptoms	Inability to rise
Tremors of hand, clumsy leg, one half of face involved affecting expression rigidity	Loss of facial expression on both sides, Decreased blinking, Speech abnormality, Truncal rigidity	Balance is affected, Unable to make involuntary and automatic adjustments	Needs assistance in activity of daily living, Can still walk and stand unassisted	Freezing prominent, Risk of fall specially while turning, Hallucination and delusion

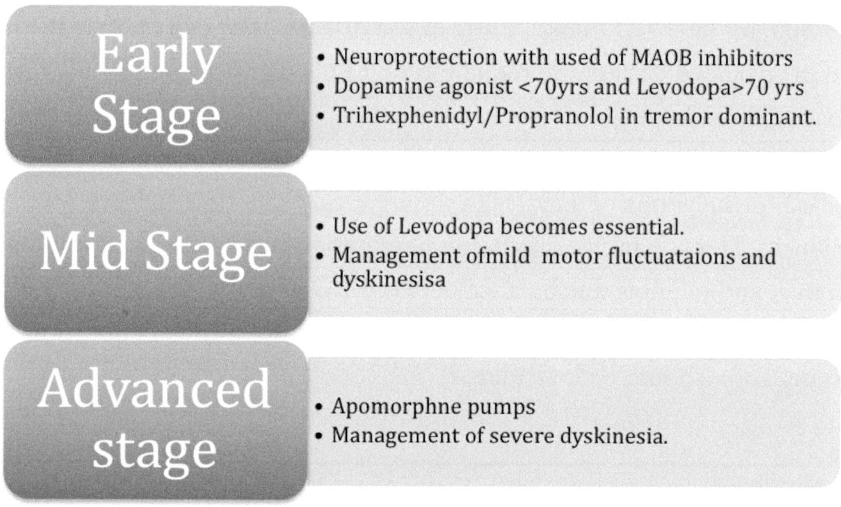

Type of motor fluctuations[6]

1. Wear off/End of dose phenomena: The patient experiences this phenomenon after three to four hours of Levodopa intake, with three to five years into the treatment. There are various methods

to overcome it. The foremost though is to check the relation of meals with LD dosing. There should be a gap of at least one to one and half hours between the meals and LD.

The dosing interval can be shortened or the dose of LD can be increased, or extended release preparation can be used or COMT inhibitors can be added to the regime. Various formulations of LD also have beneficial effects. Controlled release preparations, liquid formulations or intestinal gel delivers constant dose of infusion through percutaneous gastrojejunostomy. Subcutaneous Apomorphine can be used to overcome acute severe off periods. Apart from this, addition of Dopamine agonist and MAO B inhibitors can also help overcome this problem.

Freezing of Gait is a special type of "Off phenomena" and are important because they can result in fall increasing morbidity and impairing quality of life. A combination of pharmacological and non pharamacological measures are required to treat this difficult situation. Amantadine and subthalamic deep brain stimulation are also beneficial.

2. Unpredictable off: It implies that there is no observable relationship with LD dose. Yet, the first step in its management is to keep a home diary recording each off period and timing of LD dose to discern any relationship. Some patients have misconceptions that their off periods are unpredictable, whereas in reality. They are predictable off. This phenomenon usually occurs in advanced disease where other motor fluctuations are also prominent and the absorption of LD may become erratic due to gastroparesis or constipation. The management includes adequate spacing with meals and LD dose adjustment.
3. Failure of "On" phenomena: It is described as failure to have no symptomatic improvement after Levodopa dose. It is mainly due to poor LD absorption or prolonged off phenomena. The aim is therefore to limit the "off" state in terms of both duration and severity.
4. Peak dose dystonia/dyskinesia: Dystonia is sustained or transient twisted muscle contraction. At times they can be tremulous. They usually involve limbs and prominent craniofacial LD induced dystonia may be suggestive of atypical of Parkinsonism. Dyskinesia on the other hand are choreiform movements involving any part of body including respiratory muscles. They manifest usually 30-90 minutes later LD intake. Early in the disease, they can be managed by reducing the LD dose, but in advanced disease the reduction of LD is compensated by addition of Dopamine Agonist. Amantidine is also used to manage dyskinesia specially in advanced PD.
5. Off dystonia + Paroxysmal fluctuations: Add Dopamine agonists and COMT inhibitor and add extended release formulations of LD.
6. Bipahsic dyskineia: This is a tricky situation where dyskinesias occur both in on and off phase. It is difficult to treat and requires fine balance between LD and other dopaminergic drugs. If it does not succeed then intestinal gel infusion of levodopa-carbidopa or subcutaneous apomorphine or deep brain stimulation should be considered.

Conclusion[8]

PD is a unique disorder in the form that we have drugs to control the symptomatology and improve the quality of life- a rare characteristic in the neurodegenerative section. The management is largely individualized and depends on degree of disability. In general the younger patients (<65 years) are treated with Dopamine agonist and Levodopa is reserved for the higher age group with more advanced disease. The tremor dominant younger group may benefit from anticholinergics. Whatever treatment regime is initiated, the continuation of care to manage the adverse effects along with progression of disease pathology is of utmost importance. There is an unmet therapeutic need of neuroprotective agents for this entity, and there are many drugs in experimental trials which may become available in future. With advances in surgery, availability of internet of things (IOT), focused support from ancillary branches like physiotherapy and occupational therapist and finally the hope of better neuroprotection in future, the overall of quality of life of PD is bound to improve.

Reference

1. Parkinson J. An Essay on the Shaking Palsy. London: Sherwood, Neely, and Jones; 1817.
2. Carlsson A., Lindquist M. and Magnusson T. 3,4 Dihydroxyphenylalanine and 5-hydroxytryptophan as reserpine antagonists. Nature. 1957.
3. C.W. Olanow et al Clinical approaches to the development of neuroprotective therapy for PD. Experimental Neurology 298. 2017;246-251
4. Poewe W et al. Parkinson disease Nature Reviews. Disease Primers, 2017. Volume 3, Article 17013.
5. Jankovic et al. Principles and Practice of Movement disorder, Second Edition.
6. Ossig et al. Treatment strategies in Early and Advanced Parkinson disease. Neurol Clin. 2015;33;19–37
7. Martinez-Martin et al. Parkinson's disease severity levels and MDS-Unified Parkinson's Disease Rating scale. Parkinsonism and Related Disorder. 2015;(1);50-54.
8. Connolly et al. Pharmacological treatment of Parkinson Disease. A review. JAMA 2014,311, 1670.

Chapter 19

How Indian Doctors Treat Parkinson's Disease

Sumit Singh, Sunil Singla

Introduction

Healthcare scenario in India is unique and completely different from the healthcare systems of other parts of the world, and more so from the developed countries. There are issues with the medicine and the health delivery system of India, which are difficult to explain to health professionals of developed countries. There is strong availability of extremely qualified and competent physicians and state of the art infrastructure in the metropolitan cities and large towns. However, there is a striking contrast in the health delivery within the same metropolis, between different cities, between cities and towns, and between towns and villages. There is no uniform code of management, and there are no standardized treatment guidelines for most disorders. Even if guidelines do exist, they do not percolate to the level of general physician, and if they do, they are not strictly followed. This is the norm for most diseases, including Parkinson's disease (PD).

In addition, there are several modes of medical systems in the country, viz. allopathic, Homeopathy, Ayurvedic, naturopathy, Unani, herbal and traditional beliefs. Patients with PD have options to seek treatment from myriad sources, ranging from General Physicians, Physicians, Neurologists and Movement disorder specialists. It is not hard to imagine that the approach towards the treatment offered by each of these groups of health professionals will be grossly different from each other. The diagnostic approach would also be different between these groups. It is obviously difficult for a general physician to differentiate between PD and Parkinson's plus syndromes, and thereby, the treatment results would vary in the individuals treated. The discussion about whether and when to offer DBS to PD patients will also be markedly different in these groups. The following chapter is focused on these aspects of PD management in India. It is an unusual article, as evidence on this topic is hard to find or generate. Many thoughts communicated here are the personal approach of the authors and should be observed in a similar light.

DOI: 10.1201/9781003505488-19

Health case scenario in India

At present in India, we have 0.7 physicians per thousand population, which is much below the recommended norm, according to healthcare standards. This ratio might be extremely variable between the villages, cities and the big metropolis. The healthcare expenditure by the government is 4.2% of Gross Domestic Product (GDP), and only 20% of Indian population is covered under any kind of health insurance. Bed density is 0.9 per thousand population, i.e., 1 per 1050 (in USA, it is 1 for every 350, while in Japan it is 1 in 85). Most of the patients pay out of their own pockets, and the insurance companies (most of them) do not cover outdoor patient consultations. Investigations or medications are prescribed on an outdoor patient basis. Parkinson's disease is predominantly managed on outdoor basis, and therefore practically all patients pay out of pocket for their healthcare needs. The distribution of Neurologists is even more skewed in India.[1] Most Indians have no access to neurological centres in the country. Of all the neurologists, more than 95% live in the urban cities, and remaining in rural areas. One can easily understand that the first contact point of a PD patient will be a general physician for his diagnosis and treatment. It is extremely unlikely that a PD patient is likely to seek opinion from a neurologist for his initial symptoms. This results in the treatment of the patient to be based on the knowledge, aptitude, and preference of that particular clinician. The awareness about PD in the general population is also not adequate. In a study where elderly individuals, staying in old age homes in south India (area where the literacy rate is high), the prevalence of PD was close to 16 % (94 definite bilateral PD out of 612 patients). Only 4% of the caregivers of these patients knew that their family member had PD.[2] Therefore, the caregivers are unlikely to seek medical attention for their patients for an illness which they don't know exists.

Parkinson's disease treatment in India

Data regarding the prevalence of PD in India is far in between. According to one study, prevalence is higher in rural population compared to urban population, and is more common in men as compared to women.[3, 4] As compared to the western world, Indian patients have slightly younger age of onset. PD patients in India are usually taken care of by the caretakers, who usually stay with the patients. Because of lack of an organized appointment and referral system, the patients can "choose" their doctor. The first point of contact is usually the neighbourhood doctor, and it might not be difficult for him not to miss the diagnosis in the first attempt. The scenario in a rural setting might be completely different where a qualified medical practitioner might not be available, and patients might have to resort to "alternative" medicine, which might vary anything from Ayurveda, massage, exercises and/ or vitamins and supplements. The first contact point of a patient might be a specialist physician, or rarely a neurologist. Many patients are seen by orthopaedic surgeons, as these patients develop peri-arthritis of shoulder joint, and are later referred to a physician or a neurologist. Broadly, the medical treatment is according to the medical expertise and knowledge of the treating physician, and therefore can be described under the heading of alternative medicine, primary care, neurologist care, and a movement disorder specialist care. The treatment is likely to vary, as the level of the treating physician changes.

Ayurvedic/complementary and alternative (CAM) therapy for PD

The psyche of Indian patients is unpredictable at its best. In general, most people want the effect of the medications to be optimal but are afraid of the side effects. The availability of internet has percolated to the grass-root level, and therefore it is not unusual for patients to have half-correct information about the drugs and their side effects. Ironically, patients are concerned about the side effects first, and the effects later. Ayurveda is an ancient Indian system of medicine with most people having a huge amount of faith on its efficacy. The Indian government also promotes the alternative system of medicine a big way, and therefore the availability of these medications, even without prescription is easy. Poorly randomized and non-randomized trials have found that Ayurvedic medicines are effective in the motor and non-motor symptoms of PD. It is known that occurrence of anticholinergics are naturally present in Datura stramonium, levodopa in Mucuna pruriens and Vicia faba and dopamine agonist activity in Claviceps purpura. Alternative system of medicine like Chinese herbal medicines, massage and external application, *Panchakarma* are very commonly used along with the modern medicines. As high as 45 – 50 % percent patients, in various series report the use of these complimentary systems in PD, with variable results. In an Indian observational study done in a tertiary care hospital in India, 50% of PD patients used Complementary and Alternative Medicine (CAM). Most commonly used CAM is Ayurveda therapy. Many PD patients reported no benefits with CAM in this cross-sectional study. Though, reportedly free from side effects and cheap, there are several problems with these "Alternative" medicines. Ayurveda products are not standardized in their production, and the amount of neurotransmitters might vary from batch to batch. Most of these are in powder formulations, and therefore the dopamine delivered in each dose may be variable. While this might be of little significance in the early stages of the disease, it might create problems in advanced PD where even a small change in the dosage of dopamine replacement can cause severe motor and non-motor symptoms. These medicines have not been evaluated in large randomized trials and follow up studies are woefully lacking, thereby raising a question on their long-term efficacy and side effects.

Primary care physician and PD treatment

Primary care physician (PCP) is usually the first person consulted by the patient for his health needs. As we know PD has variable presentations, there may be patients with atypical presentations. While classical PD may not be difficult to diagnose, unusual presentations might be missed. Young onset PD might be mistaken for a psychogenic disease, while a patient with rigidity dominant PD might be confused with hemiparesis. Differentiating between PD and Parkinson plus syndromes is even more difficult for the PCP. The most common drug prescribed by the PCP is usually levodopa, in various combinations and dosage. Most physicians have little knowledge about the various types of levodopa strengths, dose adjustments needed between the controlled release and regular preparations. Levodopa is usually prescribed after meals, avoidance of protein rich supplements close to the dose are not insisted, and sustained release levodopa is often advised to be taken after being broken into half. It is a common practice in India to prescribe multivitamins as health supplements. Most of these preparations have high doses of pyridoxine, which interferes with the absorption of levodopa. There is little knowledge about when to prescribe dopamine agonists, or amantadine or rasagline. Physiotherapy and rehabilitation exercises are often not advised as an adjuvant treatment. The most important part to understand is that usually these patients are in the early stages of the disease, and might not actually need levodopa as the initial treatment. As most of them are initiated on levodopa, if they are switched over to dopamine agonists by a neurologist whom they are eventually referred to, they do not experience the same degree of response, moreover, they develop motor fluctuations sooner during the course of their disease. Interestingly, almost all these patients usually have very good response in the initial phase as they have excellent response to the medicines. Almost all patients with PD in India would eventually seek opinion from a neurologist.

A neurologist's perspective for PD treatment

With better understanding of the disease, drugs to be prescribed, drug interactions and dietary modifications, most neurologists are fairly knowledgeable in PD treatment. Most neurologists (not all) have a levodopa delaying approach, and prefer treating early or milder PD with MAO B inhibitors and dopamine agonists. The dosage of agonists is usually not taken up to the permissible levels, and there are inhibitions in this regards because of the potential side effects. Neurologists in general have to see all different varieties of patients in India, which include patients from all sub-specialities, and therefore sometimes it becomes difficult for them to focus on PD treatment alone. Usually, when the motor and non-motor complications become prominent, the neurologists consider referring the patient to a movement disorder specialists. Neurologists manage most of the aspects of PD very well, and are able to differentiate between PD and Parkinson's plus syndromes. Awareness about deep brain stimulation (DBS) is slowly growing, but as most centres where neurology training is imparted do not do DBS, most neurologist feel inhibited in referring a suitable patient for DBS. This results in the patients missing the benefits of DBS, which can be achieved by undergoing the procedure when they are likely to benefit the most. Not all neurologists may be conversant with managing PD patients motor and non-motor complications of PD as is the importance of a comprehensive management plan for advanced PD.

Movement disorders specialist and PD

There is only a hand full of movement disorders (MD) specialists in India. The approach of a MD specialist towards PD is completely different from a neurologist. Most MD specialists believe in making a clear clinical diagnosis of the exact akinetic rigid syndrome. They might consider ordering F DOPA PET (if available) or a TRODAT SPECT and PET scan brain to confirm the diagnosis before initiating treatment. Most MD specialist have a set approach towards prescription of dopaminergic drugs. Degree of dependence for activities of daily living decides which medication and the age of the patient to prescribe. If the age is older than 70 years or dependence for ADL is high, they might consider initiation of treatment with Levodopa, or might consider delaying the levodopa as far as possible if the ADL requirement is low. Dopamine agonists are usually the treatment of choice in younger individuals and Rasagline is usually added to the treatment regimen. Caution is usually taken in avoiding dopamine agonists in the night for their potential side effects. Usually the patients are counselled by the specialists about the prognosis of the disease and the fact that the medicines are to be exercised life long, which will control only the symptoms of the disease. A comprehensive plan is usually chalked out where a physical therapist, occupational therapist and a speech therapist help the patient in managing other comorbidities of PD. Levodopa is usually initiated once the severity of the disease starts to adversely affect the quality of life and activities of daily living. Most MD specialists follow the dictum of "start low and go slow". There are tertiary referral centres where dedicated movement disorders clinics are operative regularly, which cater to every possible need of the PD patients. Most such centres were in the government sector earlier, however, recently even in the private corporate hospitals the concept of Independent Practice Units (IPUs) are slowly evolving. An IPU is completely patient centric and disease specific, and a multidisciplinary approach is adapted to derive maximum benefits for a particular disease in a cost and time effective manner. Such IPUs for PD are evolving in India, including facilities for DBS, which is available for PD patients under one roof.

Conclusion

PD is a heterogenous disease with varied manifestations. treatment of this disease is equally varied, especially in a country like India. The health sector of India is extremely diverse and has a huge spectrum, where at one end we have a "healer" with little or no knowledge about the disease, to the other end where there is a MD specialist who knows the deepest intricacies of PD management. There is no fixed point where the patient "enters" this spectrum, and there is no fixed pattern which a patient will follow once he enters this system. He can choose to show to a MD specialist directly, or he can contact a traditional healer and stay with him if he wants, or deteriorate to a point where he can improve only if taken under care of by a PD specialist. The treatment will be different depending upon the physician who treats the patient first. There is little doubt that ideal treatment for PD is delivered by a MD specialist, but their number is so small that it is impossible for them to take care of all the PD patients. The burden of treating PD therefore will largely remain with the PCPs and the neurologists. It is important therefore that the awareness about PD treatment. Treatment can be offered to PD by them. Revamping of insurance sector to cover the expensive treatment of PD cannot be overemphasised.

References

1. Ganapathy K. Distribution of neurologists and neurosurgeons in India and its relevance to the adoption of telemedicine. Apollo Telemedicine Networking Foundation, Apollo Hospitals, Chennai, India. 2015;63(2);142-154
2. Gourie-Devi M, Rao VN, Prakashi R. Neuroepidemiological study in Semiurban and rural areas in south India: Pattern of Neurological disorders including motor neuron disease. Oxford and IBH Publishing Company Pvt. Ltd., 1987;11-21.
3. Razdan S, Kaul RL, Motta A, Kaul S, Bhatt RK. Epidemiology of Neurological disorders in Kashmir. *Neuroepidemiology* 1994;13;113-9.
4. Das SK, Sanyal K. Neuroepidemiology of major neurological disorders in rural Bengal. *Neurology India* 1996;44;47-58.

Chapter 20

DBS: Its Role in the Management of Parkinson's Disease

S P Gorthi, Vijay Chandran, Aprajita Chatterjee

Introduction

Parkinson's disease (PD) is clinically characterized by bradykinesia, rigidity, tremor and postural instability. PD is a neurodegenerative disease wherein deposition of Lewy bodies containing alpha synuclein is observed within neurons of various brain regions along with neuronal loss.[1] This results in abnormal function of multiple motor and non-motor neural circuits, particularly within the basal ganglia. PD patients initially show a good therapeutic response to levodopa, but with time, levodopa induced dyskinesia and motor fluctuations impair quality of life. Management of patients with dyskinesia and motor fluctuations using oral medications is quite challenging and with advancing disease, maintaining stable motor function may not be possible with oral medications alone. Before the advent of levodopa therapy, a few studies highlighting the potential role of surgical therapies in PD had been published. Dr R Meyers was the first to describe a partial caudate and pallidofugal fibre resection for the treatment of rigidity in a patient with PD.[2] This was followed by Dr Irving Cooper in 1952, while performing a mesencephalic pedunculotomy in a patient with PD; he accidentally cut the anterior choroidal artery and noted post-operative improvement in tremor and rigidity.[3] He attributed this serendipitous improvement to damage to the globus pallidus interna (GPi) and followed this case report with a surgical series.[4] The 1950's heralded the introduction of stereotactic pallidotomy by Spiegel and Wycis.[5] Svennilson et al. demonstrated that posteroventral pallidotomy using radiofrequency ablation resulted in improvement of parkinsonian symptoms in a series of patients.[6] These findings were confirmed much later in several follow up studies[7-9] and it was observed that in addition to modest improvement in bradykinesia, rigidity and tremor, there was a marked improvement in drug induced dyskinesia following this procedure. Hassler and Riechert were the first to demonstrate thalamotomy as treatment for tremor.[10] Following exploration of various targets in the thalamus for tremor, the Ventral intermediate nucleus (Vim) of the thalamus was commonly used as a target for stereotactic lesions.[11,12] The subthalamic nucleus (STN) had also been used as a target

DOI: 10.1201/9781003505488-20

for tremor control with stereotactic procedures.[13] The use of stereotactic lesions declined with the advent of effective medical therapy for PD with levodopa, only to reemerge as a treatment option with the recognition of motor complications associated with levodopa therapy. In general, stereotactic procedures are associated with significant disadvantages such as unpredictability of results, difficulties in precise identification of the target, permanence of the procedure with inability to modify outcomes and a gradual decline in efficacy.[14] Many of these drawbacks are overcome by deep brain stimulation, a technique in which one or more electrodes are inserted into the brain, targeting specific nuclei for electrical stimulation. (Fig 1)

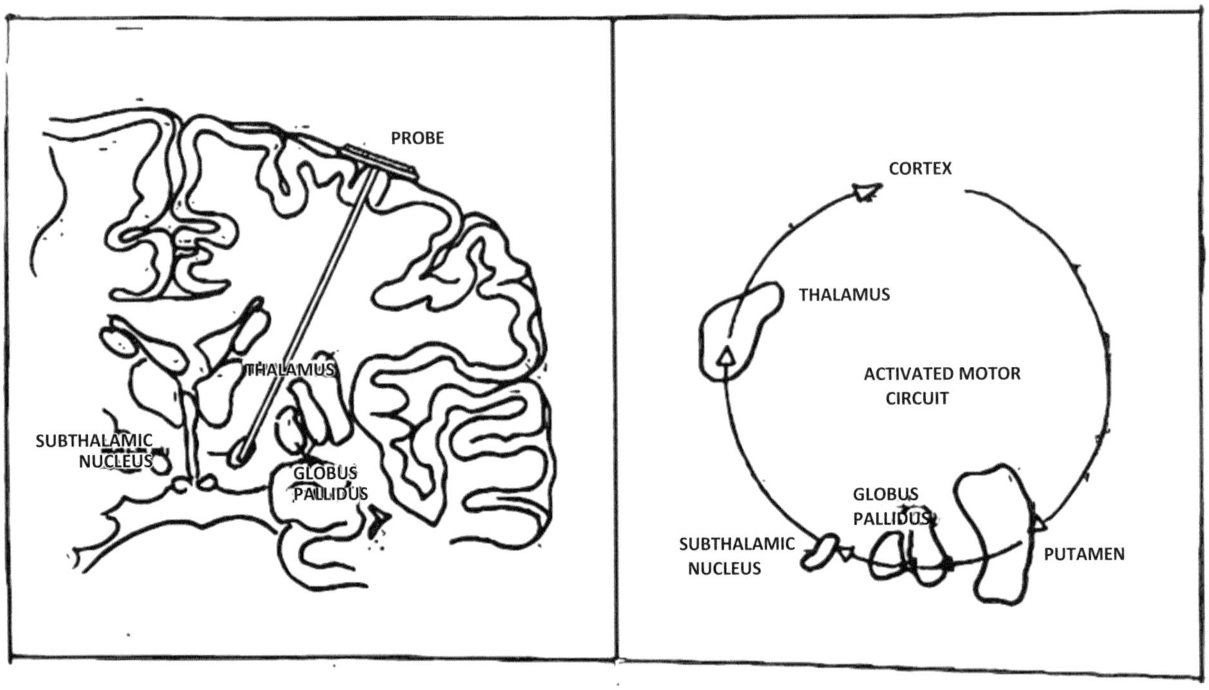

Figure 1

Stimulation of subcortical regions with implanted electrodes was initially described as a treatment for psychiatric and behavioral problems by Heath et al. in the 1950s.[15] Benabid et al. in 1987 were the first to introduce deep brain electrical stimulation of the Vim nucleus of the thalamus to treat PD tremor.[16] The improvement of tremor was found to be sustained with chronic stimulation.[17] This led to the evaluation of other targets for deep brain stimulation (DBS) namely the GPi and the STN. STN stimulation was noted to markedly improve akinesia and rigidity, thereby allowing for a reduction in oral medications,[18,19] whereas GPi stimulation, like pallidotomy, had a significantly blocking effect on dyskinesia and to lesser extent improved bradykinesia and rigidity.[19,20] The above mentioned deep nuclei continue to be the primary targets for deep brain stimulation in PD.

Mechanism of action of DBS

The exact mechanism of action of DBS is poorly understood. The time course of clinical responses varies in different diseases, moreover the time course of response of individual symptoms or signs are also variable within a disease. For example, parkinsonian tremor responds within seconds to stimulation, whereas rigidity and bradykinesia may take a few minutes to hours to show improvement.[21] This would suggest that multiple mechanisms are acting together to bring about eventual clinical benefit.[21] The current signals delivered are of very low intensity and the electrical field generated exerts most of its influence close to the electrode. However, neuronal activation depends on tissue variables; axons and dendrites are more susceptible to stimulation as compared to the soma of neurons, furthermore, axons or dendrites which are larger and oriented perpendicular to the electrical field are more likely to be excited.[21] Stimulation parameters may also play a role.[21] Since the development of DBS was linked to stereotactic ablative procedures and its effects were comparable, the net result of stimulation was thought to be reversible inhibition of the target nucleus. PD is characterized by a loss of dopaminergic neurons in the substantia nigra pars compacta (SNc), which innervates the striatum, which leads to hyperactivity of the indirect and under activity of the direct basal ganglia pathways, resulting in increased inhibition of the thalamus, consequently reducing motor cortex activity.

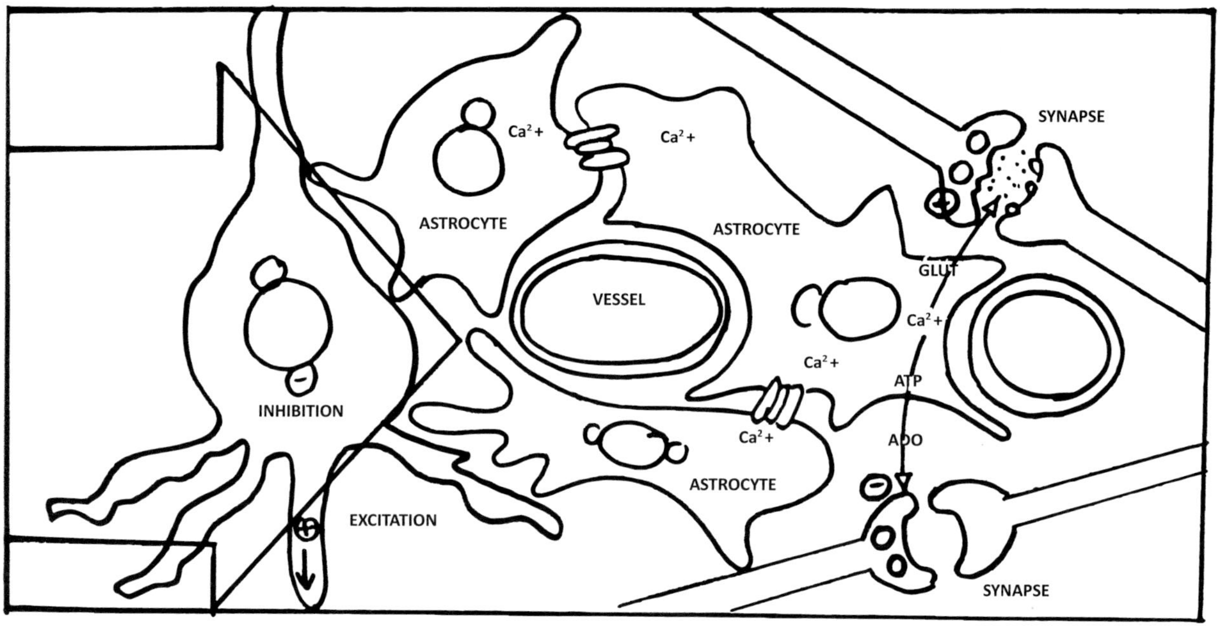

Figure 2: LOCAL EFFECTS OF DBS

Stimulation of GPi and STN has been hypothesized to result in inhibition of these nuclei, which in turn results in excitation of thalamo-cortical connections, thereby facilitating movement. Multiple oscillating loops in basal ganglia circuits are responsible for tremor and the Vim nucleus of thalamus which predominantly receives cerebellar afferents is an integral part of these circuits. DBS may interrupt rhythmic firing of this nucleus and reduce tremor. High frequency electrical stimulation is postulated to cause reversible inhibition by depolarization blockade or via activation of pre-synaptic terminals which end on neurons in the target nuclei.[21] In the case of GPi, these pre-synaptic terminals arise from the striatum and the external globus pallidus (GPe) and release inhibitory neurotransmitters, resulting in inhibition, but this may not be true for other nuclei. On the other hand, there is evidence which suggests that excitation occurs in target nuclei, primarily in axons and dendrites.[21] This excitation may, however, act like a lesion by disrupting the normal pattern of neuronal firing which conveys information. In addition to local electrical effects, DBS has been demonstrated to result in dopamine release in the globus pallidus of patients with PD.[22, 23] However, it appears unlikely that this plays a significant role in PD, in view of the following observations: DBS has been used to block the adverse effects of levodopa such as dyskinesia and furthermore in advanced PD, there are limited numbers of dopaminergic terminals to release sufficient dopamine to explain the degree of benefit seen with DBS.[21] Neural network oscillations play an important functional role in the nervous system. In normal individuals, beta band (12-30Hz) oscillations are observed in the sensorimotor network which includes the (sensorimotor) cortex, basal ganglia, thalamus and the cerebellum, during rest and tonic contraction.[21] This oscillation shifts to higher frequencies in the gamma (30-100 Hz) and high frequency oscillation (HFO) range (100-500 Hz) during movement.[21] In PD, there is an increase in the beta band oscillations as well as an increase in phase-amplitude coupling of higher frequency oscillations. Disruption of beta band oscillations may play a role in the clinical effect seen in DBS.[21] The long term effects of DBS may be mediated through synaptic plasticity, although the clinical evidence for this is limited. Long-term stimulation has revealed only minimal gliosis and moderate cell loss surrounding the stimulating electrode. DBS has been demonstrated to result in network re-organization, of note, the Parkinson's disease related pattern (PDRP) of metabolic changes seen in PD using positron emission tomography can be reversed with DBS.[24-26] Thus DBS has multiple influences on brain tissue through electrical, chemical and neural network alterations. The mechanism by which DBS leads to clinical improvement is an area of active and ongoing research.

Patient selection

In advanced PD, motor fluctuations and dyskinesia make it increasingly difficult to maintain a satisfactory "on" state. The goal of DBS is a reduction in the duration and severity of "off" time, more consistent and longer "on" time, as well as an improvement in dyskinesia. DBS is therefore not recommended in patients without motor fluctuations or dyskinesia and is unlikely to improve motor function in patients with mild disease.[27-29] DBS is usually recommended in patients with at least five years of disease duration, as this interval of time is usually required to be confident of the diagnosis of idiopathic PD, particularly since patients with atypical Parkinsonism are unlikely to benefit.[27-29] A positive levodopa challenge test assists in patient identification; the degree of improvement obtained with a supra-maximal dose of levodopa is highly predictive of the response to STN DBS and probably GPi DBS as well.[30, 31] All symptoms that respond to levodopa, including tremor, will respond to DBS. Symptoms that are not improved with levodopa include, cognitive and psychiatric problems, "on" period freezing and levodopa refractory dysarthria, dysphagia and postural instability; these are unlikely to respond to DBS.[27-29] Tremor which is refractory to levodopa therapy may be improved by targeting the Vim nucleus of thalamus. The levodopa challenge test involves evaluating potential DBS candidates with the Unified Parkinson Disease Rating Scale (UPDRS) motor component (part III) score (maximum 108) in the "off" state, usually after 12 hour overnight medication withdrawal. They are then challenged with a supra-threshold dose of dopaminergic medication and re-evaluated at their best "on" state.[32,33] Improvement of UPDRS- motor sub score by 30% or more is a strong indicator of a positive outcome. Ideally, patients would have disability restricted to motor symptoms in the "off" state, during the "on" state; patients should be independent for activities of daily living. This would be reflected in high "off" (usually >50) and low "on" (<20) scores. However, there are no recommended cut-off values for "off" and "on" scores and as in most treatment considerations in PD, decision making should be individualized based on improvements in the scores. In general, it is worth mentioning that a score of above 30 is considered to be severely disabling and usually in the "on" state, one would expect a score of less than 30.[27] Severe dyskinesia during the "on" state despite optimization of dopaminergic medications or medically refractory tremors which is impairing quality of life would be another indication for DBS.

Baseline cognitive function needs to be evaluated with at least a screening test; full neuropsychological evaluation may be needed in some patients. DBS, particularly of the STN, has a negative impact on cognition. This effect in most patients is minor in comparison to the marked benefit in motor function, but in patients with low cognitive reserve it may be significant. Patients with low cognitive function are unlikely to experience an improvement in quality of life despite improvement in motor function, may not be co-operative for the procedure and are also more likely to suffer from post-operative delirium. Therefore, level of cognitive function is an important consideration which needs to be a part of the discussion when selecting and counseling patients for DBS in PD.[27-29] Co-morbid mental health issues such as depression, anxiety and impulse control disorder would need to be addressed prior to surgery. Untreated or psychiatric illness which is refractory to treatment would be a contraindication for DBS.[27-29] DBS may have a negative impact on mood, rare instances of suicide have been documented after surgery and stimulation may occasionally drive impulse control disorders, this is especially true of STN DBS. Furthermore, patient cooperation is required if the planned surgical technique involves an awake craniotomy. DBS programming, especially during the initial few weeks, may be quite demanding, and poor cooperation or lack of motivation may preclude the ability to derive the maximum benefit from the procedure. Age and comorbid conditions also play a role in patient selection.[27-29] Increasing age and multiple comorbidities are associated with a higher incidence of peri-operative complications. However, this is an individualized decision and there should ideally be no cut-off age above which patients are excluded, although one needs to be cautious when recommending DBS in patients above 70 years of age.

In summary, characteristics of an ideal candidate for DBS in PD include a positive levodopa challenge test, motor fluctuations, disabling dyskinesia, medication-resistant tremor and reasonably preserved cognitive function.[27-29] DBS results are poor in the presence of severe dementia, unstable psychiatric disease, severe autonomic dysfunction, poor dopaminergic response, marked axial symptoms and frequent falls.[27-29] Selection of patients is a multidisciplinary process that involves a neurologist, neurosurgeon and may also require a neuropsychologist and psychiatrist for optimal results.[29]

Surgical procedure and hardware

The surgery may be performed as a single stage procedure or may be done in two stages. Dopaminergic medications are stopped 12 hours before the procedure, usually the night before. When the patient comes in for surgery the next morning, a frame with fiducial markers is mounted on the patients head followed by brain imaging using computed tomography (CT) or magnetic resonance imaging (MRI). The target for DBS is delineated by computer software and a trajectory is planned from the surface of the brain to the target avoiding important structures and vessels (Fig 3). The first part of the procedure is performed under local anesthesia; the patient remains awake during the procedure for assessment of physiological recordings and behavioral responses. A burr hole is created in the skull for the insertion of micro-electrodes and these micro-electrodes are used to obtain electrophysiological recordings and the response of the recorded signal to passive movements.

Once the surgeon is confident that the micro-electrode is in the target, the stimulation lead is placed and therapeutic stimulation can be delivered through the lead to assess responses in tremor, rigidity and bradykinesia, as well as to assess the thresholds necessary to evoke side effects (Fig 4). After the leads are positioned and tested, the electrodes are anchored to the skull.

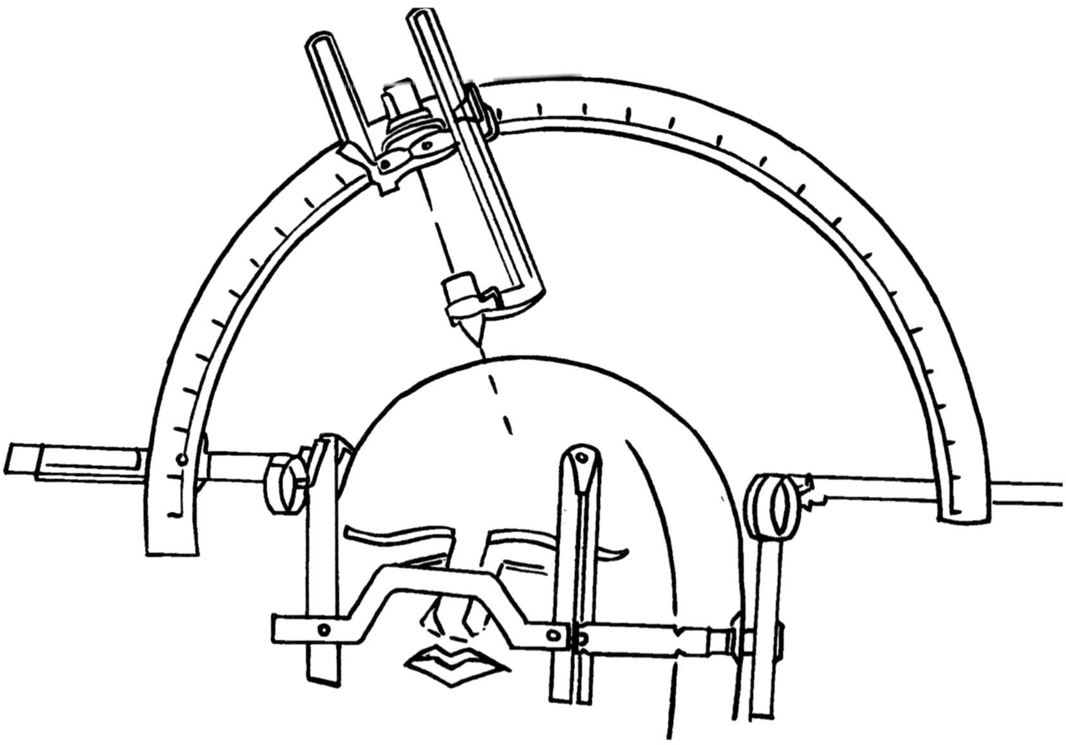

Figure 3: STEREOTACTIC FRAME - DBS

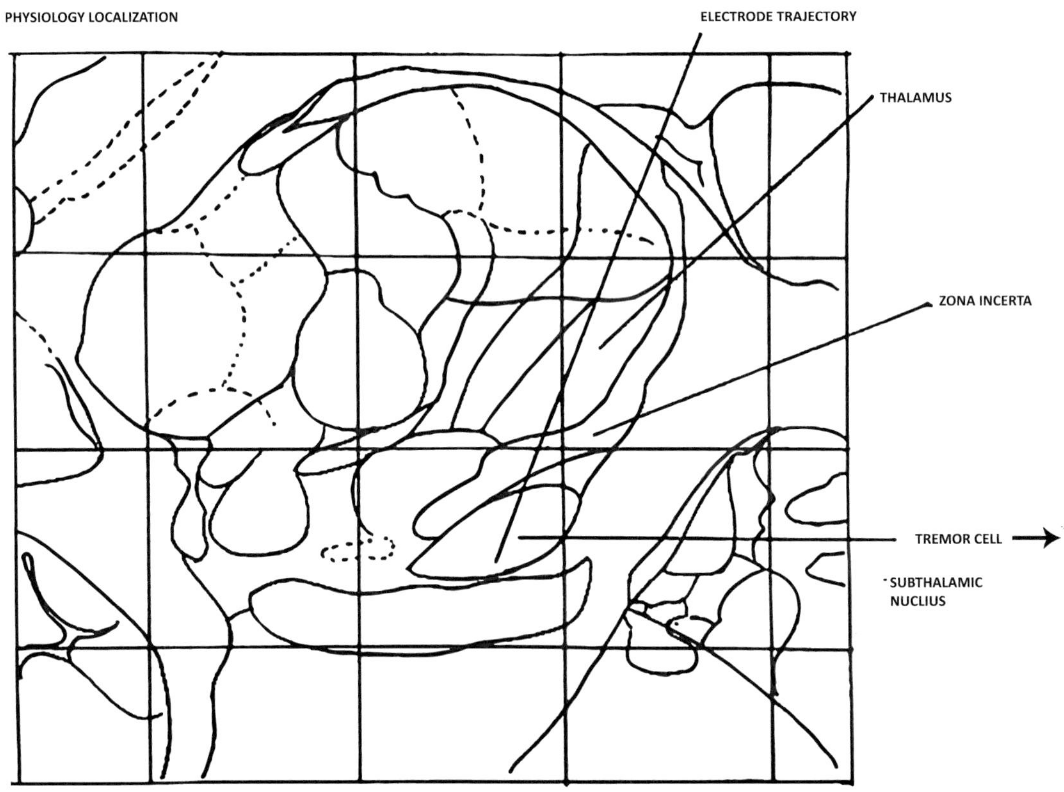

Figure 4: PARA SAGGITAL SECTION – MICROELECTRODES IN TARGET - DBS

A connecting cable is then attached to the lead and tunneled under the skin of the scalp and the neck to the anterior chest wall, where a subcutaneous pocket is created for the insertion of an impulse generator (Fig 5).

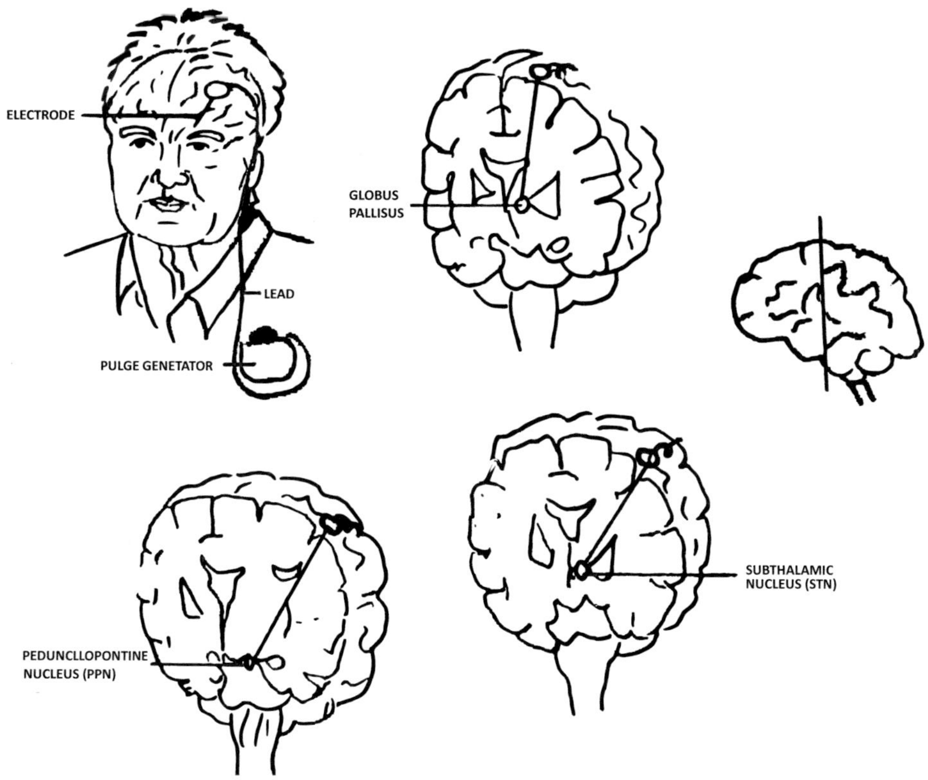

Figure 5: DBS – ELECTRODES & TARGETS

The impulse generator may be inserted in a separate procedure performed two to four weeks after electrode insertion. The conventional stimulation electrode has four contact points either 1.5 or 0.5mm apart at its distal end. It produces a spherical electrical field that may extend beyond the target area and as a result elicit side effects (Figure 6). Different types of implantable pulse generators (IPG) are available, these include a single channel IPG which is used for powering a single lead, a dual channel IPG for 2 leads, a rechargeable IPG is also available which is helpful in patients with high current settings who would otherwise require frequent replacement of the IPG.

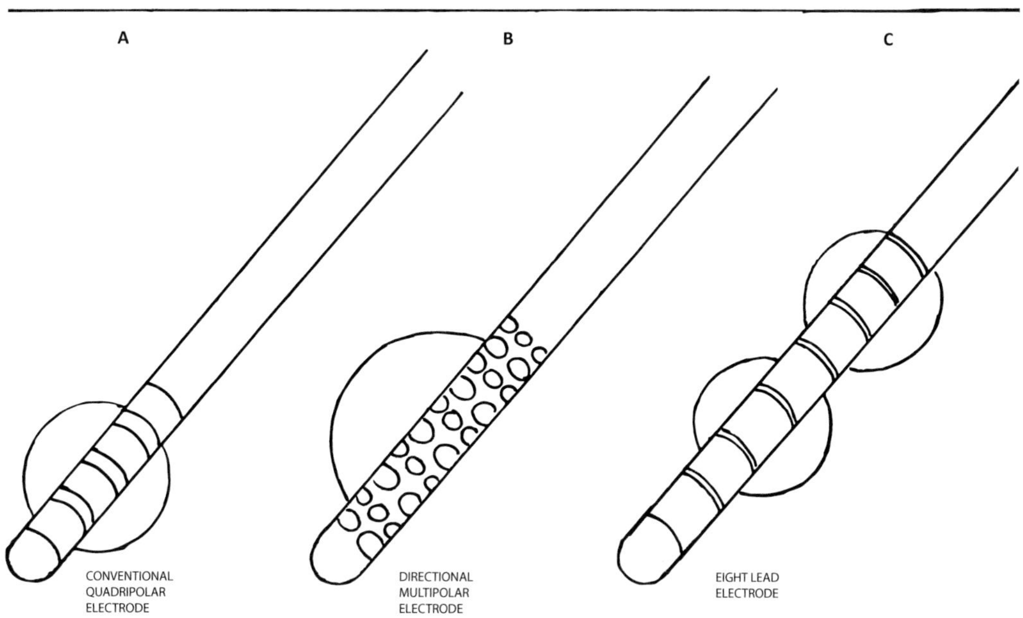

Figure 6: DIFFERENT ELECTRODES FOR DBS

Stimulation and programming (Figure 7)

Soon after the insertion of DBS electrodes, edema develops around the inserted electrode; there may also be a micro-lesional effect from the procedure which results in a transient improvement in parkinsonian symptoms that resolves over a few weeks.[34] Stimulation through the inserted electrode is therefore delayed for a few weeks to allow time for edema and the micro-lesional effect to resolve. A trans-cutaneous programming unit is used to adjust the settings. Before programming, integrity of the entire DBS system should be checked by measuring impedances at various contacts.[34] The signal delivered is continuous and may be adjusted for voltage or amplitude, pulse width and frequency. The current delivered to brain tissue will be dependent on impedance at the contact in use. Usually the threshold for effect and side effects at each contact is determined and finally the best contact to stimulate is identified. Stimulation may be programmed as unipolar or bipolar. Approaches to programming vary based on the target e.g. STN DBS usually requires medication adjustment while stimulation is increased, whereas in GPi DBS usually does not.[34] The typical pulse generator settings are as follows, frequency: 130-185 Hz, pulse width: 60-120 µsec, voltage: 2.0 to 5.0V.[34]

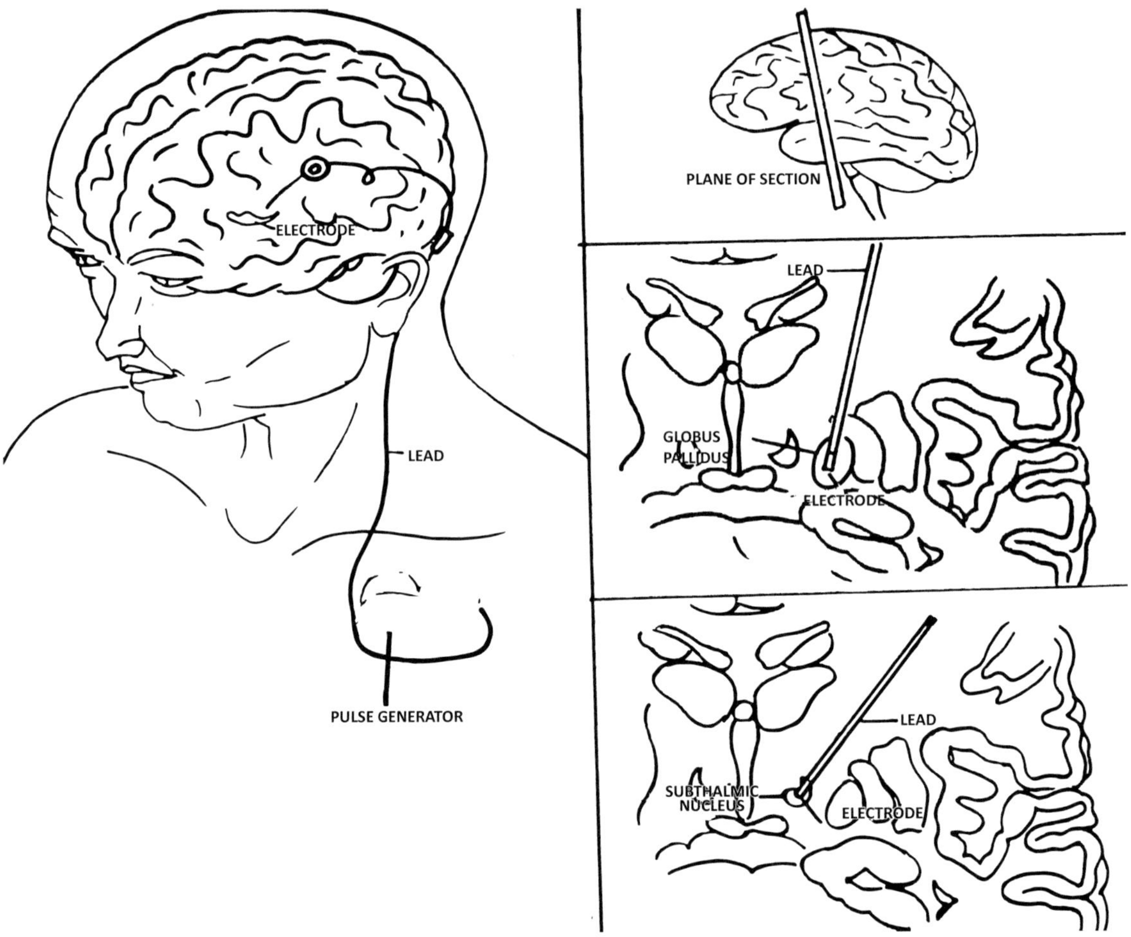

Figure 7: NEWER TARGETS FOR DBS

Complications

Complications of DBS can be classified into three groups: surgery related, hardware-related and stimulation-related.[35] Infections, intracranial hemorrhage (ICH) or infarction, infection, pneumonia, peri-operative confusion and suboptimal lead placement are important operation related complications.[35] Procedural infection rates range from 1.2% to 15.2% varying across centre's and studies. ICH is encountered in approximately 5% of patients and in 2.1%, the ICH is symptomatic. Hardware related complications include fracture of the electrode, migration of the electrode, infection of the system, and skin erosion.[35] Stimulation related side effects occur due to spread of electric field which is usually reversible by adjusting stimulation parameters or contacts and include sensory motor phenomenon and psychiatric disturbances such as mania, depression, apathy, suicidal ideation and impulse control disorder.[35] Other complications include seizures, cognitive impairment, especially decrease in verbal fluency, speech abnormalities, dysphagia, and increased falls.[35] Care must be taken following DBS when undergoing dental or surgical procedures with diathermy and while undergoing MRI. The manufacturer's recommendations are to be followed.

Evidence from clinical trials

Three randomized control trials have demonstrated that in appropriately selected patients, DBS with medication is superior to best medical management. These trials were similar in design, with some differences in specific features and in the choice of primary outcome measures. In the first of these trials conducted by Deuschl G et al.,[36] 156 patients with Parkinson's disease and persistent motor symptoms despite optimal medical therapy were randomly assigned to undergo bilateral STN DBS or to receive best medical therapy. At six months, patients assigned to DBS had significantly better mean scores on the PDQ-39 (31.8 vs. 40.2) and the UPDRS-III (28.3 vs. 46.0). Weaver FM et al.,[37] published results from a trial involving 255 patients with six months of follow-up, patients who underwent DBS of the STN or GPi gained a mean of 4.6 hours per day of "on" time versus zero hours per day for the best medical therapy group. Williams A et al.,[38] in a trial involving 366 patients treated with either DBS and best medical therapy or best medical therapy alone and one year of follow up, demonstrated an improvement in quality of life using PDQ-39 in the DBS group.

Studies have also been carried out attempting to ascertain the best target among STN and GPi. In the DBS for PD Study Group[39] electrodes were implanted bilaterally either in the STN (96 patients) or GPi (36 patients). Three months after the procedure double-blind cross over evaluations were performed, wherein the device was randomly assigned to be in the "on" or "off" state. Both STN and GPi DBS were associated with significant improvements in the motor score with the device switched on (STN: 49%, GPi: 37%). Comparing pre-operative performance with performance at six months follow up, the percentage of time during the day with good mobility increased in both groups (STN: 27% to 74%, GPi: 28% to 64%). This study suggested that both targets were almost equally effective in improving motor function in PD. Anderson et al.[40] compared GPi (n=10) and STN DBS (n=10) at one year follow up (10 in each group) using a randomized blinded design and found that off medication UPDRS motor scores (GPi vs STN: 39% vs 48%) and on medication scores were comparable. Levodopa dose reduction was more common in the STN group (38% vs 3%), but dyskinesia was reduced more often in GPi stimulation (89% vs 62%). Of note, cognitive and behavioral complications were only observed with STN stimulation. The CSP study group randomly assigned patients to either STN (n=147) or GPi (n=152) DBS and evaluated outcome at 24 months.[41] There was no difference in "off" medication UPDRS motor scores which was the primary outcome measure. However, on several secondary outcome measures, differences were noted. Patients undergoing STN DBS were more likely to have a reduction in dopaminergic medication, decline in visuomotor processing speed and increased level of depression. Based on the results, the authors felt that non-motor factors may play a significant role in patient selection. The COMPARE trial used mood (visual analog mood scale) and cognition (verbal fluency, both semantic and letter) as primary outcome measures to compare STN and GPi DBS,[42] but was unable to identify any differences when stimulation was in optimal settings. The NSTAPS study group evaluated whether GPi DBS resulted in better functional outcome as compared to STN DBS in 128 patients randomly assigned to either group.[43] On primary outcome measures, which included a functional health score and a composite score for cognitive, mood and behaviour effects, there was no difference between the two groups, however, on secondary outcome measures, STN DBS patients had a reduction in UPDRS motor scores in "off" state and were on lower dose of dopaminergic medication. These results were felt to be in favour of STN DBS.

Table 1: Comparison of DBS targets-STN and GPi

Sl. No	Feature	Subthalamic Nucleus	Globus pallidus Internus
1	Quality of life	Poor when compared to GPi	Better
2	Tremor	Good response	Less effective
3	Rigidity	Good response	Less effective
4	Axial rigidity	Improves	No effect
5	Axial motor symptoms	Less effective	More effective
6	Speech and Swallowing	Adverse prognosis	No effect
7	Motor fluctuations	Good response	Less effective
8	Dyskinesia	Indirectly reduces as medication doses decrease, however, may increase in some	Good response
9	Dystonia	Only L-Dopa responsive dystonia responds	All types of dystonia improve
10	Cognition	Greater decline	Lesser decline
11	Mood	Depression more common	Less common
12	Economic profile	Less expensive	More expensive

Newer advances

Several areas are being explored with regards to targets, surgical procedure, hardware, and stimulation parameters in DBS for PD. In addition to conventional targets such as GPi, STN and Vim nucleus of thalamus, newer targets of interest include the caudal zona incerta[44] and centromedian parafascicular nucleus of thalamaus for tremor,[45] and the pedunculopontine nucleus for gait dysfunction and axial symptoms.[46] (Fig 8) The use of multiple targets in an individual patient to provide adequate symptom relief has also been attempted.[45] In terms of surgical procedure, targeting without the help of frame i.e. frameless neuro-navigation and performing the entire procedure in a MRI suite with continuous image guidance have also been described. Newer stimulation leads are in development which would enable moulding of the generated electrical field to a shape that would maximize effect while simultaneously limiting side effects.[47] Additionally, electrodes with ability to set different stimulation parameters to each electrode contact are also being developed.[47] In terms of stimulation, constant current stimulation has been described [48] and there is a lot of potential in studying different stimulation parameters such as different frequency and pulse width settings.[47] Adaptive DBS wherein the stimulation delivered is modified based on recording of local field potentials is also being considered.[47]

Figure 8: MAIN CONNECTIONS OF THE PEDUNCULOPONTINE NUCLEUS

Conclusion

Deep brain stimulation has demonstrated substantial clinical benefit in patients with Parkinson's disease. Current understanding is that DBS activates neurons and regularizes pathologic activity and oscillations within the basal ganglia-thalamo-cortical network, which in turn improves sensorimotor processing and alleviates disease symptoms. Presently, the best results are obtained in patients with severe motor fluctuations and dyskinesias who are responsive to levodopa and in those with intractable tremors. Probably, the most important factor underlying outcome is the selection of the right candidate for stimulation.

References

1. Dickson DW. Parkinson's disease and parkinsonism: neuropathology. Cold Spring Harb Perspect Med. **2012;**2(8);pii;a009258.
2. Meyers R. Surgical experiments in the therapy of certain 'extrapyramidal' diseases: a current evaluation. Acta Psychiatr Neurol Suppl. 1951;67;1-42.
3. Cooper IS. Ligation of the anterior choroidal artery for involuntary movements; parkinsonism. Psychiatr Q. 1953;27(2);317–319.
4. Cooper IS. Surgical alleviation of Parkinsonism; effects of occlusion of the anterior choroidal artery. J Am Geriatr Soc. 1954;2(11);691-718.

5. Speigel EA, Wycis HT. Ansotomy in paralysis agitans. AMA Arch Neurol Psychiatry. 1954;71(5);598-614.
6. Svennilson E, Torvik A, Lowe R, Leksell L. Treatment of parkinsonism by stereotactic thermolesions in the pallidal region. Acta Psychiatr Scand. 1960;35;358–77.
7. Laitinen LV, Bergenheim AT, Hariz MI. Leksell's posteroventral pallidotomy in the treatment of Parkinson's disease [see comments]. J Neurosurg. 1992;76;53–61.
8. Dogali M, Fazzini E, Kolodny E, et al. Stereotactic ventral pallidotomy for Parkinson's disease. Neurology. 1995;45;753–61.
9. Lang AE, Lozano AM, Montgomery E, Duff J, Tasker R, Hutchinson W. Posteroventral medial pallidotomy in advanced Parkinson's disease [see comments]. N Engl J Med. 1997;337;15–42. Comment in: N Engl J Med 1998. 338;262–3.
10. Hassler R, Riechert T. [Indications and localization of stereotactic brain operations]. Nervenarzt. 1954;25(11);441-7.
11. Yasui N, Narabayashi **H**, Kondo T, Ohye C. Slight cerebellar signs in stereotactic thalamotomy and subthalamotomy for parkinsonism. Appl Neurophysiol. **1976-1977**;39(3-4);315-20.
12. Narabayashi H, Ohye C. Importance of microstereoencephalotomy for tremor alleviation. Appl Neurophysiol. 1980;43;222-7.
13. Hullay J, Velok J, Gombi R, Boczàn G. Subthalamotomy in Parkinson's disease. Confin Neurol. **1970;**32(2);345-8.
14. Okun MS, Vitek JL. Lesion therapy for Parkinson's disease and other movement disorders: update and controversies. Mov Disord. 2004;19(4);375–389.
15. Heath RG, Monroe RR, Wckle WA. Stimulation of the amygdaloid nucleus in a schizophrenic patient. Am J Psychiatry. 1955;111;862-63.
16. Benabid AL, Pollak P, Lovuvea A, Heanry S, de Rougermont J. Combined (thalamotomy and stimulation) stereotactic surgery of the VIM thalamic nucleus for bilateral Parkinsons disease. Appl. Neurophysiol. 1987;50;344-346.
17. Benabid AL, Pollack P, Gervason C, Hoffman D, Gao DM, Hommel M, Perret JE, De Rougemont J. Longterm suppression of tremor by chronic stimulation of the ventral intermediate thalamic nucleus. Lancet. 1991;337;403-6.
18. Limousin P, Krack P, Pollack P et al. Electrical stimulation of the subthalamic nucleus in advanced Parkinsons disease. N Engl J Med. 1998;339;1105-1111.
19. Benabid AL, Benazzouz A, Hoffmann D, Limousin P, Krack P, Pollak P. Long-term electrical inhibition of deep brain targets in movement disorders. Mov Disord. **1998;13 Suppl 3;**119-25.
20. Segfried J, Lippitz B. Bilateral chronic stimulation of ventroposterolateral pallidum; a new therapeutic approach for alleviating all parkinsonism symptoms. Neurosurgery.1994;35;1126-1129.
21. Herrington TM, Cheng JJ, Eskandar EN. Mechanisms of deep brain stimulation. Neurophysiol. 2016;115(1);19-38.
22. Zsigmond P, Dernroth N, Kullman A, Augustinsson LE, Dizdar N. Stereotactic microdialysis of the basal ganglia in Parkinson's disease. J Neurosci Methods. **2012;207(1);**17-22.

23. Martinez RC, Hamani C, de Carvalho MC, et al. Intraoperative dopamine release during globus pallidus internus stimulation in Parkinson's disease. Mov Disord. **2013;28(14);**2027-32.
24. Fukuda M, Mentis MJ, Ma Y, et al. Networks mediating the clinical effects of pallidal brain stimulation for Parkinson's disease: a PET study of resting-state glucose metabolism. Brain. **2001;124(Pt 8);**1601-9.
25. Asanuma K, Tang C, Ma Y, et al. Network modulation in the treatment of Parkinson's disease. Brain. **2006;129(Pt 10);**2667-78.
26. Trost M, Su S, Su P, et al. Network modulation by the subthalamic nucleus in the treatment of Parkinson's disease. Neuroimage. 2006;31(1);301-7.
27. Lang AE, Houeto JL, Krack P, et al. Deep brain stimulation: preoperative issues. Mov Disord. 2006;21;S171–S196.
28. Rodriguez RL, Fernandez HH, Haq I, et al. Pearls in patient selection for deep brain stimulation. Neurologist. 2007;13;253–260.
29. Bronstein JM, Tagliati M, Alterman RL, et al. Deep brain stimulation for Parkinson disease: an expert consensus and review of key issues. Arch Neurol. **2011;68(2);**165.
30. Charles PD, Van Blercom N, Krack P, et al. Predictors of effective bilateral subthalamic nucleus stimulation for PD. Neurology. 2002;59(6);932–934.
31. Kleiner-Fisman G, Herzog J, Fisman DN, et al. Subthalamic nucleus deep brain stimulation: summary and meta-analysis of outcomes. Mov Disord. 2006;21(14);S290–S304.
32. Defer GL, Widner H, Marie RM, et al. Core assessment program for surgical interventional therapies in Parkinson's disease (CAPSIT-PD) Mov Disord. 1999;14;572–584.
33. Merello M, Nouzeilles MI, Arce GP, Leiquadra R. Accuracy of acute levodopa challenge for clinical prediction of sustained long-term levodopa response as a major criterion for idiopathic Parkinsons Disease diagnosis. Mov Disord 2002;17;4;795-8.
34. Gunther Deuschl, Jan Herzog, Galit Kleiner-Fisman, et al. Deep Brain Stimulation: Postoperative Issues. Movement Disorders. 2006; 21(14);S219–S237.
35. Jose Fidel Baizabal Carvallo, MD, MSc, Richard Simpson, MD, PhD, and Joseph Jankovic, Diagnosis and Treatment of Complications Related to Deep Brain Stimulation Hardware. MD1 Movement Disorders; 26(8);2011.
36. Deuschl G, Schade-Brittinger C, Krack P, et al. A randomized trial of deep-brain stimulation for Parkinson's disease. N Engl J Med. 2006;355;896-908.
37. Weaver FM, Follett K, Stern M, et al. Bilateral deep brain stimulation vs best medical therapy for patients with advanced Parkinson disease: a randomized controlled trial. JAMA. 2009;301;63-73.
38. Williams A, Gill S, Varma T, et al. Deep brain stimulation plus best medical therapy versus best medical therapy alone for advanced Parkinson's disease (PD SURG trial): a randomised, open-label trial. Lancet Neurol. 2010;9:581-91.
39. Deep-brain stimulation of the subthalamic nucleus or the pars interna of the Globus Pallidus in Parkinson's disease. The Deep-Brain Stimulation for Parkinson's Disease Study Group N Engl J Med 2001;345;956-63.

40. Anderson VC, Burchiel KJ, Hogarth P, Favre J, Hammerstad JP. Pallidal vs subthalamic nucleus deep brain stimulation in Parkinson disease. Arch Neurol. **2005;62(4);**554-60.

41. Follett KA, Weaver FM, Stern M, et al. Pallidal versus subthalamic deep-brain stimulation for Parkinson's disease. N Engl J Med. **2010;362(22);**2077-91.

42. Okun MS, Fernandez HH, Wu SS, et al. Cognition and mood in Parkinson's disease in subthalamic nucleus versus globus pallidus interna deep brain stimulation: the COMPARE trial. Ann Neurol. **2009;65(5);**586-95.

43. Odekerken VJ, van Laar T, Staal MJ, et al. Subthalamic nucleus versus globus pallidus bilateral deep brain stimulation for advanced Parkinson's disease (NSTAPS study): a randomised controlled trial. Lancet Neurol. **2013;12(1);**37-44.

44. Blomstedt P, Fytagoridis A, Åström M, Linder J, Forsgren L, Hariz MI. Unilateral caudal zona incerta deep brain stimulation for Parkinsonian tremor. Parkinsonism Relat Disord. 2012;18(10);1062-6.

45. Stefani A, Peppe A, Pierantozzi M, et al. Multi-target strategy for Parkinsonian patients: the role of deep brain stimulation in the centromedian-parafascicularis complex. Brain Res Bull. 2009;78(2-3);113-8.

46. Hamani C, Lozano AM, Mazzone PA, et al. Pedunculopontine Nucleus Region Deep Brain Stimulation in Parkinson Disease: Surgical Techniques, Side Effects, and Postoperative Imaging. Stereotact Funct Neurosurg. **2016;94(5);**307-319.

47. Fang JY, Tolleson C. The role of deep brain stimulation in Parkinson's disease: an overview and update on new developments. Neuropsychiatr Dis Treat. **2017;13;**723-732.

48. Okun MS, Gallo BV, Mandybur G, et al. Subthalamic deep brain stimulation with a constant-current device in Parkinson's disease: an open-label randomised controlled trial. Lancet Neurol. 2012;11(2);140-9.

Chapter 21

Complementary and Alternative Therapy in Parkinson's Disease

M Behari

Parkinson's disease (PD) is characterized by progressive loss of dopaminergic nigro-strial neurons resulting in the characteristic symptoms of PD. As it is a progressive disorder with no cure, no drugs are available which can prevent the occurrence of PD or slow its progress.[1] This creates a lacuna in the armamentarium of therapy of PD. Since the 1960's, after the discovery of levodopa, other drugs discovered so far only provide a replacement to deficient dopamine. Dopamine replacement therapy (DRT) with levodopa is still the best symptomatic therapy and is known to extend life expectancy.[2] Over the years of use of levodopa, the therapeutic efficacy of levodopa wanes, both in terms of benefit in motor symptoms and in duration of effect. A large number of patients experience disabling side effects in the form of sleep attacks and psychosis, with incomplete relief of symptoms.[3-6] A large number experience levodopa induced motor fluctuations as the disease advances. This necessitates modification of the dosage, change in the formulation of levodopa, and combine it with DA agonists or other drugs to control the adverse symptoms. Complementary and Alternative Medicine (CAM) is sought when pharmacotherapy fails to meet the requirements of PD patients. However, a small percent use CAM even in the early stage of illness when there are no problems with pharmacotherapy or as the sole therapy. Complementary and alternative medicine (CAM) are a group of management practices which are not part of western medicine.[7] The term "complementary" is used when these management practices are used along with conventional medicines.[8] On the other hand the term "alternative" is used as they are solely used in the absence of conventional medicines. CAM's use varies from 40-76%.[9,10] Though, CAM has been used by PD patients in all parts of the world, it is rather traditional in some parts of countries like China, India, Korea and the Amazon-region, where the use of traditional medicines are very strong. It varies from preparations of herbs containing anticholinergic agents, levodopa and MAO-B inhibitors. Different forms of CAM which are generally used, consist of oral ingestion or application of these preparations on skin surface as ointments or pastes. They also include practices such as yoga, meditation, dance, music, exercise etc. Many patients describe reasons for using CAM for the "feel good factor", belief that it is free from side effect, or as

DOI: 10.1201/9781003505488-21

a replacement of western medicines, as these are considered to have side effects.[10] Commonly used CAMs include Ayurvedic medicines, homeopathic medicines and acupuncture.

Epidemiology

While CAM is used by PD patients all over the world, it is more common in countries where traditional medicine has ancient history and is strongly imbibed in the thought of masses. In Asian countries like India, China and Korea, frequency of CAM use is 46%, 61% and 76% in India, Singapore and in Korea, respectively.[11-13] In the western countries, its use ranges from 34% to 54%.[14,15-17] The spectrum of CAM used for PD is diverse; acupuncture is the intervention most often used worldwide.[18] CAMs which used more often are massages, ingestion of herbs acupressure and ingestion of vitamins/health supplements. Predictors of CAM use are cultural background and countries with ancient systems of medicines.[18] Among the US PD patients, natural products such as vitamins, health supplements and herbs are the most commonly used CAM,[14] while mind and body practices such as acupuncture, yoga and massage are common in Sweden,[16] Argentina[15] and the UK.[17] In Asian countries, traditional and oriental medicine is the most common form of CAM.[12,13] Certain social factors, which determine CAM use: are higher income,[12] rural living,[13] perceived poor health,[16] female gender,[15] younger age[17] and higher education level.[11,14,15] Authors from Korea found that patients with lower educational level used traditional oriental medicine, while those with a higher educational level opted for traditional medicine and massage therapies more often. Certain disease-specific factors which govern CAM use are younger age at onset of PD,[14,15] higher levodopa dose,[13] higher severity of symptoms[12] and longer duration of PD.[12, 13]

Types of CAM[10]

CAM treatments can be categorized in to three main types:[11]

1) Natural products including a variety of products such as herbs, vitamins, minerals and probiotics, which may be ingested or applied on the outer surface of body. In this group, products which are ingested includes: Ayurvedic medicines [*Macuna pruriens* (natural source of levodopa), *Ashwagandha, Brahmi;* (derived from Bacopa Monnieri family: Scrophulariaceae), *Jeeva mala,* Wheat grass, *Amla* (Indian gooseberry) juice (*Phyllanthus emblica* or *Emblica officinalis*), *Balwach, Sumanta, Kampavatras, Mochpak, Swarasta rist, Til* oil (Sesame seeds oil), Aloe vera, known as *Ghrit Kumari* in Sanskrit, a member of lily family, *Vicia faba* , *Ashwagandha* (*Withania somnifera,* family Solanaceae known as Indian Winter cherry or Indian Ginseng, or *Sattvic Kapha Rasayana*), babies tear, Herpestis monniera, water hyssop which contains Sitoindosides and Withaferin-A and acyl-steryl-glucosides]. Of about 240 species, *Aloe Vera Barbadenis,* is used for therapy in several formulations (external application and oral preparations as well). *Amla* juice is rich in vitamin C and is extensively used in Ayurveda. One of the patients in our study reported oral consumption of placental cord blood for treatment of PD.[11] Some people also use homeopathic medicines.

   There is another group of products, which are used on surface of body. In India, in the Ayurvedic practice *Panchkarma (Pancha* = five in Sanskrit and *Karma* = therapeutic activities, therefore *Panchakarma* means five types of therapeutic activities) is utilized. In this method of treatment, the body is cleansed of harmful *dosha* (toxic materials) by five activities, which include various herbal massages, fomentations such as steam, external oil treatments, *Basti* (medicated enemas), *Virechana* (purgation through herbs), *Vamana* (emesis through herbs), *Nasya* (nasal administration of oils) etc. Mud application over whole body (used in Naturo-therapy).

2) Other procedures which work through mind-body axis, affects body through acupuncture, acupressure, massage, meditation, movement therapies, relaxation techniques, spinal manipulation such as chiropractic and osteopathic manipulations, Tai Chi, Qi Gong, Reiki Yoga, healing touch and hypnotherapy. In this group, therapies included are: physiotherapy, music therapy, dance therapy, water therapy, Yoga, massage, transcutaneous electric nerve stimulation (TENS), Tai Chi, acupressure, acupuncture, magnet therapy, cow dung application over whole body, exercise, physiotherapy and foot reflexology. Reflexology or zone therapy, is an alternative medicine in which the practitioner applies pressure various parts of body like feet, hands, or ears using thumb, fingers, and other hand techniques without use of oil or lotions.
3) The next category does not consist of either use of herbs or pastes or any procedures or practices of traditional healers, Ayurvedic medicine, traditional Chinese medicine, homeopathy, naturopathy, etc.[12]

Scientific comments on some of the CAMs

- **Exercise**

In the recent times, exercise has become an important form of CAM for PD by neurologists due to benefits observed in both motor and non-motor symptoms of PD. Some studies also reported its possible neuro-protective effect as well. It is believed to work by improving mitochondrial function and stimulating neuroplasticity. It therefore, improves both motor and non-motor symptoms.[19] In a recent study it was shown that exercise increases DA and glutamate transporters levels, and suppresses inflammation.[20] Exercise is an effective promising complimentary therapy. Long-term and follow-up studies are needed to evaluate its efficacy. Some studies on mouse model of PD have shown that regular exercise improves dopamine levels in brains and increases neurotropic factors.

- **Conventional physical exercises**

Aerobic exercises, including aerobic walking and stretching, has been shown to improve several motor functions such as gait, balance, physical performance and some non-motor functions such as depression, fatigue and cognition.[20, 21] However, it does not prevent falls in PD patients. Intensive training exercises help to strengthen muscles and improve stiffness in muscles as well as also improve mobility.[22, 23]

- **Tai Chi and Qi Gong**

Tai Chi, a traditional Chinese exercise combines deep breathing and slow movements. It has been observed to improve axial symptoms like balance impairment, and prevent falling tendencies.[24-26] Tai Chi improves quality of life and is safe and feasible. It is recommended in PD patients with not very severe disease. Qi Gong is an exercise similar to Tai Chi practiced in traditional form in China. It works by transferring internal energy to the muscles and motor functions. Though not proven in class 1 studies, one randomized controlled trial has suggested that Qi Gong improves motor functions of PD and non-motor symptoms as well.[27] However, on the other hand a smaller RCT did not confirm the findings of the larger study.[28] Therefore, larger well-planned studies are required to prove the efficacy of Qi Gong in PD patients.

A systemic-analysis on Tai Chi in PD patients in 2014 by Yang et al. has shown an overall increased benefit with Tai Chi as compared to no intervention as well as add on therapy with active medications.

- **Yoga**

Yoga as a discipline originated in India by sage Patanjali. It is good for improving flexibility, strength, gait, stamina and quality of life. Some pilot studies have shown usefulness of yoga in improving UPDRS scores, tremor and some other activities.[29] On the others hand, other studies have shown an improvement in some functions such as sit-and-reach, single-leg balance and on some non-motor symptoms such as depression.[30] Larger well-planned studies may confirm or refute the efficiency of yoga for PD patients.

- **Dance**

Several types of dances have been suggested. These improve both motor and non-motor symptoms. A recent meta-analysis suggests that short-term practice of dance is better than no intervention in improving UPDRS scores, balance and gait.[31] Of all the dance forms, Tango has been reported to improve motor function and balance, as compared to common exercise.[32]

- **Massage**

Massage is a common practice in India for different types of pain, injury and paralyses. It is a common form of complementary therapy in PD too. A small study including 10 patients using Japanese massage technique for two months showed good effect in stiffness of shoulders, muscle pain and fatigue.[33] Another study on 40-minute Anma massage showed improvement in movements and general well-being.[34]

- **Acupuncture**

Acupuncture originated as CAM in traditional Chinese medicine. Bee venom acupuncture is supposed to have anti-inflammatory effect.[35] An eight week intervention with both acupuncture and bee venom acupuncture has been shown by a randomized controlled trial to improve UPDRS scores of PD patients.[35] In another randomized controlled trial, patients were divided into acupuncture, covert placebo and overt placebo groups to study the efficacy of acupuncture and placebo and showed superiority of acupuncture over placebo in improving motor function.[36] Several experimental studies suggest that acupuncture may work by releasing neuroprotective factors[44,45] as well as having symptomatic benefits[44,45] in PD animal models. It is suggested to reduce oxidative stress,[46,47] up-regulation of neurotrophic factors,[48,49] reorganizing neural network in basal ganglion, [46,48,49] reducing activity of microglial,[50,51] modulation of apoptosis[52] and neuroinflammation which help in PD. Despite these hypotheses, the precise mechanisms of the neuroprotective effect and beneficial effects of acupuncture in improving symptoms are still unclear. In Chinese centers, many types of acupuncture are practiced while in Korea venom acupuncture is preferred.

Systematic reviews studying usefulness of acupuncture in PD published so far[53-56] have concluded that efficacy of acupuncture in PD is not established because of the poor quality of the trials. Most of the studies are flawed due to poor methodology, rigour, poor controls, different acupoints and methodology of performing acupuncture (type, parameter and schedule) in different studies. Also, acupuncture parameters such as stimulation frequency and duration, retention time and treatment sessions were variable between trials.

- **Ayurveda**

Of the Ayurvedic medicines trials, only four clinical trials are available using Ayurvedic medicines in PD. Additionally, there are no systematic review or meta-analysis. Of these four trials, three were open clinical trials[35-37] and only one was a pilot study which was randomized double-blind cross-over RCT .[38] Among Ayurvedic medicines, *Mucuna pruriens* is the medicine tested in clinical trials. All these clinical trials were open label and suggested some beneficial effect of Ayurvedic treatment. These studies were flawed as they did not include a placebo arm.[37,39] In the pilot study, patients on *Mucuna pruriens* had longer period with drug on-time and higher peak levodopa concentrations, compared with an equivalent dose of L-dopa, however, there was no significant difference in UPDRS motor score and dyskinesia score.[40] This study was a single dose administration without long term effect, and the sample size was also small. For measuring strength of any RCT it is important to have large sample size, blinding of participant as well as the evaluator and presence of a control arm; none of the studies conducted on Ayurvedic medicines stand to this rigour of good clinical trial.

- **Herbal medicines**

Three systematic reviews analyzed RCTs for efficacy of herbal medicines in PD.[41-43] Though a large number of trials have been conducted using Chinese herbal medicine in PD, not many stand to the rigour of good RCT. In a systematic review, nine RCTs were analyzed. However, in each RCT, a different herbal medicine was tried.[41] In four of these studies the herbal medicine was used as monotherapy. In another study which tested *Mucuna pruriens*, similar or superior efficacy was observed as compared to conventional drugs on the Webster rating scale, UPDRS motor score, Abnormal Involuntary Movement Scale or Goetz scale.[42] Herbal medicines were also tested as adjunct therapy in five studies which showed better outcome as compared to the sole use of either herbal medicine or conventional drugs on Webster rating scale or UPDRS scores. Trials conducted in China (seven out of nine); had major flaws of study design. Another meta-analysis included 19 studies, assessing clinical efficacy of Chinese herbal medicine for PD.[42] This meta-analysis suggested improvement of UPDRS I, II, III, IV when herbal medicine was used as an adjunctive therapy compared with conventional western treatment alone. It was also observed that there was no uniformity in composition of herbal medicines, dosage used, preparation and method of manufacturing.

A systematic review conducted by Kim et al., which included 2432 relevant publications reviewed 64 studies to evaluate the evidence of efficacy of herbal medicine in PD.[43] All studies used combination of several different herbal medicines. Combination therapy showed improvement in PD related symptoms and a decrease in the dose of conventional antiparkinsonian drugs.[43]

Efficacy of CAM

In a study from Singapore, 40% of CAM users reported some improvement, 58% reported no improvement and 2% reported worsening of their motor symptoms.[12] A similar study on Korean patients reported improvement in 41.6% (37/89), of which a substantial number that is 83.8% reported only a mild beneficial effect and 9.6% reported side effects of CAM. However, 57.4% of CAM users planned to continue using CAM.[13] In another study from Sweden, efficacy depended on type of CAM used.[16] 50% of acupuncture users reported no improvement, while 50% reported some improvement. Of those who used herbal medicines and other drugs, "good improvement", "some improvement" and "no effect" was equally divided among the group. However, most PD patients reported no benefit with other CAMs.[16] In a study from India from our centre, 66% patients reported no benefit with CAM and 10% reported worsening of symptoms with CAM (when taken alone without allopathic medicines).[10]

Cost of CAM

Globally, the use of CAM costs a lot as it is not covered by insurance. Hence, most users have to pay for it from their own pocket. Over the years, the cost of CAM is increasing due to high demand. In Korea, the cost of traditional medical treatment is rising at the rate of 7.7% per year. According to a study by Kim et al., the cost of CAM paid by an average patient from his pocket was 102.3 US Dollars (USD) per month, while on the other hand, the cost of western medical treatment for PD paid by patients averaged at 72.8 USD per month in the same study.[13] Though, CAM relieved the symptoms only partially, its cost was 12.4 times higher than that of western therapy.[57] In Singapore, the average cost of CAM was 43.53 USD/person/type of CAM/month,[12] while in Sweden, 20% of CAM users reported spending an amount of 50 USD on CAM in the previous six months.[16] According to our study,[10] mean cost of CAM per month in Indian Rupees (INR) was 2211 (about 44USD, in 2010) and median was INR 1000 (20USD) per month (range INR 0-25000). While per capita income of our cohort was mean ± SD INR 12,202.70±9073 (range INR 1200 - 50000). Therefore, on average, our PD patients used almost 20-40% of their monthly income on CAM.

Conclusion

PD patients are using complementary and alternative medicines (CAM) frequently globally at a fairly high cost and side effects. Our knowledge about efficacy and safety is not very clear, despite the fact that a large number of PD patients are using it. The half-baked knowledge available on internet and some on print media attracts lots of gullible patients to use it without discussing it with their neurologist, resulting in side effects and loss of money. Scientific basis for their use is not robust. Most of the studies using these methods are flawed due to poor design and dose and preparation used. Most patients do not inform their doctors that they are using CAMs, and this results in drug interaction, over-medication and other side effects. In most countries the cost of CAM is out of pocket expenditure by most patients, and hence the patients have to bear their cost as well.

References

1. Lohle M, Reichmann H. Clinical neuroprotection in Parkinson's disease—still waiting for the break-through. J Neurol Sci. 2010;289;104–114.
2. Cao XB, Guan Q, Xu Y, Wang L, Sun SG. Mechanism of over-activation in direct pathway mediated by dopamine D (1) receptor in rats with levodopa-induced dyskinesia. Neurosci Bull. 2006;22;159–164.

3. 6-9 Fahn S, Oakes D, Shoulson I, Kieburtz K, Rudolph A, Lang A, et al. Levodopa and the progression of Parkinson's disease. N Engl J Med. 2004; 351;2498–2508.
4. Pahwa R, Factor SA, Lyons KE, Ondo WG, Gronseth G, Bronte-Stewart H, et al. Practice Parameter: treatment of Parkinson disease with motor fluctuations and dyskinesia (an evidence-based review): re-port of the Quality Standards Subcommittee of the American Academy of Neurology. Neurology. 2006; 66;983–995.
5. Hobson DE, Lang AE, Martin WR, Razmy A, Rivest J, Fleming J. Excessive daytime sleepiness and sudden-onset sleep in Parkinson disease: a survey by the Canadian Movement Disorders Group. JAMA. 2002;287;455–463.
6. Comella CL. Daytime sleepiness, agonist therapy, and driving in Parkinson disease. JAMA. 2002;287;509–511.
7. Pearson NJ and Chesney MA. The CAM Education Program of the National Center for Complementary and Alternative Medicine: an overview, Acad Med J Assoc Am Med Coll 82. 2007;921–926.
8. Ferry P, Johnson M and Wallis P. Use of complementary therapies and non-prescribed medication in patients with Parkinson's disease, Postgrad Med J 78. 2002;612–614.
9. Kim SR, Lee TY, Kim MS, Lee MC and Chung SJ. Use of complementary and alternative medicine by Korean patients with Parkinson's disease, Clin Neurol Neurosurg 111. 2009;156–160.
10. Pandit AK, Vibha D, Srivastava AK, Shukla G, Goyal V, Behari M.

Complementary and alternative medicine in Indian Parkinson's disease patients. J Tradit Complement Med. 2016;6(4);377–382

11. Tan LC, Lau PN, Jamora RD, Chan ES. Use of complementary therapies in patients with Parkinson's disease in Singapore. Mov Disord. 2006; 21; 86-89. Pecci C, Rivas MJ, Moretti C, Raina G, Ramirez CZ, D´ıaz S, Uribe Roca C, Micheli FE. Use of complementary and alternative therapies in outpatients with Parkinson's disease in Argentina. Mov Disord. 2010;25;2094-2098.
12. Kim SR, Lee TY, Kim MS, Lee MC, Chung SJ. Use of complementary and alternative medicine by Korean patients with Parkinson's disease. Clin Neurol Neurosurg. 2009;111;156-160.
13. Rajendran PR, Thompson RE, Reich SG. The use of alternative therapies by patients with Parkinson's disease. Neurology. 2011;57;790-794.
14. Pecci C, Rivas MJ, Moretti C, Raina G, Ramirez CZ, D´ıaz S, Uribe Roca C, Micheli FE. Use of complementary and alternative therapies in outpatients with Parkinson's disease in Argentina. Mov Disord. 2011;25;2094-2098.
15. Lokk J, Nilsson M. Frequency, type and factors associated with the use of complementary and alternative medicine in patients with Parkinson's disease at a neurological outpatient clinic. Parkinsonism Relat Disord. 210;16;540-544.
16. Ferry P, Johnson M, Wallis P. Use of complementary therapies and non-prescribed medication in patients with Parkinson's disease. Postgrad Med J. 2002;78;612-614.
17. Wang Y, Xie CL, Wang WW, Lu L, Fu DL, Wang XT, Zheng GQ. Epidemiology of complementary alternative medicine use in patients with Parkinson's disease. J Clin Neurosci. 2013;20;1062-1067.

18. Burini D, Farabollini B, Iacucci S, Rimatori C, Riccardi G, Capecci M, Provinciali L, Ceravolo MG. A Randomised Controlled Cross-over Trial of Aerobic Training versus Qigong in Advanced Parkinson's Disease. Eura. Medicophys. 2006;42(3);231-238.

19. Uc EY, Doerschug KC, Magnotta V, Dawson JD, Thomsen TR, Kline JN, Rizzo M, Newman SR, Mehta S, Grabowski TJ, Bruss J, Blanchette DR, Anderson SW, Voss MW, Kramer AF, Darling WG. Phase I/II Randomized Trial of Aerobic Exercise in Parkinson Disease in a Community Setting. Neurology. 2014;83(5);413-425.

20. Canning CG, Sherrington C, Lord SR, Close JCT, Heritier S, Heller GZ, Howard K, Allen NE, Latt MD, Murray SM, O'Rourke SD, Paul SS, Song J, Fung VSC. Exercise for Falls Prevention in Parkinson Disease: A Randomized Controlled Trial. Neurology. 2015;84(3);304-312.

21. Cruickshank TM, Reyes AR, Ziman MR. A Systematic Review and Meta-Analysis of Strength Training in Individuals With Multiple Sclerosis Or Parkinson Disease. Medicine (Baltimore). 2015;94(4).

22. Uhrbrand A, Stenager E, Pedersen MS, Dalgas U. Parkinson's Disease and Intensive Exercise Therapy - a Systematic Review and Meta-Analysis of Randomized Controlled Trials. J. Neurol. Sci. 2015;353(1-2);9-19.

23. Li F, Harmer P, Fitzgerald K, Eckstrom E, Stock R, Galver J, Maddalozzo G, Batya SS. Tai Chi and Postural Stability in Patients with Parkinson's Disease. N. Engl. J. Med. 2012;366(6);511-519.

24. Tsang WWN. Tai Chi Training is Effective in Reducing Balance Impairments and Falls in Patients with Parkinson's Disease. A recent meta-analysis has shown Tai Chi to have positive effects in motor function and balance, but not in gait velocity, step length and gait endurance J. Physiother. 2013;59(1);55.

25. Yang Y, Li XY, Gong L, Zhu YL, Hao YL. Tai Chi for Improvement of Motor Function, Balance and Gait in Parkinson's Disease: A Systematic Review and Meta-Analysis. PLoS One. 2014;9(7).

26. Schmitz-Hübsch T, Pyfer D, Kielwein K, Fimmers R, Klockgether T, Wüllner U. Qigong exercise for the symptoms of Parkinson's disease: a randomized, controlled pilot study. Mov. Disord. 2006;21(4);543-548.

27. Burini D, Farabollini B, Iacucci S, Rimatori C, Riccardi G, Capecci M, Provinciali L, Ceravolo MG. A Randomised Controlled Cross-over Trial of Aerobic Training versus Qigong in Advanced Parkinson's Disease. Eura. Medicophys. 2006;42(3);231-238.

28. Colgrove Y, Sharma N, Robbins K, Wagner K. A Randomized Controlled Pilot Study of the Therapeutic Effects of Yoga in People with Parkinson's Disease. Int. J. Yoga. 2015;8(1);74.

29. Boulgarides LK, Barakatt E, Coleman SB. Measuring the Effect of an Eight-Week Adaptive Yoga Program on the Physical and Psychological Status of Individuals with Parkinson's Disease. A Pilot Study. Int. J. Yoga Ther. 2014;24;31-41.

30. Sharp K, Hewitt J. Dance as an Intervention for People with Parkinson's Disease: A Systematic Review and Meta-Analysis. Neurosci. Biobehav. Rev. 2014;47;445-456.

31. Hackney ME, Kantorovich S, Levin R, Earhart GM. Effects of Tango on Functional Mobility in Parkinson's Disease: A Preliminary Study. J. Neurol. Phys. Ther. 2007;31(4);173-179.

32. Donoyama N, Ohkoshi N. Effects of Traditional Japanese Massage Therapy on Various Symptoms in Patients with Parkinson's Disease: A Case-Series Study. J. Altern. Complement. Med. 2012;18(3);294-299.

33. Donoyama N, Suoh S, Ohkoshi N. Effectiveness of Anma Massage Therapy in Alleviating Physical Symptoms in Outpatients with Parkinson's Disease: A before-after Study. Complement. Ther. Clin. Pract. 2014;20(4);251-261.

34. Cho SY, Shim SR, Rhee HY, Park HJ, Jung WS, Moon SK, Park JM, Ko CN, Cho KH, Park SU. Effectiveness of Acupuncture and Bee Venom Acupuncture in Idiopathic Parkinson's Disease. Parkinsonism Relat. Disord. 2012;18(8);948-952.

35. Chae Y, Lee H, Kim H, Kim CH, Chang DI, Kim KM, Park HJ. Parsing Brain Activity Associated with Acupuncture Treatment in Parkinson's Diseases. Mov. Disord. 2009;24(12);1794-1802.

36. HP-200 in Parkinson's Disease Study Group. An alternative medicine treatment for Parkinson's disease: Results of a multicenter clinical trial. J Altern Complement Med.1995;1;249-255.

37. Nagashayana N, Sankarankutty P, Nampoothir MR, Mohan PK, Mohanakumar KP. Association of L-dopa with recovery followingAyurveda medication in Parkinson's disease. J Neurol Sci. 2000;176;124-127.

38. Vaidya AB, Rajagopalan TG, Mankodi NA, Antarkar DS, Tathed PS, Purohit AV, Wadia NH. Treatment of Parkinson's disease with the cowhage plant-Mucuna pruriens Bak. Neurol India. 1978;26;171-176.

39. Katzenschlager R, Evans A, Manson A, Patsalos PN, Ratnaraj N, Watt H, Timmermann L, Van der Giessen R, Lees AJ. Mucuna pruriens in Parkinson's disease: A double blind clinical and pharmacological study. J Neurol Neurosurg Psychiatry. 2004;75;1672-1677.

40. Chung V, Liu L, Bian Z, Zhao Z, Leuk Fong W, Kum WF, Gao J, Li M. Efficacy and safety of herbal medicines for idiopathic Parkinson's disease: A systematic review. Mov Disord. 2006;21;1709-1715.

41. Wang Y, Xie CL, Lu L, Fu DL, Zheng GQ. Chinese herbal medicine paratherapy for Parkinson's disease: A meta-analysis of 19 randomized controlled trials. Evid Based Complement Alternat Med. 2012;534861.

42. Kim TH, Cho KH, Jung WS, Lee MS. Herbal medicines for Parkinson's disease: A systematic review of randomized controlled trials. PLoS One. 2012;7, e35695.

43. Kim SN, Kim ST, Doo AR, Park JY, Moon W, Chae Y, Yin CS, Lee H, Park HJ. Phosphatidylinositol 3-kinase/Akt signaling pathway mediates acupuncture induced dopaminergic neuron protection and motor function improvement in a mouse model of Parkinson's disease. Int J Neurosci. 2011; 121;562-569.

44. Wang H, Pan Y, Xue B, Wang X, Zhao F, Jia J, Liang X, Wang X. The antioxidative effect of electroacupuncture in a mouse model of Parkinson's disease. PLoS One. 2011;6;e19790.

45. Khalil WK, Assaf N, ElShebiney SA, Salem NA. Neuroprotective effects of bee venom acupuncture therapy against rotenone-induced oxidative stress and apoptosis. Neurochem Int. 2014;80;79-86.

46. Yu YP, Ju WP, Li ZG, Wang DZ, Wang YC, Xie AM. Acupuncture inhibits oxidative stress and rotational behavior in 6-hydroxydopamine lesioned rat. Brain Res. 2010;1336;58-65.

47. Liang XB, Liu XY, Li FQ, Luo Y, Lu J, Zhang WM, Wang XM, Han JS. Long-term high-frequency electro-acupuncture stimulation prevents neuronal degeneration and up-regulates BDNF mRNA in the substantia nigra and ventral tegmental area following medial forebrain bundle axotomy. Brain Res Mol Brain Res. 2002;108;51-59.

48. Liang XB, Luo Y, Liu XY, Lu J, Li FQ, Wang Q, Wang XM, Han JS. Electro-acupuncture improves behavior and upregulates GDNF mRNA in MFB transected rats. Neuroreport. 2003;14;1177-1181.

49. Kang JM, Park HJ, Choi YG, Choe IH, Park JH, Kim YS, Lim S. Acupuncture inhibits microglial activation and inflammatory events in the MPTP-induced mouse model. Brain Res. 2007;1131;211-219.

50. Liu XY, Zhou HF, Pan YL, Liang XB, Niu DB, Xue B, Li FQ, He QH, Wang XH, Wang XM. Electroacupuncture stimulation protects dopaminergic neurons from inflammation-mediated damage in medial forebrain bundle-transected rats. Exp Neurol. 2004;189;189-196.

51. Doo AR, Kim ST, Kim SN, MoonW, Yin CS, Chae Y, Park HK, Lee H, Park HJ. Neuroprotective effects of bee venom pharmaceutical acupuncture in acute 1-methyl- 4-phenyl-1,2,3,6-tetrahydropyridine-induced mouse model of Parkinson's disease. Neurol Res. 2010;32;88-91.

52. Lee MS, Shin BC, Kong JC, Ernst E. Effectiveness of acupuncture for Parkinson's disease: Asystematic review. Mov Disord. 2008;23;1505-1515.

53. Lam YC, Kum WF, Durairajan SS, Lu JH, Man SC, Xu M, Zhang XF, Huang XZ, Li M. Efficacy and safety of acupuncture for idiopathic Parkinson's disease: A systematic review. J Altern Complement Med. 2008;14;663-671.

54. Lee HS, Park HL, Lee SJ, Shin BC, Choi JY, Lee MS. Scalp acupuncture for Parkinson's disease: A systematic review of randomized controlled trials. Chin J Integr Med. 2003;19;297-306.

55. Kim HJ, Jeon BS. Is acupuncture efficacious therapy in Parkinson's disease? J Neurol Sci. 2014;341;1-7.

56. Kim JY, Kim HJ, Jeon BS. Prevalence and characteristics of non-motor symptoms in Korean Parkinson's disease patients and its relationship with experience of alternative therapies. J Korean Neurol Assoc. 2013;31;8-14.

Chapter 22

Stem Cell Therapy in Parkinson's Disease

Bhavna Kaul, Jasmine Parihar

Introduction

By definition, cells that can self-renew and produce progenies as well as differentiate into multiple cell lineages are termed as stem cells.[1] Essentially, these are undifferentiated cells that are able to differentiate into multiple specialized cell types. Since stem cells have the potential to replace or restore lost cells, they have been evaluated and considered as potential therapeutic agents in neurological diseases.

Parkinson's disease (PD) is the second most common neurodegenerative disorder of the central nervous system (CNS) and typically presents around the age of 70 years. PD causes degeneration of midbrain dopaminergic neurons controlling movements and gait. It is currently incurable and leads to progressive disability, despite responding well to dopaminergic (DA) replacement therapies early on in its clinical course. While the condition is defined by the loss of the dopaminergic nigrostriatal neurons and the formation of α-synuclein Lewy bodies, it is now recognized that the pathology and clinical features extend well beyond this pathway. A combination of influence of protein aggregates and participation of other neural cells in the degeneration of dopaminergic neurons contribute to the underlying pathology in PD.

Many of the motor symptoms can be treated for several years with L-DOPA, which is readily converted to dopamine by the remaining dopamine neurons. Eventually, nearly all of dopaminergic neurons succumb to the degenerative process and patients become unresponsive to L-DOPA. Because of the important role of dopaminergic neurons, there has been extensive interest in the idea that replacement of diseased cells may help in ameliorating the symptoms of Parkinson's disease.

Current status of medical and surgical treatment in Parkinson's disease–

Current therapies focus on extrinsically increasing levels of dopamine in the brain or simulating its effects by way of dopamine agonists and deep brain stimulation surgery. These measures effectively

DOI: 10.1201/9781003505488-22

alleviate disease symptoms for a few years, but do not retard the disease progression or restore neuronal function. Moreover, the side effects of pharmacological treatment are more severe in the aged population. These include confusion, hallucinations, fatigue and orthostatic hypotension.

Due to possible adverse side effects such as brain haemorrhage, infarction, seizures and even death, invasive surgical methods are not recommended for elderly patients. Cell based therapies for Parkinson's disease which stimulate endogenous dopamine production present a tempting prospect of disease modifying effect by means of retardation of stem cell degeneration along with a more physiological release of dopamine in the brain.

Basic approach of stem cell treatment in Parkinson's disease involves transplants of dopaminergic cells grafted into the brain to replace those lost to the disease process. This approach is also not designed to cure PD, but to treat the dopamine responsive symptoms of the disease. For stem cell therapy to be incorporated into the mainstream, it should not only be as efficacious in alleviating symptoms as the existing conventional therapies, but also have advantages over these existing agents–either practically, therapeutically in the long-term, economically or a combination thereof.

To achieve a successful cell-based therapy in PD, some criteria for cell transplantation are generally suggested[2,3]- (1) The cells should possess molecular, morphological and electrophysiological properties of DA neurons in substantia nigra; (2) grafts should be able to reverse the motor deficits of PD; (3) therapy should enable 1,00,000 or more DA neurons to survive long term in human putamen; (4) grafted cells should re-establish dense terminal network throughout the striatum to functionally integrate into host neural circuitries.

In addition to displaying these functional characteristics, the cells also need to be safe which means they should not present any risk of forming masses/tumors; migrate away from the site of grafting and integrate into unaffected neural circuits; be able to de-differentiate back to the cell of origin.

Sources of stem cells

Stem cells for DA production can be derived from the following sources[4-5]

1. Human fetal ventral midbrain (hfVM) cells
2. Human embryonic stem cells (hESCs)
3. Human neural stem/precursor/progenitor cells (hNSCs/hNPCs)
4. Human mesenchymal stem cells (hMSCs)
5. Human induced neural stem cells (hiNSCS)
6. Human induced pleuripotent stem cells (hiPSCs)
7. Human retinal pigment epithelial cells (hRPECs)

Human fetal ventral midbrain (hfVM)

Human fetal ventral mid-brain-cells are derived from fetal midbrain, but they have poor ability to generate dopaminergic neurons. Also because of inadequate source, supply, technical and ethical issues in obtaining sufficient fetal brain tissue its clinical application is limited.

Human embryonic stem cells (hESCs)

These are derived from inner mass of developing blastocyst and have capability to produce cells of all three primary germ layers i.e. ectoderm, mesoderm and endoderm. Using several morphogens (like all trans-retinoic acid, fibroblast growth factor and epidermal growth factor), feeder cell co-culture, growth or neurotropic factor and genetic manipulation the hESCs are induced to produce DA neurons. The advantages of hESCs are their ability of totipotent differentiation and preserved pluripotency even after in-vitro expansion. Thus, they can be virtually limitless source of DA precursor cells. Several human and animal studies have shown considerable graft survival, dopamine release and improvement in motor symptoms with hESCs. hESCs have the disadvantage of having tumerogenic potential leading to teratoma formation in intended zone of transplantation and immunorejection.[6] Also, there are ethical concerns in obtaining source tissue leading to limited tissue availability.

Human neural stem/precursor/progenitor cells (hNSCs/hNPCs)

Unlike ESCs, NSCs are committed to neural lineage differentiation and form neurons, oligodendriocytes and astrocytes in vivo. Neural stem cells can be endogenous, or stem cell derived. As compared to ESCs, NSCs have less tumorigenic potential and immunological rejection. Though NSCs appear perfect therapeutic candidate for neurological disease, limited tissue availability is a concern.

Human mesenchymal stem cells (hMSCs)

MSCs are non-haematopoietic, multipotent cells derived from stromal structures of bone marrow, umbilical cord, dermis and peripheral blood. These cells can differentiate into osteocytes, chondrocytes and adipocytes or they can trans-differentiate into other non-mesenchymal cell types especially neurogenic cells. There are two types of MSCs: naieve MSCs and neurally induced MSCs. Differentiation of MSC into DA neurons can be induced through combination of chemicals growth factors and signalling molecules. MSCs are capable of proliferation, migration and differentiation into endogenous NSCs. MSCs are most promising option in regenerative therapy as these can be easily produced and expanded. Also, there are no ethical concerns and risk of immunorejection associated with MNCs, but like ESCs and NSCs there is risk of tumor formation with these cells.

Induced pluripotent stem cells (iPSCs)

ESC-like pleuripotent cells could be derived from embryonic and adult mouse fibroblasts by means of virally introduced OSKM (Oct3/4, Sox2, Klf4, and c-Myc) gene transcription factors known as iPSCs.[7] Viral vectors used can be adenovirus, lentivirus and retrovirus. [8,9,10] In PD patients, the iPSC can be used to generate patient-specific midbrain DA neurons while using the autologous donor cells reducing the risk of graft rejection significantly. But since the process involves integrating viruses, there is risk of introduction of oncogenes especially c-Myc and Klf4 leading to tumorogenesis. Besides, only limited quantity of genome can be carried be viral vector, the host can develop resistance to repeated infections, the procedure is complicated and there can be unforeseen genome alterations.[11]

These problems have been circumvented in several studies by application of non-integrating viruses[14,15] and by use of non-viral delivery vectors such as expressing plasmids,[14] episomal vectors,[14] minicircle vectors,[14] piggyBac transposon[15] and synthetic modified RNAs.[14] Though many animal studies have been carried out on application of iPSCs in PD patients, the practical application of this technique is still lagging much behind its availability.

Thus, iPSCs, NSCs and MSCs are the most likely sources of stem cells for PD therapy. In all cases, autografts may be used as demonstrated by several successes in rodent and primate PD models. However, the diverse differentiation methods, low production rate and low survival rate after transplantation are still obstacles that need to be overcome before clinical use.

What do we know so far?

Allograft transplants of hfVM tissue contain the developing midbrain dopaminergic neurons and have been shown to provide clinical benefits lasting well over a decade in some patients with normalization of dopaminergic innervation of the transplanted striatum.[14] However, other studies have not shown such clear benefits and even reported significant side effects, such as graft-induced dyskinesias.[14,15]

Several studies have shown that DAergic neurons derived from ESCs or iPSCs grafted into PD model systems can both survive and integrate into the host network with remarkable functional improvements.[14] Thus, the use of iPSCs and induced dopamine neurons offer suitable approach for autologous cell-based therapy in PD.[14]

Transplantation of autologous bone marrow mesenchymal stem cells causes marginal amelioration of symptoms, but nevertheless improves quality of life in PD patients, in the absence of significant adverse effects.[14]

Human retinal pigment epithelial cells (hRPECs)

Human retinal pigment epithelial (hRPE) cells, an alternative to human fetal tissue transplants, possess the features of NPCs that permit differentiation into DA neurons. Due to this property, they have been used for transplanting in animal models as well as human PD patients. In a double blind randomized controlled trial, there was statistical difference between those patients who received retinal pigment epithelial cells transplant and those who underwent sham surgery.[14] Overall, outcomes of the stem cell transplantation in PD have been quite diverse and adverse events such as dyskinesia have diminished the therapeutic effect. [12] Behari et al. treated 6-OHDA induced mice model of parkinsonism with hRPEC transplanted in striatum and observed significant improvement in parkinsonism motor behaviour tests after 16 weeks (unpublished data).

Graft

It is currently unknown whether symptomatic relief would be best achieved by implanting a pure population of DA neurons or a graft containing a portion of glial cells. Several studies support the necessary role of astrocytes for neural differentiation during embryonic development, implying that glial cells are important for fate determination of precursors during implantation.[13]

Another key issue in performing grafts for PD treatment is that implanting the most suitable sub-type of DA neurons is also critical for the outcome of transplantation. DA rich-ventral mesencephalic grafts contain two types of DA neuron progenitors, including A9 Substantia Nigra neurons and A10 dopamine neurons of the Ventral Tegmental Area.[14] The former is beneficial for PD treatment as only these send innervated axons into the striatum in rats.

Patient selection

Younger patients who are levodopa responsive are more likely to benefit from stem cell transplantation surgery. In addition, patients with DA neuron loss restricted to the caudate-putamen will also receive more symptomatic benefit after transplantation.

How to apply stem cells?

Stem cells may be administered for the treatment of central nervous system (CNS) disorders in four ways: intracerebrally, intrathecally/intraventricularly, intranasally, or intravascularly. Intracerebral application of stem cells involves applying stem cells directly to the injured or disease-affected tissue. In an unpublished report from AIIMS, New Delhi (Behari et al. 2014), 12 PD patients with advanced PD received atulogous bone marrow derived MSC implanted stereotactically in putamen bilatrally. Outcome was measured by change in total UPDRS scale and Tc SPECT scan done 3, 6 and 12 months after implant. The authors also looked for short term and long-term side effects (up to 12 months). The results showed that there was significant improvement in UPDRS score and also in SPECT scan which showed increase in DaT uptake. One patient developed dyskinesia 3 months following transplant.

Intrathecal/intraventricular application of stem cells involves applying stem cells directly into the cerebral spinal fluid. Intranasal application of stem cells involves applying stem cells into the brain via the olfactory epithelium and fila olfactoria. Finally, intravascular application involves applying stem cells into the blood vessels. Adult stem cells and especially MSCs are capable of migrating toward injured or damaged tissue by a process called homing. However, treating diseases of the CNS may require directed application because the blood–brain barrier may hamper CNS penetration of the stem cells. One of the major challenges to the use of stem cells may be ensuring that the highest number of stem cells reaches the lesion.

Intracerebral injection of stem cells requires fewer stem cells and is a precise delivery route and is therefore, a good method of administration of neural stem cells. However, due to invasive procedures and severe adverse effects, such as seizures, motor exacerbations, syncope, chronic subdural haematoma and tumorigenicity, direct stereotaxic injection has been used rarely in clinical trials. [15] Intracerebroventricular (ICV) injection is relatively less invasive than intracerebral injection and has not shown cell-related tumorigenicity. However, only a small fraction of transplanted cells migrate to the target site and ICV injection may cause higher risk of fever, infection and meningitis. [16]

Issues after grafting

- *Graft induced dyskinesia*

Freed et al. described development of "graft induced dyskinesia (GID)" in 15% of transplanted patients one year after transplantation.[17] This incidence rate was 56% in study by Olanow. [18] Immunosuppression may also be an important factor since dyskinesia did not develop until immunosuppression withdrawal in several reports. [19,20] The other possible issue is that heterogeneous grafts were found in ventral putamen, containing serotonergic neurons. Disease grafts led to islands of re-innervation and abnormal production of DA. [21,22] Therefore, transplanting sufficient number of cells containing pure population of DA neurons in basal ganglion with immunosuppression is suggested to avoid the development of GID.

- *Grafts affected by PD pathological process*

Evidence that PD pathology may propagate from host to grafts is emerging. [23,24] Presence of Lewy bodies and Lewy neurites in grafted DA neurons were generally observed 11–16 years after human fetal mesencephalic transplantation. [25] The exact reasons for this negative outcome remain unresolved. It was suggested that Lewy bodies behaved like an infective agent, similar to prion. Hence, suggesting that PD may indeed be a prion disease and spreads from diseased cell to healthy donor cells. Therefore, although the spread of PD pathology may occur after transplantation, the period of beneficial effects from transplantation is still longer than that of current medications, suggesting stem cells still could be a potential clinical therapy.

- *Timing of therapy*

In practice, when is the best time to institute this therapy is an unanswered question. It has been found that Clinical Randomized Trials (CRT) will give best results when instituted in early stage of disease. But starting CRT in early stages, which is expensive, complicated and more invasive compared to present therapies, cannot be justified. Besides, clinical improvement after CRT does not show immediate result. Also withholding other therapies in a symptomatic patient is a matter of ethical concern. Theoretically, if started in preclinical stage of PD, CRT can abort the development of clinical phase of PD.

To summarize, stem cell therapy is a promising treatment option for Parkinson's disease. The results found in animal studies have not been convincingly duplicated in human subjects. This may be due to combination of factors, such as patient selection, amount and mode of tissue engraftment and the level of immunosuppression. Although graft induced dyskinesia is a major concern, the graft tissue was found infected by the disease pathology as long as 10 years after transplantation. Hence, this therapy continues to hold promise for the future. The major advantage of this strategy is the restorative and trophic abilities of the grafted cells which reach far beyond drugs prescribed in current practice.

References

1. Lunn JS, Sakowski S A, Hur J, and Feldman EL (2011). Stem cell technology for neurodegenerative diseases. Ann Neurol 70, 353–361. doi: 10.1002/ana.22487
2. Lindvall O, Hagell P (2000) Clinical observations after neural transplantation in Parkinson's disease. Prog Brain Res 127:299–320.
3. Lindvall O, Kokaia Z (2006). Stem cells for the treatment of neurological disorders. Nature 441:1094–1096.
4. Aggarwal S, Pittenger MF. Human mesenchymal stem cells modulate allogeneic immune cell responses. Blood. 2005;105(4):1815-1822. doi:10.1182/blood-2004-04-1559.
5. Kriks S, Shim J-W, Piao J, et al. Dopamine neurons derived from human ES cells efficiently engraft in animal models of Parkinson's disease. Nature. 2011;480(7378):547-551. doi:10.1038/nature10648.

6. Brederlau A, Correia AS, Anisimov S V, et al. Transplantation of Human Embryonic Stem Cell-Derived Cells to a Rat Model of Parkinson's Disease: Effect of In Vitro Differentiation on Graft Survival and Teratoma Formation. Stem Cells. 2006;24(6):1433-1440. doi:10.1634/stemcells.2005-0393.
7. Takahashi K, Yamanaka S. Induction of Pluripotent Stem Cells from Mouse Embryonic and Adult Fibroblast Cultures by Defined Factors. Cell. 2006;126(4):663-676. doi:10.1016/j.cell.2006.07.024.
8. Zhou H, Wu S, Joo JY, et al. Generation of Induced Pluripotent Stem Cells Using Recombinant Proteins. Cell Stem Cell. 2009;4(5):381-384. doi:10.1016/j.stem.2009.04.005.
9. Yu J, Vodyanik MA, Smuga-Otto K, et al. Induced pluripotent stem cell lines derived from human somatic cells. Sci (New York, NY). 2007;318(5858):1917-1920. doi:10.1126/science.1151526.
10. Huangfu D, Osafune K, Maehr R, et al. Induction of pluripotent stem cells from primary human fibroblasts with only Oct4 and Sox2. Nat Biotechnol. 2008;26(11):1269-1275. doi:10.1038/nbt.1502.
11. Thomas CE, Ehrhardt A, Kay M. Progress and problems with the use of viral vectors for gene therapy. Nat Rev Genet. 2003;4(5):346-358. doi:10.1038/nrg1066.
12. Karussis D, Petrou P, Kassis I, 2013. Clinical experience with stem cells and other cell therapies in neurological diseases. J. Neurol. Sci. 324, 1–9.
13. Song H, Stevens CF, Gage FH (2002), Astroglia induce neurogenesis from adult neural stem cells. Nature 417:39–44.
14. Thompson L, Barraud P, Andersson E, Kirik D, Bjorklund A (2005), Identification of dopaminergic neurons of nigral and ventral tegmental area subtypes in grafts of fetal ventral mesencephalon based on cell morphology, protein expression, and efferent projections. J Neurosci 25:6467–6477.
15. Savitz SI, Dinsmore J, Wu J, Henderson GV, Stieg P, Caplan LR, 2005. Neurotransplantation of fetal porcine cells in patients with basal ganglia infarcts: a preliminary safety and feasibility study. Cerebrovasc. Dis. (Basel Switzerland) 20, 101–107.
16. Rabinovich SS, Seledtsov VI, Banul NV, et al., 2005. Cell therapy of brain stroke. Bull. Exp. Biol. Med. 139, 126–128.
17. Freed CR, Greene PE, Breeze RE et al. (2001), Transplantation of embryonic dopamine neurons for severe Parkinson's disease. N Engl J Med 344:710–719.
18. Olanow CW, Goetz CG, Kordower JH et al. (2003), A double blind controlled trial of bilateral fetal nigral transplantation in Parkinson's disease. Ann Neurol 54:403–414.
19. Piccini P, Lindvall O, Bjorklund A et al. (2000), Delayed recovery of movement related cortical function in Parkinson's disease after striatal dopaminergic grafts. Ann Neurol 48:689–695.
20. Lane EL, Soulet D, Vercammen L, Cenci MA, Brundin P (2008), Neuroinflammation in the generation of post-transplantation dyskinesia in Parkinson's disease. Neurobiol Dis 32:220–228.
21. Ma Y, Feigin A, Dhawan V et al. (2002), Dyskinesia after fetal cell transplantation for parkinsonism: a PET study. Ann Neurol 52:628–634.
22. Carlsson T, Carta M, Munoz A et al. (2009), Impact of grafted serotonin and dopamine neurons on development of L-DOPA induced dyskinesias in parkinsonian rats is determined by the extent of dopamine neuron degeneration. Brain 132:319–335.

23. Visanji NP, Brooks PL, Hazrati LN, Lang AE (2013), The prion hypothesis in Parkinson's disease: Braak to the future. Acta Neuropathologica Communications 1:2.
24. Kordower JH, Chu Y, Hauser RA, Freeman TB, Olanow CW (2008a), Lewy body like pathology in longterm embryonic nigral transplants in Parkinson's disease. Nat Med 14:504–506.
25. Li JY, Englund E, Holton JL et al. (2008), Lewy bodies in grafted neurons in subjects with Parkinson's disease suggest host to graft disease propagation. Nat Med 14:501–503.

Chapter 23

Quality of Life in Parkinson's Disease

Rajeev Aggarwal, Madhuri Behari

Parkinson's disease (PD) is a progressive neuro-degenerative disease characterized by loss of pigmented dopaminergic nigro-striatal neurons leading to motor symptoms such as rest tremor, rigidity and bradykinesia and several non-motor symptoms. With improving medical facilities and life expectancy, the number of persons with Parkinson's disease (PWP) is expected to rise consistently throughout the world. In western world, median age of onset of PD is around 60 years, but among Indians it is around 50 years. The lower age at onset of symptoms by about one decade[1] results in large number of people affected by PD in the peak of their career affecting their earning capacity with catastrophic effects when sole bread earner of the family is affected.

PD is a complex and challenging disease. Initially easy to manage, but as the disease progresses multiple factors crop up making management more complicated (Table 1).

Table 1: Factors and their sequel which make the life of PWP more troublesome

Factors	Sequels
Clinical	Emergence of new motor and/or non-motor symptoms
Treatment related	Side effects of medications
Psychological	Anxiety and depression
Social	Social isolation
Financial	Direct cost as well as indirect cost of treatment

In chronic diseases like PD where treatment aims at amelioration of symptoms, evaluation of patients' perspective about their disease is of paramount importance, in addition to various clinical tests to assess severity of disease. Patient reported outcome (PRO) measures are excellent tools to know the patients' perspective about their disease and its implication. PRO measures are usually questionnaire-based scales where a person fills it without any interference by clinician or other person. PRO measures are commonly used to assess symptoms and their severity, physical capacity/disability and quality of life (QoL).

DOI: 10.1201/9781003505488-23

In last few decades QoL has become a very popular term used by clinicians, socialists, politicians and human rights people. QoL is a term which is used rather vaguely without any numerical measurable parameters. This is talked about very commonly, many people not knowing what it is exactly about.[2]

Historical aspect

Historical development of concept of QoL can be traced back to the Nichomachean Ethics, in which Aristotle (384-322 BC) stated that, "Both the multitude and persons of refinement . . . conceive 'the good life' or 'doing well' to be the same thing as 'being happy'. "However, happiness is difficult to define, it may mean different thing to different people at different times. Although, the term "Quality of life" did not exist in the Greek language 2000 years ago, Aristotle agreed that QoL does not mean different things to different people and it changes according to status of the person.

Scientific advancements in health sector has increased longevity of people creating a new realm of inquisitiveness for achieving "goodness" of life or life satisfaction in extended years of life. The Oxford's Dictionary defines quality as "degree of excellence of a characteristic", but there is lack of consensus on definition of QoL.

The common areas that affect QoL of a person ranges from personal, psychological, economic, social (neighbourhood), national, international and environmental factors. To evaluate QoL in an individual suffering from PD demographic issues, violence, crime, social justice, housing issues etc., remain unaltered, whereas physical fitness, psychological, social, family and financial issues become of paramount importance.

As per WHOQOL Group 1995, *"Quality of life encompasses individuals' perception of their position in life in the context of the culture and value systems in which they live and in relation to their goals, expectations, standards and concerns. It is a broad ranging concept affected in a complex way by the person's physical health, psychological state, level of independence, social relationships, personal beliefs and their relationship to salient features of their environment."*[3] This definition is too broad for medical perspectives as health professionals are usually concerned about impact of disease on patients' lives and its management. To meet this expectation, Health Related Quality of Life (HRQoL) term is used to describe the subset of QoL related to health and disease.

Recently Aggarwal et al.[4] conceptualized the essential components of HRQoL in PD and postulated that *"Health related quality of life in Parkinson's disease is defined as multi-dimensional concept consisting of patient's self-evaluation of the impact of motor and non-motor symptoms, effect of treatment, psychological state, spirituality, social and financial implication and overall health satisfaction."*[4]

HRQoL scales in PD

HRQoL is measured by two types of scales: 1) Generic scales measures QoL broadly and 2) Disease specific QoL scales which measures QoL as influenced by that specific disease. One needs to evaluate both generic and disease specific HRQoL scales to globally evaluate HRQoL in PD.

Generic HRQoL scales for PD patients

Generic HRQoL scales measure the perception of an individual to his general health. WHOQoL-BREF, Nottingham health profile (NHP), Sickness impact profile (SIP), 36 item short form questionnaire (SF36), Schedule for the evaluation of Individual Quality of life–Direct Weighting (SEIQoL–DW), EuroQoL (EQ-5D) and 15D questionnaire have been used as generic HRQoL scales in PD.[5,6] Out of these scales, Movement Disorder Society (MDS) has recommended SF36, NHP, SIP and EQ-5D for use in PD patients. EQ-5D is a five item utility scale for econometric issues. NHP measures health status and consists of 38 items in eight domains that has substantial floor effects in PD, while SIP has 136 items health status questionnaire that has 12 domains and two dimensions i.e., Physical and psychosocial. It is limited in use for PWP due to its length. SF36 is a 36-item questionnaire with eight domains. These domains are physical function (PF), role physical (RP), body pain (BP), general health (GH), vitality (VT), social function (SF), role emotion (RE) and mental health (MH). In addition, SF36 provides two summary score viz. physical component summary (PCS) and mental component summary (MCS). It is most commonly used generic scale in PD. It shows good reliability, responsiveness and discriminative validity in PD.[5–8] These questionnaires are useful for comparing the impact of disease across the number of diseases and to compare the burden of various diseases on the society. There are normative values for the general population for SIP, NHP, SF36 and EQ-5D which provides a baseline for comparison of results. However, these questionnaires lack specific questions pertaining to PD, hence are poor in evaluating the progression of disease. On the other hand, these scales are sensitive to general influence of a disease on overall QoL.

PD specific HRQoL scales

Parkinson's disease specific HRQoL scales are developed to assess QoL in PWP. These scales contain questions pertaining to PD and their implications on lives of PWP. These scales are more appropriate, sensitive, responsive and have good discriminative validity for PWP. These scales cannot be used for healthy population or persons of other diseases; therefore, comparing the impact of various diseases is not feasible. BELA-P-k[9] is the first PD specific scale on psychosocial aspects, which was developed in German language in 1993. This scale is not very popular due to its non-availability in other languages. In 1995, Peto et al. developed PD specific scale as 39 item Parkinson's disease questionnaire (PDQ39),[10] since then this is the most commonly used scale in PD as PRO measure. It has been translated in more than 20 languages. Since 1995, many PD specific scales have been developed in various countries [11–14] (Table 2). In 2002, first review article on PD specific HRQoL scales, published by Marinus et al. included PDQ39, PDQL, PIMS and PLQ; and reported good internal consistency of all scales, however, content and construct validity of PIMS was poor.[23] In 2007, two systematic review articles were published by Oudsten et al.[6,24] on patient-based measures in PD. They included generic as well as PD specific scales and criticized the inappropriate use of PDQ39, PDQL, PIMS and PDQUALIF as HRQoL scales when these are actually health status (HS) scales. The entire literature has created a complicated situation, as the fundamental nature of QoL measurement that it should reflect the patients' perception is missing in most of these scales. Questions on life satisfaction-Movement disorder and Deep brain stimulation modules (QLS-MD and QLS-DBS) are the only scales which meet the criteria as HRQoL scale, but this scale has hardly been used in literature. QLS-MD and QLS-DBS are the supplement modules to original generic questionnaire "Questions on life satisfaction" (QLS) scale.[15] As health status influences and predicts QoL and due to lack of any PD specific HRQoL scale; MDS recommended PDQ39, PDQ8, PDQL, PIMS and SCOPA-PS for use as PD specific HRQoL scales. Apart from these scales PDQUALIF is suggested as a measure of QoL, though PLQ, PPS and BELA-P-k have also been listed.[5]

In spite of large number of PD specific HRQoL scales, there is a serious threat to the validity of these scales while adopting any of these scales for low income countries like India due to following reasons:

i. All of these instruments were developed in western countries that have distinctly different education system, social structure, economic status, environmental factors, treatment options and social security system compared to Indian sub-continent.

ii. Most of these instruments were developed 15-20 years back and do not cover non-motor symptoms (NMS) comprehensively. Recently, NMS in PD have been increasingly recognized and considered to be major determinant of QoL;[16–19] and are described to be present several years, even decades before motor symptoms appear.[20,21]

iii. PD has significant impact on financial status of the patient. Treatment of PD is associated with lots of expenditure, not only direct cost such as expenditure on medicines, investigations, doctor's fee, travel to hospital etc. but also indirect cost, such indirect cost/ hidden costs in the form of lost wages, services from informal care giver and changing job to a less paying job are substantial. Their magnitude is even more important considering that the family generally has fixed income and when caregiver is a spouse equally old.[22] This problem further aggravates when the sole bread earner of the family is affected by PD and becomes dependent on the family for his treatment. In spite of being a major stress in life, financial implication of the disease on QoL has not been considered in PDQ39, PDQL and SCOPA-PS which are much commonly used instruments as compared to PDQUALIF and PIMS. PDQUALIF has kept a single item on finance in "self-image and sexuality" domain, whereas PIMS has included an item on finance in "financial security" domain.

iv. All of these instruments although intend to measure a single construct, i.e., PD, specific HRQoL include different domains and items.[23] In fact, none of these instruments cover all relevant domains for HRQoL in PD and do not assess QoL in holistic manner.[24]

To address these issues Aggarwal et al. (Unpublished work) developed a new PD specific HRQoL scale in Hindi language called as "Quality of life in Parkinson's disease" (QLPD) instrument. QLPD is a 45-item scale with one item on global QoL. The internal consistency of QLPD is 0.94 and test-retest reliability (Intra-class correlation coefficient) is 0.977. QLPD has nine domains with internal consistency ranging from 0.42 to 0.91. QLPD has shown excellent construct validity, concurrent validity, precision of scores, interpretability of scores and discriminatory property.

Following is a list of all scales that have been developed to evaluate HRQoL in PD and their description (Table 2).

Table 2: Available Health Related Quality of Scales for PD in use

Scale (Year)	Author and country	Total no of items	Domains (No of items)
BELA-p-k (1993)	Ellgring et al.[9] Germany	19	Achievement capability/physical (5), Fear/emotional (4), Social functioning (5), Partner-bonding/family (5)
PDQ39 (1995)	Peto et al.[10] UK	39	Mobility (10), ADL (6), Emotional well-being (6), Stigma (4), Social support (3), Cognition (4), Communication (3), Bodily discomfort (3)
PDQ8* (1998)	Peto et al.[25] UK	8	
PDQL (1996)	De Boer et al.[11] Netherland	37	Parkinsonian symptoms (14), Systemic symptoms (7), Emotional functioning (9), Social functioning (7)
PIMS (1996)	Calne et al.[12] Canada	10	Psychological (5), Social (2), Physical (2), Financial (1)
SCOPA-PS (2003)	Marinus et al.[14]Netherland	11	Psychosocial (11)
PD-QUA-LIF (2003)	Welsh et al.[13]U-SA	32 plus 1	Social and role function (9), Self-image and sexuality (7), Sleep (3), Outlook (4), Physical functioning (5), Independence (2), Urinary functions (2)
QLPD (2013)	Aggarwal et al. (Unpublished) India	45+1	ADL (6), Mobility (4), Psychological (4), Fear (4), Social (4), Family (3), Treatment (3), Finance (3), Non-Motor Symptoms (14)

PDQ–39, Thirty nine item Parkinson's Disease Questionnaire; PDQL, Parkinson's Disease Quality of Life questionnaire; BELA–p –k, Belastungsfragebogen Parkinson Kurzversion; PDQ–8*, Short Form Parkinson's Disease Questionnaire (PDQ-8 is a short version of PDQ-39 and contains one item in every domain), PIMS, Parkinson's IMpact Scale; PLQ, Fragebogen Parkinson Lebens Qualitat; PDQUALIF, Parkinson's Disease QUAlity of LIFe scale; QLPD, Quality of Life in Parkinson's Disease; SCOPA–PS, SCale for Outcome in Parkinson's disease- Psychological.

Predictors of QoL in PD

Large numbers of studies have delineated various factors as predictors or determinants of QoL in PD, which is significantly affected in PD due to progressive nature of the disease.[26–35]In a study on 278 Indian PWP Behari et al.[29] reported that female PD patients, presence of depression, higher levodopa dose (>400 mg/day), higher degree of dependence degree and higher UPDRS (ADL) score were associated with poor QoL. The main predictors and determinants of HRQoL in PD are as follow:

i. Depression

Depression is prevalent in 2.7-70% of all PD patients.[29,36–38] Depression may pre-date motor symptoms in a large majority of patients[36]and is the most significant contributor of QoL[14,29,38] after adjusting other variables. The impact of depression on QoL is more than double compared to motor symptoms.[39] Severity of depression does not correlate with the clinical severity of motor symptoms of PD. Depression is reported in higher percent of PD patients early in the disease, when the diagnosis is made and disclosed to the patients.[40] Symptoms of depression have been reported to be much higher in young onset PD patients,[34,35] which could be attributed to higher family responsibilities and jeopardy of active lifestyle.

ii. Anxiety

The prevalence of anxiety in PD is up to 40%,[21] whereas up to 29% have a circumscribed anxiety disorder.[21] Seven to 30% suffer from panic disorders and up to 15% from some type of phobia.[41] Anxiety has been reported in patients even before motor symptoms appear due to neurotransmitter (serotonin and dopamine) disturbances and is a main determinant of QoL in PD.[42]

iii. Financial status

QoL in PD is associated with financial status of the patient. The direct costs due to expenditure on medications, consultations and visits to the hospital as well as indirect costs in the form of reduced income and need of caregiver are substantial.[22] Monthly income has been correlated with QoL in PD.[43] A secured income is more conducive for good quality of treatment and in turn good QoL. Pre-mature retirement and complex benefit system adversely affectQoL[44] as many health insurance agencies adopt multiple screening before dispensing medicines or reimbursement for medicines putting strain on incapacitated PD patients.

iv. Treatment efficacy and satisfaction

After initial few years of successful management of Parkinsonian symptoms with dopaminergic therapy, the medications may lead to motor complications e.g. dyskinesia, inadequate response or psychosis; causing a feeling of lack of control.[45] Maintaining an optimal level of functioning with consideration of other factors like age, occupation, long-term side effects of the drug and patient's satisfaction is an arduous task. Successful management of the symptoms and meeting the patients' expectations has direct relationship with patients' QoL.[46] Higher levodopa dose (>400 mg/day) is associated with poor HRQoL.[29] Physiotherapaeutic exercises [47] are positively correlated with HRQoL in PD.

v. Social life

PD affects the social behavior and social functioning profoundly.[48] Personality changes associated with frontal lobe dysfunction, mood and cognitive dysfunction, depression, apathy and impulse control disorder may all affect social life of the patient. Communication difficulties and masked facies may further worsen the situation.[49] The social stigma associated with mask facies has significant influence on a patient's life. Facial expressions are usually first sign for social interaction; in fact, a person without facial expressions is considered less trustworthy and his smile without upper facial movements may be considered as deceptive smile.[50]According to WHO, social life is an integral part of health and significantly contributes to HRQoL in PD.

vi. Family

PD has significant burden on family and caregivers.[51] Effect of PD on family relationship and support varies with social and cultural structure of a nation. Lack of social security and free medical treatment put a major burden on the finances. Patients who have a supportive caregiver have been reported to be twice as satisfied as those who do not.[52] Younger patients experience more family disruption compared to their elderly counterparts.[34]Female experience poorer QoL compared to male counterparts.[29]

vii. Non-motor symptoms (NMS)

PD has been associated with large number of NMS that has been quoted as main determinant of QoL. The prevalence of whole spectrum of NMS has been reported up to 100% and contributes more variability in QoL than UPDRS,[16,53] confusing movement disorder term for PD.NMS constitute large spectrum of symptoms and have been studied individually or in group for their effect on QoL (Table 3). A community-based study explored the patients, who survived 20 years of PD diagnosis (their average age at follow up was 74 years) and reported high prevalence of NMS; falls were experienced by 87%, dysarthria of moderate dysarthria in 81%, dementia in 83%, hallucinations (visual) in 74%, excessive daytime sleepiness in 70%, a urinary incontinence in 71%, choking in 48% and symptomatic postural hypotension in 48%.[54] Table 3 shows the prevalence of non-motor symptoms and how they impact quality of life.

Table 3: *Non-motor symptoms and quality of life*

Non-motor symptom	Prevalence in PD	Effect on QoL
Constipation	Three times more common than age matched control	Significantly worse QoL [55]
Fatigue	33% to 58% [56]	Affects routine daily activities, work, leisure activities and social activities affecting overall QoL.[57,58]
Urinary dysfunction	38-71%	Nocturia, urgency, daytime frequency and incontinence are common urinary dysfunctions associated with PD. Cause social embarrassment and significantly impair QoL.[59]
Sleep disorders	Night time sleep problems: 18-98% Daytime sleepiness problem: 11-84% [60]	Sleep problems are associated with pain, aches and depression in PD. Sleep disorder is an independent predictor of poor QoL.[27,60,61]
RBD	Approx. 20%	Vibha et al.[62] reported significant association of RBD with other sleep disorders that can significantly affect QoL of the patient as well as his/her bed partner.
RLS		Patients with RLS show psychological problems in multiple domains which has substantial effect on QoL.[63]
Pain	40-85%	Affects all domains of QoL.
Sialorrhea	70-80%	Associated with social embarrassment, isolation and worsening of depressive symptoms.[64]
Dysphagia	90% develop dysphagia during their disease course	Leading cause of death as aspiration pneumonia. Dysphagic people experience anxiety or panic attacks during meal and avoid eating with others that leads to social isolation, frustration and reduction in mental health.[65]
Sexual dysfunction	54-79% male PD patients: Erectile dysfunction.[66] Compulsive sexual behaviour: 3.5 % [67]	YOPD patients have poor sexual satisfactions compared to age and sex matched control group.[68]
Visual symptoms		Visual acuity, contrast sensitivity, colour perception and motor perception are impaired in PD that might lead to problem of complex visual hallucination.[69] Visual dysfunction significantly affects QoL.

RBD= Rapid eye movement sleep behaviour disorder; RLS= Restless leg syndrome *viii. Postural instability and gait disorder*

Postural instability and gait disorders are amongst the cardinal symptoms of PD and have predominant effect on QoL.[26] Fear of fall and dependence for mobility are the most important contributory factors which negatively affect QoL in PD.[27] Patients with axial symptoms, postural and gait disorder compared to tremor dominant patients have significantly worse QoL.[70]

ix. Freezing

Freezing is characteristic symptom of PD associated with one third of patients mainly in advanced stage of the disease. Apart from effect on gait and mobility, freezing has profound effect on QoL of PD patients. Freezing is associated with increased number of falls, social isolations, cognitive, emotional and communication problems in addition to its effect on mobility, bodily discomfort and adverse effect on ADL.[71]

x. Functional independence

Maintaining functional independence is of paramount importance to every person. Dependency for ADL is a predictor for worsening QoL.[26]

xi. Age

Age at onset of symptoms has no linear relationship with QoL in PD. QoL is dependent on the consequences of diseases and the disruption of life. Young age onset PD patients (YOPD) have significantly worse QoL as younger patients are more likely to be employed and have more family responsibilities. YOPD have longer disease duration and suffer from more physical, emotional and financial consequences [35] in addition to higher prevalence of dystonia and dyskinesia.[72] Younger patients are also associated with higher depression level which is a strong predictor of poor QoL. YOPD patients have two-fold increased odd ratio of worse QoL as measured by PDQ39 in a large population study.[35] After retirement, age does not have strong relationship with QoL.

xii. Education

PD patients with higher level of education have better QoL, independent of other factors.[35,73] Education is one of the most important indicators of life span and health during old age.[74] Higher level of education helps people to understand the disease better and handle difficult situations during management of disease. Higher number of years of education is associated with better cognitive functions, lower degree of depression and better QoL in PD patients.[75]

xiii. Personality traits

Higher score in extraversion (behaviour where a person enjoys being around people) is associated with better emotional health in male PWP, but opposite for female PWP. Neuroticism (a long-term tendency to be in a negative emotional state) is associated with poor QoL.[57]

Conclusion

PD severely affects QoL in people affected by it especially as it is a progressive disorder. QoL in PD is mainly assessed by physical, emotional and social functioning of a person. Non-motor symptoms, financial implication and impact of treatment on PWP are newer domains that profoundly affect

QoL in PWP. In QoL assessment patients' perception and judgment about the disease, symptoms and management are crucial. Available generic and PD specific QoL scales to assess QoL in persons with Parkinson have methodological limitations and unsuitable for adoption in low income countries. Newly developed QLPD scale addresses these limitations of PD specific scales and may prove as a useful scale for assessment of QoL in persons with Parkinson.

References

1. Vinish M, Prabhakar S, Khullar M, Verma I, Anand A. Genetic screening reveals high frequency of PARK2 mutations and reduced Parkin expression conferring risk for Parkinsonism in North West India. J Neurol Neurosurg Psychiatr. 2010;81(2);166–70.
2. Campbell A. Subjective measures of well-being. Am Psychol. 1976;31(2);117–24.
3. The World Health Organization Quality of Life assessment (WHOQOL): position paper from the World Health Organization. Soc Sci Med. 1995;41(10);1403–9.
4. Aggarwal R, Goyal V, Pandey RM, Kumar N, Singh S, Shukla G, et al. What Should Constitute a Health Related Quality of Life Scale for Parkinson's Disease? Journal of Clinical and Diagnostic Research. 2016;10(9);OC35-OC39.
5. Martinez-Martin P, Jeukens-Visser M, Lyons KE, Rodriguez-Blazquez C, Selai C, Siderowf A, et al. Health-related quality-of-life scales in Parkinson's disease: critique and recommendations. Mov Disord. 2011;26(13);2371–80.
6. Den Oudsten BL, Van Heck GL, De Vries J. The suitability of patient-based measures in the field of Parkinson's disease: a systematic review. Mov Disord. 2007;22(10);1390–401.
7. Hagell P, Törnqvist AL, Hobart J. Testing the SF-36 in Parkinson's disease. Implications for reporting rating scale data. J Neurol. 2008;255(2);246–54.
8. Brown CA, Cheng EM, Hays RD, Vassar SD, Vickrey BG. SF-36 includes less Parkinson Disease (PD)-targeted content but is more responsive to change than two PD-targeted health-related quality of life measures. Qual Life Res. 2009;18(9);1219–37.
9. Ellgring H, Seiler S, Perleth B, Frings W, Gasser T, Oertel W. Psychosocial aspects of Parkinson's disease. Neurology. 1993;43(12);S41-44.
10. Peto V, Jenkinson C, Fitzpatrick R, Greenhall R. The development and validation of a short measure of functioning and well being for individuals with Parkinson's disease. Qual Life Res. 1995;4(3);241–8.
11. De Boer AG, Wijker W, Speelman JD, De Haes JC. Quality of life in patients with Parkinson's disease: development of a questionnaire. Journal of Neurology, Neurosurgery & Psychiatry. 1996;61(1);70–74.
12. Calne S, Schulzer M, Mak E, Guyette C, Rohs G, Hatchard S, et al. Validating a quality of life rating scale for idiopathic parkinsonism: Parkinson's Impact Scale (PIMS). Parkinsonism Relat Disord. 1996;2(2);55–61.
13. Welsh M, McDermott MP, Holloway RG, Plumb S, Pfeiffer R, Hubble J. Development and testing of the Parkinson's disease quality of life scale. Mov Disord. 2003;18(6);637–45.

14. Marinus J, Visser M, Martinez-Martin P, Van Hilten JJ, Stigggelbout AM. A short psychological questionnaire for patients with Parkinson's disease: the SCOPA-PS. Journal of Clinical Epidemiology. 2003;56(1);61-7

15. Kuehler A, Henrich G, Schroeder U, Conrad B, Herschbach P, Ceballos-Baumann A. A novel quality of life instrument for deep brain stimulation in movement disorders. Journal of Neurol, Neurosurg Psychiatry. 2003;74(8);1023–1030.

16. Gallagher DA, Lees AJ, Schrag A. What are the most important nonmotor symptoms in patients with Parkinson's disease and are we missing them? Mov Disord. 2010;15;25(15);2493–500.

17. Pandya M, Kubu CS, Giroux ML. Parkinson disease: Not just a movement disorder. Cleveland Clinic J of Medicine. 2008;75(12);856–64.

18. Krishnan PR, Bhatia M, Behari M. Restless legs syndrome in Parkinson's disease: A case-controlled study. Mov Disord. 2003;18(2);181–185.

19. Kumar S, Bhatia M, Behari M. Sleep disorders in Parkinson's disease. Mov Disord. 2002;(4);775–81.

20. Lim S-Y, Lang AE. The nonmotor symptoms of Parkinson's disease—An overview. Mov Disord. 2010;25(S1);S123–S130.

21. Richard IH. Anxiety disorders in Parkinson's disease. Adv Neurol. 2005;96;42–55.

22. Whetten-Goldstein K, Sloan F, Kulas E, Cutson T, Schenkman M. The burden of Parkinson's disease on society, family, and the individual. J Am Geriatr Soc. 1997;45(7);844–9.

23. Marinus J, Ramaker C, Hilten JJ van, Stiggelbout AM. Health related quality of life in Parkinson's disease: a systematic review of disease specific instruments. J Neurol Neurosurg Psychiatry. 2002;72(2);241–8.

24. Den Oudsten BL, Van Heck GL, De Vries J. Quality of life and related concepts in Parkinson's disease: A systematic review. Mov Disord. 2007;22(11);1528–37.

25. Peto V, Jenkinson C, Fitzpatrick R. PDQ-39: a review of the development, validation and application of a Parkinson's disease quality of life questionnaire and its associated measures. J Neurol. 1998;245;10–14.

26. Marras C, McDermott MP, Rochon PA, Tanner CM, Naglie G, Lang AE. Predictors of deterioration in health-related quality of life in Parkinson's disease: Results from the DATATOP trial. Mov Disord. 2008;23(5);653–659.

27. Forsaa EB, Larsen JP, Wentzel-Larsen T, Herlofson K, Alves G. Predictors and course of health-related quality of life in Parkinson's disease. Mov Disord. 2008;23(10);1420–1427.

28. Schiehser DM, Han SD, Lessig S, Song DD, Zizak V, Filoteo JV. Predictors of health status in nondepressed and nondemented individuals with Parkinson's disease. Arch Clin Neuropsychol. 2009;24(7);699–709.

29. Behari M, Srivastava AK, Pandey RM. Quality of life in patients with Parkinson's disease. Parkinsonism Relat Disord. 2005;11(4);221–6.

30. Chapuis S, Ouchchane L, Metz O, Gerbaud L, Durif F. Impact of the motor complications of Parkinson's disease on the quality of life. Mov Disord. 2005;20(2);224–30.

31. Dowding CH, Shenton CL, Salek SS. A Review of the Health-Related Quality of Life and Economic Impact of Parkinson's Disease. Drugs & Aging. 2006;23(9);693–721.
32. Fitzsimmons B, Bunting LK. Parkinson's disease. Quality of life issues. Nurs Clin North Am. 1993;28(4);807–18.
33. Martinez-Martin P, Rodriguez-Blazquez C, Kurtis MM, Chaudhuri KR, on Behalf of the NMSS Validation Group. The impact of non-motor symptoms on health-related quality of life of patients with Parkinson's disease. Mov Disord. 2011;26(3);399–406.
34. Schrag A, Hovris A, Morley D, Quinn N, Jahanshahi M. Young- versus older-onset Parkinson's disease: Impact of disease and psychosocial consequences. Mov Disord. 2003;18(11);1250–1256.
35. Knipe MDW, Wickremaratchi MM, Wyatt-Haines E, Morris HR, Ben-Shlomo Y. Quality of life in young- compared with late-onset Parkinson's disease. Mov Disord. 2011;26(11);2011–2018.
36. Lemke MR, Fuchs G, Gemende I, Herting B, Oehlwein C, Reichmann H, et al. Depression and Parkinson's disease. J Neurol. 2004;251(Suppl 6);24-27.
37. Burn DJ. Depression in Parkinson's disease. Eur J Neurol. 2002;9(Suppl 3);44–54.
38. Costa FH de R, Rosso ALZ, Maultasch H, Nicaretta DH, Vincent MB. Depression in Parkinson's disease: diagnosis and treatment. Arquivos de Neuro-Psiquiatria. 2012;70(8);617–20.
39. Hinnell C, Hurt CS, Landau S, Brown RG, Samuel M, on behalf of the PROMS-PD Study Group. Nonmotor versus motor symptoms: How much do they matter to health status in Parkinson's disease? Movement Disorders. 2012;27(2);236–41.
40. Ravina B, Camicioli R, Como PG, Marsh L, Jankovic J, Weintraub D, et al. The impact of depressive symptoms in early Parkinson disease. Neurology. 2007;69(4);342–7.
41. Leentjens AFG, Dujardin K, Marsh L, Richard IH, Starkstein SE, Martinez-Martin P. Anxiety rating scales in Parkinson's disease: a validation study of the Hamilton anxiety rating scale, the Beck anxiety inventory, and the hospital anxiety and depression scale. Mov Disord. 2011;26(3);407–15.
42. Carod-Artal FJ, Ziomkowski S, Mourão Mesquita H, Martínez-Martin P. Anxiety and depression: main determinants of health-related quality of life in Brazilian patients with Parkinson's disease. Parkinsonism Relat Disord. 2008;14(2);102–8.
43. Ray J, Das SK, Gangopadhya PK, Roy T. Quality of life in Parkinson's disease--Indian scenario. J Assoc Physicians India. 2006;54;17–21.
44. Clarke CE, Zobkiw RM, Gullaksen E. Quality of life and care in Parkinson's disease. Br J Clin Pract. 1995;49(6);288–93.
45. Denny AP, Behari M. Motor fluctuations in Parkinson's disease. J Neurol Sci. 1999;165(1);18–23.
46. Uitti RJ. Treatment of Parkinson's disease: focus on quality of life issues. Parkinsonism Relat Disord. 2012;18;S34–6.
47. Goodwin VA, Richards SH, Taylor RS, Taylor AH, Campbell JL. The effectiveness of exercise interventions for people with Parkinson's disease: a systematic review and meta-analysis. Mov Disord. 2008;23(5);631–40.
48. Karlsen KH, Tandberg E, Årsland D, Larsen JP. Health related quality of life in Parkinson's disease: a prospective longitudinal study. J Neurol Neurosurg Psychiatry. 2000;69(5);584–9.

49. McNamara P, Stavitsky K, Durso R, Harris E. The Impact of Clinical and Cognitive Variables on Social Functioning in Parkinson's Disease: Patient versus Examiner Estimates. Parkinsons Dis. https://doi.org/10.4061/2010/263083.

50. Pitcairn TK, Clemie S, Gray JM, Pentland B. Non-verbal cues in the self-presentation of parkinsonian patients. Br J Clin Psychol. 1990;29(2);177–84.

51. Agrawal V, Goyal V, Shukla G, Behari M. Predictors of caregivers' burden of Parkinson's disease in India: experience of a tertiary care centre in India. Journal of Parkinsonism and Restless Leg Syndrome. 2012;(2);59–65.

52. Fujii C, Aoshima T, Sato S, Morita T, Ohkoshi N, Nakamura T, et al. Quality of life for patients with intractable diseases. Subjective satisfaction of patients with Parkinson's disease. Kango Kenkyu. 1997;30(4);11–21.

53. Li H, Zhang M, Chen L, Zhang J, Pei Z, Hu A, et al. Nonmotor symptoms are independently associated with impaired health-related quality of life in Chinese patients with Parkinson's disease. Mov Disord. 2010;25(16);2740–6.

54. Hely MA, Reid WGJ, Adena MA, Halliday GM, Morris JGL. The Sydney multicenter study of Parkinson's disease: the inevitability of dementia at 20 years. Mov Disord. 2008;23(6);837–44.

55. Kaye J, Gage H, Kimber A, Storey L, Trend P. Excess burden of constipation in Parkinson's disease: A pilot study. Mov Disord. 2006;21(8);1270–3.

56. Friedman JH, Brown RG, Comella C, Garber CE, Krupp LB, Lou J-S, et al. Fatigue in Parkinson's disease: A review. Mov Disord. 2007;22(3);297–308.

57. Dubayova T, Nagyova I, Havlikova E, Rosenberger J, Gdovinova Z, Middel B, et al. Neuroticism and extraversion in association with quality of life in patients with Parkinson's disease. Qual Life Res. 2009;18(1);33–42.

58. Beiske AG, Loge JH, Hjermstad MJ, Svensson PhD E. Fatigue in Parkinson's disease: Prevalence and associated factors. Mov Disord. 2010;25(14);2456–60.

59. Sakakibara R, Uchiyama T, Yamanishi T, Kishi M. Genitourinary dysfunction in Parkinson's disease. Mov Disord. 2010;25(1);2–12.

60. Caap-Ahlgren M, Dehlin O. Insomnia and depressive symptoms in patients with Parkinson's disease. Relationship to health-related quality of life. An interview study of patients living at home. Arch Gerontol Geriatr. 2001;32(1);23–33.

61. Gómez-Esteban JC, Tijero B, Somme J, Ciordia R, Berganzo K, Rouco I, et al. Impact of psychiatric symptoms and sleep disorders on the quality of life of patients with Parkinson's disease. J Neurol. 2011;258(3);494–9.

62. Vibha D, Shukla G, Goyal V, Singh S, Srivastava AK, Behari M. RBD in Parkinson's disease: a clinical case control study from North India. Clin Neurol Neurosurg. 2011;113(6);472–6.

63. Scholz H, Benes H, Happe S, Bengel J, Kohnen R, Hornyak M. Psychological distress of patients suffering from restless legs syndrome: a cross-sectional study. Health and quality of life outcomes. 2011;9(1);73.

64. Chou KL, Evatt M, Hinson V, Kompoliti K. Sialorrhea in Parkinson's disease: A review. Mov Disord. 2007;22(16);2306–13.

65. Plowman-Prine EK, Sapienza CM, Okun MS, Pollock SL, Jacobson C, Wu SS, et al. The relationship between quality of life and swallowing in Parkinson's disease. Mov Disord. 2009;24(9);1352–8.
66. Bronner G, Royter V, Korczyn AD, Giladi N. Sexual dysfunction in Parkinson's disease. J Sex Marital Ther. 2004;30(2);95–105.
67. Weintraub D, Koester J, Potenza MN, Siderowf AD, Stacy M, Voon V, et al. Impulse control disorders in Parkinson disease: a cross-sectional study of 3090 patients. Arch Neurol. 2010;67(5);589–95.
68. Jacobs H. Sexuality in young patients with Parkinson's disease: a population based comparison with healthy controls. J Neurol, Neurosurg Psychiatry. 2000;69(4);550–2.
69. Archibald NK, Clarke MP, Mosimann UP, Burn DJ. Visual symptoms in Parkinson's disease and Parkinson's disease dementia. Mov Disord. 2011;26(13);2387–95.
70. Hariz GM, Forsgren L. Activities of daily living and quality of life in persons with newly diagnosed Parkinson's disease according to subtype of disease, and in comparison to healthy controls. Acta Neurologica Scandinavica. 2011;123(1);20–27.
71. Moore O, Peretz C, Giladi N. Freezing of gait affects quality of life of peoples with Parkinson's disease beyond its relationships with mobility and gait. Mov Disord. 2007;22(15);2192–5.
72. Wickremaratchi MM, Ben-Shlomo Y, Morris HR. The effect of onset age on the clinical features of Parkinson's disease. Eur J Neurol. 2009;16(4);450–6.
73. Carod-Artal FJ, Vargas AP, Martinez-Martin P. Determinants of quality of life in Brazilian patients with Parkinson's disease. Mov Disord. 2007;22(10);1408–15.
74. Constantine D. Sick and Can't Go to School? Or, Sick Because You Didn't Go? The New York Times [Internet]. 2007 Jan 3 [cited 2013 Jun 28]; Available from: http://www.nytimes.com/2007/01/03/science/20070103_AGING_GRAPHIC.html
75. Klepac N, Trkulja V. Education effect on depression and quality of life in non demented Parkinson's disease patients. J Neuropsychiatry Clin Neurosci. 2009;21(3);314–22.

Chapter 24

Caregiver burden in Parkinson's disease

Vikas Agarwal

Introduction

Parkinson's disease (PD) is a common neuro-degenerative disesase globally.[1] Prevalence of PD is 41 per every 100,000 persons worldwide, 1900 per 100, 000 in those aged above 80 years while in India the studies show large variation from 6-328 per 100,000 population. Men are more likely (1.5 times higher) to have PD than women.[2,3] Classical clinical features of PD are rigidity, imbalance, bradykinesia and resting tremor. Other non-motor symptoms, like sleep disorder, affective, cognitive, autonomic dysfunction and psychiatric disorders are also present in a good majority of PD patients.[1] PD patients require long-term support, particularly from their family members or caregivers (e.g., spouse, children, siblings and other family members or someone assisting them).[4] Chronic illnesses caregivers have a better outcome in terms of morbidity, mortality, and quality of life. However, with increasing duration of disease, disease progression and aging of PD patients can increase the work and stress levels of caregivers.[2] Caregiver burden is the adverse effect of the disease on the physical, mental socioeconomic wellbeing of the patient's primary caretaker. Caregiver burden is an important predictor for recurrent hospitalization of patients.[4]

Caregiver burden

Caregivers of patients with PD take up many responsibilities like patient's personal care (bathing, combing, dressing and walking), giving regular medications, talking to and encouraging their loved ones, constant monitoring, preventing falls, putting up with behavioural changes as well as cognitive problems, keeping a watch during night time for sleep behavioural disorders and to accompany them for visits to bathroom. Over a period of time, the patient may become completely dependent on them for various aspects of their personal, functional, occupational and social needs.[4] These activities result in caregivers having a heavy load or burden. If the caregiver is either spouse, child or a parent they may also have to accept the disease as well as adjust to role changes in a long-term relationship. Parkinson disease is incurable and burden on the caregiver are likely to progressively increase during advancing stages of the disease.[6]

DOI: 10.1201/9781003505488-24

Caregiver burden can result in burnout, increased stress and takes a major toll on of the emotional, physical, psychological and functional health of caregivers. This in return can impact the care given to PD patients and have shown a change in the patients' quality of life, incidence of falls and depression.[5] Recent studies have implicated caregiver burden to be significantly associated with neuropsychiatric symptoms, behavioural symptoms, sleep disturbances and dementia in PD patients.[7] It is important to assess the caregiver burden and suggest measures for improving it as that will comprehensively improve the health of both patients and caregivers and can lead to lesser stress on health resources.

There are different scales for assessing caregiver burden. Zarit Caregiver Burden inventory (ZBI) is a widely used worldwide and acceptable measure for caregiver burden. The scale used since 1991 has 22 domains with two subscales and has been adapted for assessing care giver burden in PD patients.[8,9] The other scales commonly used are the World Health Organization's quality of life scale (WHO-QOL-BREF) and Short Form 36 (SF36) which is a eight domain scale, including functional health and mental health.[6]

Factors association with caregiver burden

Caregivers' burden can be divided broadly into two groups: 1) Self-related burden 2) PD patients related burden.

Caregiver burden

Age

Age is important factor for caregiver burden in PD patients. Studies have found that younger caregivers have lesser burden compared to older age group.[4,5,10-12] Apart from the lack of concomitant health issues related to age (osteoarthritis, reduced endurance, etc.) the younger group tend to be more observant, can easily adopt new methods and access various modalities of social support, contributing to significantly reduction in their burden.[10] However, few studies have inferred that there is no significant association of caregiver burden with age after adjustment of disease duration.[13-18]

Gender

Gender may play a role in burden of caregivers, especially in neurodegenerative diseases. The impact is also based on the customs and the socioeconomic status of the population. Although the concept is changing, still in many countries, women are expected to be the primary caregivers of their relatives and may increase their burden while men are more likely to seek support from others. Existing studies have observed the impact of gender on caregivers and results are variable.[19-21] Few studies have found that female caregivers perceive higher burden and have increased chronic health problems.[5] Oguh et al., reported in his study that caregivers of male patients have higher caregiver burden than female patients.[19] Contrary to this report a recent study showed male gender to have more burden than female gender (MCSI score 19.9 vs 16.4, $p < 0.0001$).[20] Few studies have shown no difference between genders after adjustment of disease duration.[21] The variation as has been mentioned earlier may depend on the ethnicity, cultural and social customs prevalent in the population studied.

Socioeconomic status

Various studies have identified socioeconomic status to have a major role on caregiver burden in PD.[4, 5, 10] Caregivers with lower socioeconomic status have higher burden than high socioeconomic group.

Compared to those working in offices, more burden was seen among house-wife caregivers.[22] Jost et al. noted in his study that care of PD patients requires substantial financial support for purchase of anti-PD medications, hospital follow-ups, other miscellaneous costs, and burden can be higher among caregivers (spouses) who are not employed as it further reduces available funds[23] However, some studies have not found any difference in caregiver burden between low and high socioeconomic status.[5, 10]

Number of caregivers: single or multiple

The number of caregivers attending PD patients is inversely related to caregiver burden and in association a positive correlation with feature of life of PD patients with single caregiver having more depression.[4,24] Although a study by Wandrekar et al showed contrary results,[5] sharing the burden seems to reduce the stress in most studies.

Other miscellaneous

Various other factors can contribute to caregiver burden. Particularly in PD, lack of knowledge regarding Parkinsonian symptoms and social isolation due to the stigma associated with PD can play a major role in increasing the caregiver burden.[25]

Caregiver spouse or children and others

Studies have examined the relationship between caregiver burden and caregivers relationship to patients and the results are controversial. Orguh et al, studied 2476 US patients with PD and demonstrated that caregivers who were spouses had higher burden compared to non-spouses,[19] a similar study from Turkey also found spouses who were caregiver had significantly higher burden than non-spouses.[26] However, few studies have found that no significant difference between caregiver spouses and non-spouses.[21,27] Shin et al., noted that there was no difference in the burden scores, but there were different factors contributing to caregiver burden in spouses and their children.[28]

Patient's related burden to caregivers

Anxiety

Anxiety is a major problem in PD patients and contributes to caregiver burden.[18, 29] PD patients may present with various phobia, fear of falling, feeling of stress[4] and may insist on the presence of their caregivers being around all the time, thus reducing the ability of caregivers to maintain social network.[30]

Apathy

A few studies have estimated that apathy is prevalent in around 40% of PD patients.[31] Apathy is loss of motivation and emotional feelings in PD patients. This can lead to a blunting of effect and reduced ability of patients to have normal social and occupational life and in severe cases can cause impairment in their activities of daily living. Some studies inferred that apathy in PD patients was associated with a greater caregiver burden.[14, 32] Clinical symptoms also include lack of warmth and reduced emotional feeling and may overlap with depression and dementia.[6] These symptoms may reduce the positive emotional feedback reported from caregiver. With severe apathy, caregivers may require to give a lot of time and attention for even maintaining patients' daily routine activities of living such as brushing of teeth, washing, and administration of medication as well as maintaining hygiene.[6]

Depression

Numbers of studies have found depressive symptoms in PD patients increase caregiver burden.[33-35] These symptoms present as sadness, hopelessness, suicidal ideation, sense of failure and lack of enjoyment. As is the case with apathy, depression reduces the natural responses of the patients, reduces their interaction with the caregivers and can reduce the warmth and happiness in their relationship with the caregivers. Depression can increase fatigue in PD patients as well as the drive to perform and can make them dependant on their caregivers. This makes the task of caregivers more demanding with lesser psychological feedback and can give them a feeling of isolation. Suicidal tendencies can severely impair the social movements of the caregivers and they can become preoccupied to ensure the safety of the patient.[6]

Psychosis

Psychosis can occur in advanced stages of PD patients, especially on multiple drugs. Psychosis is manifested as delusions and visual hallucinations. Studies have shown psychosis with PD patient is associated with caregiver burden.[6,36] Hallucinations frequently present at night time and cause sleep disturbances to caregivers. Delusions such as Othello syndrome (delusion that the spouse is unfaithful) as well as hallucinations can make the patient irritable, angry, unresponsive and may harm the caregiver in their mistaken beliefs.[6] Presence of psychosis can lead to unexpected problems in patients' management and restrict movement of the caregivers. The caregivers can face various social stigmas secondary to this and can lead to heightened burden and stress.

Impulse-control disorders (ICD)

Symptoms of impulse-control disorder (ICD) can be seen in PD patients on dopamine agonists and though usually reversible can persist in a subset of patients. ICD, as the name suggests are disorders in which patient is unable to resist an impulse such as eating, shopping, gambling, sex desire, etc. ICD is defined as a failure to resist a temptation, an urge, an impulse, or the inability to not speak on a thought and may worsen the burden on caregivers.[6,37] The common ICDs of pathological gambling can impair the economic aspect of the family, while excessive eating and dopamine dysregulation syndrome can lead to impaired motor control and physical limitations in patients causing more dependence on the caregivers.[38] Hyper-sexuality, if associated with PD patients may lead to inappropriate sexual behaviour with others leading to caregivers isolation and increased burden of caregivers or households.

Sleep

Sleep disturbance is common in PD patients and cause worsening of caregiver burden.[18,39,40] PD patients may have sleep disturbances preceding their motor illness with REM sleep behavioral disorders, nocturnal akinesia, pain, hallucinations or restless legs syndrome. Caregivers who sleep with the patients or near their bed may have insufficient sleep. Several studies have identified that 95% of caregivers complain of sleep disturbance and is a risk factor for poor health of caregivers.[39, 41]

Deep Brain Stimulation (DBS)

Deep brain stimulation (DBS) of bilateral subthalamic nuclei or globus pallidus interna, is an approved therapy for PD and shows improvement of motor symptoms comparative to medical therapy, but caregiver responsibilities are still persistent.[42,43] Studies have shown occasional worsening of patients transiently or progressively after surgery, with non-motor symptoms like mood changes, cognitive impairment or ICD and anxiety. These symptoms can increase the caregiver burden.[44] However, recent studies have shown no significant changes in caregiver burden with or without DBS.[42,45]

Cognitive impairment

Cognitive impairment in PD patient's has a strong association with caregiver burden.[46,47] PD patients may have impairment in visuospatial orientation, memory and language domains as well as executive functions. These impairments lead to high dependence on the caregivers. In addition, majority of patients with PD have psychological symptoms like vocalization, agitation, wandering, behavioural changes and may have increased physical aggression.[48] This can affect regular activities and medication administration. Some PD patients have reduced awareness of their deficits and can be overconfident of their executive functions and may sometimes feel that they are performing well. A different assessment by caregiver and measures of safety may lead to conflict and increased caregiver burden.[49]

Duration of PD

Several studies have established strong association between duration of PD and caregivers burden.[4,19,50,51] Kim et al. found in their study that caregiver burden increases linearly with duration of PD (from 1- 5 years).[12] Tanji et al. conducted two cohort studies, one from US and second from Japan, US cohort had increased caregiver burden with increasing disease duration, but Japanese cohort did not show any difference.[17] Caregivers who were more optimistic about the disease seemed to have slightly lesser burden but it did not reach statistical significance. Caregivers seem to have the highest burden around five years after disease onset as their role becomes more significant and they still have to adjust to the requirements demanded from them. However, few studies have found no correlation between caregiver burden and PD duration.[52]

Hoehn & Yahr stages and disease progression

Studies have established all stages of illness (Hoehn & Yahr Stages 1,2,3,4 and 5) were significantly associated with caregiver burden.[27, 53] Carter et al., observed in their study strong association of disease stages with caregiver burden.[54] Soon et al found that at an early stage of disease (Hoehn and Yahr stage 1 and 2), caregivers had lesser burden, this burden increased in areas of frustration, worry, direct care and overall stress at moderate disease (modified Hoehn and Yahr stage 2 and 3), while there was increased burden at Hoehn and Yahr 4 and 5.[10] One of the studies found that increasing the stages of PD was associated with increase in the caregivers social, physical, psychological burden.[55]

Similarly disease severities including motor symptoms, non-motor symptoms as well as complications of therapy [all four sub-scores of Unified Parkinson's Disease Rating Scale (UPDRS)] were significantly associated with caregiver burden.[4,54,56] As the disease progressed, caregiver burden and depression increased, and their quality of life worsened.[4]

Disease progression leads to increasing burden for caregivers, as disease advances, physical and mental independence of patients come down and dependence on caregivers increases.[57] However, Wandrekaer et al. noted in their study no significant association between stages of illness and caregiver burden, but small number of patients may have led to contradictory result.[5]

Management

Identifying caregiver stress is important. Caregiver stress may manifest as impairment of physical health (headaches, fatigue, more visits to hospital), changes in mood (depression, confusion), reduced interaction with other members and dejection and anxiety.

The management has to be individualised. The caregivers should be counselled frequently with the regular assessment of their burden, inform them regarding the disease, complications of therapy and include them in the planning of patient care. Concern regarding the problems of caregivers' should be shown and possible coping mechanisms can be discussed with the entire family to ensure smooth management of patients. Information regarding resources such as day care centres, support groups, use of physiotherapy, occupational therapy and psychological support should be given and encouraged to use.

Conclusion

PD patients have motor and non-motor symptoms and contribute to significant caregiver burden. Health professionals should be attentive to the frequently persistent caregiver burden and detailed knowledge of the disease and coping mechanisms should be provided to them. Social support can improve the physical and mental health of caregivers and patients.

References

1. Rukmini Mridula K, Borgohain R, Jabeen SA, Padmaja G, Bandaru VS, Ankathi P, Kanikannan MA, Ali Khan MS. Comparison of frequencies of non motor symptoms in Indian Parkinson's disease patients on medical management versus deep brain stimulation: A case-control study. Iran J Neurol. 2015;14;86-93.
2. https://parkinsonsnewstoday.com/parkinsons-disease-statistics accessed date 10.03.2018.
3. Mridula KR, Borgohain R, Chandrasekhar Reddy V, Bandaru VCh, Suryaprabha T. Association of Helicobacter pylori with Parkinson's Disease. J Clin Neurol. 2017;13; 181–186.
4. Agrawal V, Goyal V, Shukla G, Behari M. Predictors of caregivers' burden of Parkinson's disease in India: experience of a tertiary care center in India. Journal of Parkinsonism & Restless Legs Syndrome. 2012;2;59–65.
5. Wandrekar J, D'Souza N, Barretto M. Factors related to Caregiver Burden in Caregivers of Patients with Parkinson's disease in Mumbai, India. The International Journal of Indian Psychology. 2014;2:Id: B00231V2I12014.
6. Mosley PE, Moodie R, Dissanayaka N. Caregiver Burden in Parkinson Disease: A Critical Review of Recent Literature. J Geriatr Psychiatry Neurol. 2017;30;235-252.
7. D'Amelio M, Terruso V, Palmeri B, Di Benedetto N, Famoso G, Cottone P, Aridon P, Ragonese P, Savettieri G. Predictors of caregiver burden in partners of patients with Parkinson's disease. Neurol Sci. 2009;30;171-174.
8. Ballesteros J, Santos B, Gonzalez-Fraile E, Munoz-Hermoso P, Domínguez-Panchón AI, Martín-Carrasco M. Unidimensional 12-item Zarit Caregiver Burden Interview for the assessment of dementia caregivers' burden obtained by item response theory. Value Health. 2012;15;1141-1147.
9. Bachner YG, O'Rourke N. Reliability generalization of responses by care providers to the Zarit Burden Interview. Aging Mental Health. 2007;11;678–85.

10. Soon KM, Ja KB, Hee K, Myoung-Ae C, Yi M, Hah YS, Chung SJ, Kwon SH. Subjective and Objective Caregiver Burden in Parkinson's Disease. Journal of Korean Academy of Nursing. 2007;37;242-248.

11. Carter, J.H., Lyons, K.S., Stewart, B.J., et al. Does Age Make a Difference in Caregiver Strain? Comparison of Young versus Older Caregivers in Early-Stage Parkinson's Disease. Movement Disorders. 2010;25;724-730.

12. Kim, K.S., Kim, B.J., Kim, K.H., et al. Subjective and Objective Caregiver Burden in Parkinson's Disease. Daehan Ganho Haghoeji. 2007;37;242-248.

13. Kudlicka, A., Clare, L. and Hindle, J.V. Quality of Life, Health Status and Caregiver Burden in Parkinson's Disease: Relationship to Executive Functioning. International Journal of Geriatric Psychiatry. 2014;29;68-76.

14. Leroi, I., Harbishettar, V., Andrews, M., et al. Carer Burden in Apathy and Impulse Control Disorders in Parkinson's Disease. International Journal of Geriatric Psychiatry. 2012;27;160-166.

15. Leroi, I., McDonald, K., Pantula, H. and Harbishettar, V. Cognitive Impairment in Parkinson Disease: Impact on Quality of Life, Disability, and Caregiver Burden. Journal of Geriatric Psychiatry and Neurology. 2012;25;208-214.

16. Kelly, D.H., McGinley, J.L., Huxham, F.E., et al. Health-Related Quality of Life and Strain in Caregivers of Australians with Parkinson's Disease: An Observational Study. BMC Neurology. 2012;12;57.

17. Tanji, H., Koyama, S., Wada, M., et al. Comparison of Caregiver Strain in Parkinson's Disease between Yamagata, Japan, and Maryland, The United States. Parkinsonism & Related Disorders. 2013;19;628-633.

18. Carod-Artal, F.J., Mesquita, H.M., Ziomkowski, S. and Martinez-Martin, P. Burden and Health-Related Quality of Life among Caregivers of Brazilian Parkinson's Disease Patients. Parkinsonism & Related Disorders. 2013;19;943-948.

19. Oguh, O., Kwasny, M., Carter, J., et al. Caregiver Strain in Parkinson's Disease: National Parkinson Foundation Quality Initiative Study. Parkinsonism & Related Disorders. 2013;19;975-979.

20. Dahodwala N, Shah K, He Y, Wu SS, Schmidt P, Cubillos F, Willis AW. Sex disparities in access to caregiving in Parkinson disease. Neurology. 20182;90:e48-e54.

21. Schrag A, Hovris A, Morley D, Quinn N, Jahanshahi M.Caregiver-burden in Parkinson's Disease is closely associated with psychiatric symptoms, falls, and disabilityParkinsonism Relat Disord. 2006;12;35-41

22. https://www.gmjournal.co.uk/media/21766/gm2may2011p22.pdf accessed date 10.04.2018

23. Jost WH Costs in the treatment of parkinsonism. JNeurol. 2000;247;S31-S33.

24. Kim, H., Chang, M., Rose, K., Kim, S. Predictors of caregiver burden incaregivers of individuals with dementia. Journal of Advanced Nursing. 2011;68;846-855.

25. https://www.healthplexus.net/article/identifying-and-managing-caregiver-burden-among-spouses-individuals-parkinsons-disease accessed date 03.04.2018

26. Calder, S.A., Ebmeier, K.P., Stewart, L., et al. The Prediction of Stress in Carers: The Role of Behaviour, Reported Self-Care and Dementia in Patients with Idiopathic Parkinson's Disease. International Journal of Geriatric Psychiatry. 1991;6;737-742.

27. Razali, R., Ahmad, F., Rahman, F.N.A., et al. Burden of Care among Caregivers of Patients with Parkinson Disease: A Cross-Sectional Study. Clinical Neurology and Neurosurgery. 2011;113;639-643.
28. Shin, H., Lee, J.Y., Youn, J., et al. Factors Contributing to Spousal and Offspring Caregiver Burden in Parkinson's Disease. Movement Disorders.2012;27;S311-S311
29. Ozdilek, B. and Gunal, D.I. Motor and Non-Motor Symptoms in Turkish Patients with Parkinson's Disease Affecting Family Caregiver Burden and Quality of Life. Journal of Neuropsychiatry and Clinical Neurosciences. 2012;24;478-483,
30. Dissanayaka NN, White E, O'Sullivan JD, Marsh R, Pachana NA,Byrne GJ. The clinical spectrum of anxiety in Parkinson's disease.Mov Disord. 2014;29;967-975.
31. den Brok MG, van Dalen JW, van Gool WA, Moll van Charante EP, de Bie RM, Richard E. Apathy in Parkinson's disease: A systematic review and meta-analysis. Mov Disord. 2015;30;759-69.
32. Leiknes, I., Tysnes, O.-B., Aarsland, D. and Larsen, J.P. Caregiver Distress Associated with Neuropsychiatric Problems in Patients with Early Parkinson's Disease: The Norwegian ParkWest Study. Acta Neurologica Scandinavica. 2010;122;418-424.
33. Zhong M, Peppard R, Velakoulis D, Evans AH. The relationship between specific cognitive defects and burden of care in Parkinson's disease. Int Psychogeriatr. 2016;28;275-281
34. Martinez-Martin P, Rodriguez-Blazquez C, Forjaz MJ, et al.Neuropsychiatric symptoms and caregiver's burden in Parkinson's disease. Parkinsonism Relat Disord. 2015;21;629-634.
35. Santos-Garcia D, de la Fuente-Fernandez R. Factors contributing to caregivers' stress and burden in Parkinson's disease. Acta Neurol Scand. 2015;131;203-210.
36. Aarsland D, Bronnick K, Ehrt U, et al. Neuropsychiatricsymptoms in patients with Parkinson's disease and dementia: frequency, profile and associated care giver stress. J Neurol Neurosurg Psychiatry. 2007;78;36-42.
37. Harbishettar V, Andrews M, McDonald K, Byrne EJ, Burns A. Carer burden in apathy and impulse control disorders in Parkinson's disease. Int J Geriatr Psychiatry. 2012;27;160-166.
38. Weintraub D, Koester J, Potenza MN, et al. Impulse control disordersin Parkinson disease: a cross-sectional study of 3090 patients. Arch Neurol. 2010;67;589-595.
39. Cupidi C, Realmuto S, Lo Coco G, et al. Sleep quality in caregivers of patients with Alzheimer's disease and Parkinson's disease and its relationship to quality of life. Int Psychogeriatr. 2012;24;1827-1835.
40. .Viwattanakulvanid P, Kaewwilai L, Jitkritsadakul O, et al. The impact of the nocturnal disabilities of Parkinson's disease on caregivers' burden: implications for interventions. J Neural Transm (Vienna). 2014;121(suppl 1);S15-S24.
41. Pal PK, Thennarasu K, Fleming J, Schulzer M, Brown T, Calne SM. Nocturnal sleep disturbances and daytime dysfunction in patients with Parkinson's disease and in their caregivers. Parkinsonism Relat Disord. 2004;10;157-168.
42. Lewis CJ, Maier F, Horstkotter N, et al. The impact of subthalamic deep brain stimulation on caregivers of Parkinson's disease patients: an exploratory study. J Neurol. 2015;262;337-345.
43. Lewis CJ, Maier F, Horstko¨tter N, et al. Subjectively perceived personality and mood changes associated with subthalamic stimulation in patients with Parkinson's disease. Psychol Med.2015;45;73-78.

44. Mosley PE, Marsh R. The psychiatric and neuropsychiatric symptoms after subthalamic stimulation for Parkinson's disease. J Neuropsychiatry Clin Neurosci. 2015;27;19-26.
45. Mosley PE, Breakspear M, Coyne T, Silburn P, Smith D. Caregiver burden and caregiver appraisal of psychiatric symptoms are not modulated by subthalamic deep brain stimulation for Parkinson's disease.NPJ Parkinsons Dis. 2018;4;12.
46. Coelho M, Marti MJ, Sampaio C, et al. Dementia and severity of parkinsonism determines the handicap of patients in late-stage Parkinson's disease: the Barcelona-Lisbon cohort. Eur J Neurol. 2015;22;305-312.
47. Carter JH, Lyons KS, Lindauer A, Malcom J. Pre-death grief in Parkinson's caregivers: a pilot survey-based study. Parkinsonism Relat Disord. 2012; 18(suppl 3);S15-S18.
48. Bruno V, Mancini D, Ghoche R, Arshinoff R, Miyasaki JM. High prevalence of physical and sexual aggression to caregivers in advanced Parkinson's disease. Experience in the palliative care program. Parkinsonism Relat Disord. 2016;24;141-142
49. Peters, M., Fitzpatrick, R., Doll, H., et al. Does Self-Reported Well-Being of Patients with Parkinson's Disease Influence Caregiver Strain and Quality of Life? Parkinsonism & Related Disorders. 2011;17;348-352.
50. Lyons, K.S., Stewart, B.J., Archbold, P.G. and Carter, J.H. Optimism, Pessimism, Mutuality, and Gender: Predicting 10-Year Role Strain in Parkinson's Disease Spouses. Gerontologist. 2009;49;378-387.
51. Peters, M., Fitzpatrick, R., Doll, H., et al. Does Self-Reported Well-Being of Patients with Parkinson's Disease Influence Caregiver Strain and Quality of Life? Parkinsonism & Related Disorders. 2011;17;348-352.
52. Martinez-Martin, P., Forjaz, M.J., Frades-Payo, B., et al. Caregiver Burden in Parkinson's Disease. Movement Disorders. 2007;22;924-931.
53. Shin, H., Lee, J.Y., Youn, J., Kim, J.S., Cho, J.W. Factors contributing to spousal and offspring caregiver burden in Parkinson's disease. European Neurology. 2012;67;292-296.
54. Carter, J. H., Stewart, B. J., Archbold, P. G., Inoue, I., Jaglin, J.,Lannon, M., Rost-Ruffner, E., Tennis, M., McDermott, M. P.,Amyot, D., Barter, R., Cornelius, L., Demong, C., Dobson, J.,Duff, J., Erickson, P., Gardiner, N., Gauger, L., Gray, P.,Kanigan, B., Kiryluk, B., Lewis, P., Mistura, K., Malapira, T., Zoog, K. Living with a person who has Parkinson's disease: the spouse's perspective by stage of disease. Mov Disord. 1998;13;20-28
55. Berry, R. A., Murphy, J. F. Well-being of caregivers of spouses with Parkinson's disease. Clin Nurs Res. 1995;4;373-386.
56. Thommessen B, Aarsland D, Braekhus A, Oksengaard AR, Engedal K, Laake K. The psychosocial burden on spouses of the elderly with stroke, dementia and Parkinson's disease. Int J Geriatr Psychiatry. 2002;17;78–84
57. Goldsworthy B, Knowles S. Caregiving for Parkinson's disease patients: an exploration of a stress-appraisal model for quality of life and burden. J Gerontol B Psychol Sci Soc Sci. 2008;63;P372-6

Chapter 25

Health Economics of Parkinson's Disease in India

Vipin Ola, Madhuri Behari

Introduction

Parkinson disease (PD) is a slowly progressing neurodegenerative disorder with significant health and economic implications. With improving life expectancy in developing countries, elderly population affected with degenerative disorders is also increasing. It has significant economic impact on the society as a whole. Measuring morbidity and mortality are important for estimating the burden any disease. However, only focusing on morbidity and mortality provides an incomplete picture of the adverse impact of ill health on individual and society.

People with PD have direct and indirect economic implications. First, in absence of universal health insurance, most of the patients have to pay out of pocket for drugs, investigations, travel and stay in another city for healthcare, posing significant economic burden on the family. Due to the disease one may lose his/her job or may become less efficient, resulting in economic losses. They may either need caregiver support, or someone from family may have to quit job to take care causing indirect economic losses to the family.

Health economics

Economic impact of any disease can be estimated at macroeconomic level (societal perspective) and microeconomic level (individual, firm and government). When a person is affected with chronic disease, it affects by reducing the productivity, leading to reduced earning and economic losses. Moreover, it increases consumption of health services (in the form of expenditure on hospital visits, admissions, investigations, travel, stay, expenditure on drugs). It results in decreased non-health consumption like food, education, clothes, savings, etc. causing a lot of impact on family as well as society (Flow chart 1).

DOI: 10.1201/9781003505488-25

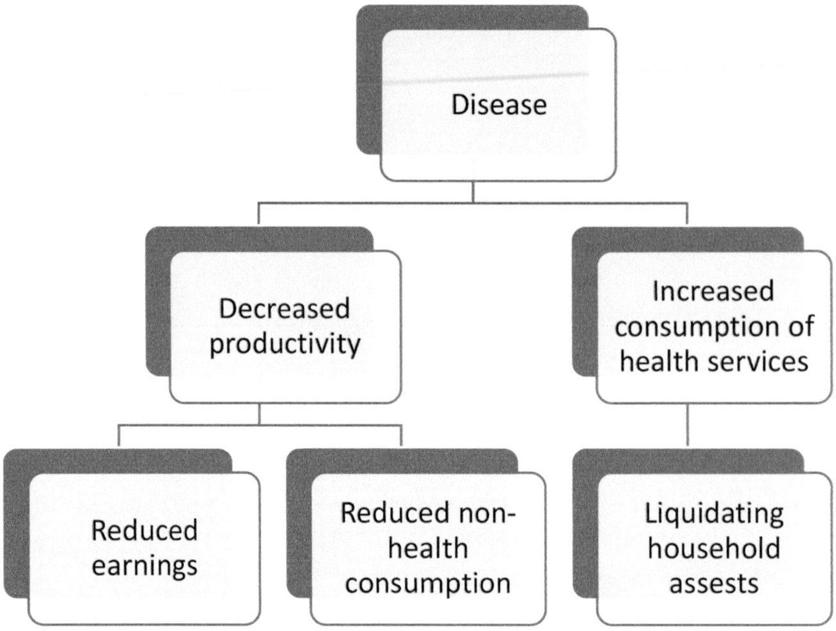

Figure 1: Flowchart 1 (Economic impact of a disease)

Economic impact of PD may be understood with more clarity in terms of direct cost and indirect cost.

Table 1: Direct and indirect costs in the treatment of PD

Direct cost	Indirect cost
Drugs	Loss of production
Outpatient cost consultation	Loss due to job loss
Inpatient cost	Time devoted by family members
Investigations	
Transportation	
Home care	
Paid care giver	

Another important concept in chronic diseases is that of Disability Adjusted Life Years (DALY). It is a measure of disease burden, reflecting the number of years lost due to ill health, disability or early death. It takes into account years of life lost due to illness (YLL) and Years Lost due to Disability (YLD).

Magnitude of problem

Age adjusted prevalence of PD in Indian population is 52.8 per 1,00,000 population.[1] In a study from India[2] with 175 PD patients, average direct cost per year was estimated to be 8,853 INR. Total annual income of about half of the patients was less than 50,000 Rupees. Average PD patient spent about 16% to 47% of average Indian gross national income on purchasing drugs. Only a small number of people (3%) had health insurance. Cost of treating PD was substantial according to average gross national income of Indian population. This study highlighted the state of affairs in India with disappointingly low health insurance.

Our unpublished work done at a tertiary care center in India, with 200 patients with PD showed that most of them belonged to Hoehn and Yahr (H&Y) stages 2 and 3 of PD, forming 73 % of the total study population. Median yearly income of patients was INR 240000. This figure is more than the projected per capita income of an Indian. Median direct cost on disease was INR 25817.6, while median yearly cost on drugs was INR 18691.2, which formed 75.50% of the total cost on treatment. It increased with progression of disease. The percentage of total income spent on drugs also increased with increasing stage of disease, with patients in stages 3 and 4 spending more than one-third of their income on drugs. Most patients paid for their medicines, which amounted to almost 6.72% (median) of their annual income. Surprisingly, expenditure on drugs was highest in H&Y stage 2.5, possibly due to use of dopamine agonist in high doses in stage 2, whereas patients in more advanced stage received Levodopa/carbi-dopa combination more often and in higher doses, which is less expensive as compared to dopamine agonist. Mean expenditure on travel and stay was INR 1750. It was 38% of the total expenditure on treatment in stage 1. However, there was no correlation between stage of disease and money spent on travel.

Mean years lived with disability (YLD) was 15.88± 4.79(yrs.) in our PD population. The YLD remained significantly unaffected irrespective of progression of disease and stage of disease, since almost all patients, irrespective of H&Y stage of disease in which they presented to us, the age of onset of symptoms was similar.

Twenty-one (10.5%) patients had to quit their jobs or incurred financial loss due to the disease. Median loss due to this was INR 60,000. This increased with worsening disease. Seven (3.5%) patients had employed paid caregivers (in stages 3 and 4). Mean expenditure on this item was INR 6000/year. No one from family members changed or had to quit their jobs.

Six (3%) patients underwent Deep Brain Stimulation (DBS) surgery for PD. The total median cost was INR 525000, which included hardware, operation charges and charges for MRI to localize target and stay in general ward. Only 10 patients out of 200 patients recruited in the study had medical insurance (5%).

Most of people in India do not have health insurance. Expenditure on a chronic disease imposes significant direct and indirect impact on family, starting a vicious cycle. Drugs form the major part of direct cost.[3,4] Out of the drugs, dopamine agonists are the costliest.[2] Cost of drugs increases with worsening stage of disease. Number of drugs used per patient increases with advancing disease. Furthermore, Levodopa Equivalent Daily Dose (LEDD) increases, thereby adding to the cost.[3]

Inpatient care also forms a significant proportion of the cost. As the disease advances, PD patients are more likely to be admitted for related complications like freezing, motor fluctuations,[3] aspiration pneumonia, resulting in further increase in costs. Cost of additional drugs for managing non-motor complications of PD further add to the cost.

With the advent of DBS as a mode of therapy, management of complications has improved. However, the cost of DBS is significantly more than the medical management. It is beyond the reach of most of the patients. Only people having some kind of insurance, either personal or from state, can afford it. Treatment with DBS exponentially increases the economic burden of disease.

Transportation for the treatment contributes to direct costs. In developing countries, in the absence of robust primary healthcare, patients are forced to visit higher centers to seek care. These are usually tertiary care centers, usually situated in major cities far away from places of residence of most people. PD patients are physically disabled in advanced stages with limited mobility, further complicating the transportation process. Hence, it becomes difficult to seek healthcare using public transport and therefore, they are forced to hire personal transport, adding to the cost of treatment significantly. In addition, one or two members of family must accompany the patient, especially with advancing disease. Accompanying person loses the days' wages as well, contributing to indirect cost.

Many times, patients have to stay outside the healthcare facility before an appointment or in between visits or while waiting for results of some tests during outpatient appointment at the hospital. It further increases the economic burden. Number of outpatient visits usually increases with increasing duration of disease, causing manifold increase in the cost of transportation and stay at every visit.

Investigations for management also contribute to economic burden, especially in the early stage of disease, when the main emphasis is to establish the diagnosis. Home care, both while using services of a family member or an employed healthcare provider adds to the economic burden indirectly as well as directly respectively. Indirect costs such as money lost due to early retirement, change of job, especially choosing part time jobs instead of full-time job/opting for less paying jobs contribute to economic burden as well. Often, PD patients are forced to opt for jobs or curtail work hours due to general fatigue, leading to loss of production or wages contributing to indirect cost. They are more likely to take sick leaves.

A Spanish study assessed it in PD cases and found 67,582 DALY in 2000.[5] Many patients with PD belong to age at which they are usefully employed, especially in India where the median age at onset of PD symptoms is about one decade earlier than that observed in western countries. Loss of production due to disease is important from the perspective of society. Patients with PD usually work for fewer hours per week or month and tend to take more sick leaves, adding to the overall burden on society. A Swedish study assessed the burden of lost production across the stages. It did not find any correlation between lost production and stage of disease.[6]

Help provided, and time spent by family members on PD patient in caring for them cannot be assigned any monetary value, nevertheless, it is has important bearing. In addition, care provided by family member has more personal touch, and keeps in mind the likes and dislikes of the person with PD (PwP). In a study from India on PwP in 2012, 51% of the caregivers were spouses (of these 84% were wives).[7] Most caregivers admitted having stress and burden of care giving. Caregivers burden correlated positively with presence of autonomic disturbance (urinary frequency/urgency/incontinence), duration of disease, presence of depression, psychosis in patient and caregivers' age and monthly income and inversely with number of caregiver.

Conclusion

Burden of PD is considerable in our country, and the magnitude of problem increases in view of paucity of universal healthcare system and lack of primary healthcare at grass root level as well as lack of specialists who can diagnose and take care of the patients locally, hence necessitating avoidable expenses on frequent travelling and staying in a city outside their own. As PD is not curable and no treatment is available to stop the progression, it has lifelong economic implications for the patient, family and society.

References

1. Das SK, Misra AK, Ray BK, Hazra A, Ghosal MK, Chaudhuri A, et al. Epidemiology of Parkinson disease in the city of Kolkata, India: a community-based study. Neurology. 2010;75(15);1362–9.
2. Ragothaman M, Govindappa ST, Rattihalli R, Subbakrishna DK, Muthane UB. Direct costs of managing Parkinson's disease in India: concerns in a developing country. Mov Disord Off J Mov Disord Soc. 2006;21(10);1755–8.
3. Wang G, Cheng Q, Zheng R, Tan Y-Y, Sun X-K, Zhou H-Y, et al. Economic burden of Parkinson's disease in a developing country: a retrospective cost analysis in Shanghai, China. Mov Disord Off J Mov Disord Soc. 2006;21(9);1439–43.
4. Hagell P, Nordling S, Reimer J, Grabowski M, Persson U. Resource use and costs in a Swedish cohort of patients with Parkinson's disease. Mov Disord Off J Mov Disord Soc. 2002;17(6);1213–20.
5. Cubo E, Alvarez E, Morant C, de Pedro Cuesta J, Martínez Martín P, Génova R, et al. Burden of disease related to Parkinson's disease in Spain in the year 2000. Mov Disord Off J Mov Disord Soc. 2005;20(11);1481–7.
6. Hagell P, Nordling S, Reimer J, Grabowski M, Persson U. Resource use and costs in a Swedish cohort of patients with Parkinson's disease. Mov Disord Off J Mov Disord Soc. 2002;17(6);1213–20.
7. Agrawal V, Goyal V, Shukla G, Behari M. Predictors of caregivers' burden of Parkinson's disease in India: experience of a tertiary care centre in India. Journal of Parkinsonism and restless leg syndrome.2012:2;59-65.

Chapter 26

Education and Empowerment of Patient's with Parkinson's disease

SP Gorthi, Vijay Chandran, Arvind Prabhu

Introduction

Empowerment is a concept which one frequently encounters in various health-related fields. Despite being frequently used, the context and what it means is quite often unclear and may vary among authors, probably because of difficulty in defining the concept. In this chapter, we approach empowerment based on the conceptual analysis of Tengland,[1] wherein empowerment is discussed as a process by which a person (or group) A acts towards another person (or group) B in order to support the latter in improving their control over the determinants of their quality of life and this process leads to a reduction in A's influence over B with regards to goal formulation and decision making, and B gains some control over the situation, process or goal. For most individuals, the important determinants of their quality of life include: Health, both physical and mental, home or living space, work and income, close relationships, leisure and values. As one reads through these determinants, one quickly realizes that in Parkinson's disease (PD) many of these areas can be significantly affected. Healthcare providers can play an important role in the process of empowering patients with PD.

One of the initial roles played by a healthcare provider is informing the diagnosis to patients with PD. Awareness or knowledge about the disease one is suffering from can be considered the first step towards empowerment. This is evidently not an easy conversation to have and most neurologists have unfortunately received very little, or no training, in the skill of breaking bad news.[2,3] Moreover, in the case of PD, the lack of a diagnostic test and the resulting diagnostic uncertainty makes the situation even more complex. To our knowledge, there is no published literature which studies the best approach to revealing the diagnosis of PD. However, there is literature available from other medical scenarios such as cancer which may help guide clinicians.[4] The acronym SPIKES summarizes a protocol used in oncology, which recommends the following: **S**etting up the interview involves arranging for privacy and dedicated time for revealing the diagnosis. It would be better if the interview

DOI: 10.1201/9781003505488-26

is carried out in the presence of a significant other and without interruptions. Assessing patient's perception as an initial part of the interview may be a useful starting point and enables the physician to correct any misconceptions and reveal the bad news in the most appropriate manner. At this stage, obtaining patients intimation regarding how much information they would like to hear would be appropriate and would direct the next step of the process. Providing knowledge and information to the patient should be at a level that the patient understands, using non-technical words. Assessment of patient's emotions and addressing them with empathetic responses would help in providing support and help in calming patients down. Patients who have a plan for the future are likely to be less anxious and in the case of PD the availability of medications and surgical strategies which improve symptoms and quality of life may help in reassuring patients. If the patient would like to have a second opinion with regards to diagnosis, this should also be arranged.

In chronic conditions, there has been an increasing emphasis on empowering patients and there is specific literature available with regards to this in patients with diabetes,[5] however, we were unable to identify any published information in the specific context of PD. In general, policies for patient empowerment in chronic conditions, stress upon active patient participation and self- management.[6] Patient participation is described as an interaction between the patient and the health care provider wherein the patient actively provides information which assists in diagnosis and problem solving, expresses preferences with regards to treatment, thereby directing care in alignment with their personal preferences.[7] In PD, patient's environment, which enables patient participation, should be encouraged. Since treatment should be individualized, physicians should identify symptoms which are impairing functioning and thereby impacting quality of life. The risks and benefits of any medication or surgical treatment directed at troublesome symptoms should be explained to the patient and the final decision to initiate therapy should be based upon patient choice. Medication should not be prescribed with a paternalistic attitude wherein the physician ignores individual choice. This may entail redefining the traditional patient-physician relationship. The extent of the healthcare provider's role may vary depending upon the clinical situation; for example, in patients with cognitive impairment, significant depression or impulse control disorder, the health care provider and family members or friends may have to play a larger role. Self management is defined as 'the care taken by individuals towards their own health and well-being; it comprises the actions they take to lead a healthy lifestyle; to meet their social, emotional and psychological needs; to care for their long-term condition and to prevent further illness or accidents.[8] Although there is some overlap of this concept with that of patient participation, self management is more focussed patient's health related behaviour, once a plan has been arrived at following patient participation. In terms of leading a healthy lifestyle, patients with PD may benefit from information with regards to sleep hygiene, regular exercise, dietary advice and strategies to aid with cognitive function and fall prevention. Patient's with PD must be advised to keep mentally active and should be encouraged to interact socially as well, as these may help maintain mental health and cognitive function. In patients with depression and apathy, increasing self-management may prove to be difficult, unless these mental health issues are adequately addressed.

An individual's ability to work and earn a living is an important determinant of quality of life. Patients with PD retire 4-7 years earlier than expected and 23%-75% of the time, patients attribute early retirement to the disease, with slowness and fatigue being the symptoms that interfere with work.[9]

Jobs which are physical or mental may both be difficult to carry out. A supportive work environment may enable patient's with PD to work longer, however quite often employers are not willing to make the necessary adjustments. Health care providers may advocate for their patients if needed, they may also help ease the transition. Notably, some patients observe an overall improvement in their health after they stop working, as they are less stressed and able to devote more of their time to managing their disease.

Social media and the internet are playing an ever-increasing role in people's lives, including patients with PD. They provide a platform for patients to share their experiences, address educational needs, support decision making, and thereby improve patient participation and self- management.[10] However, there are few drawbacks, including potential violation of privacy and incorrect information, which may mislead patients.[10] Specific to patients with PD are the risks of gambling online and internet addiction, which may be manifestations of impulse control disorder.[11,12] Despite these negatives, if patients are directed towards the appropriate resources online and monitored, they may benefit significantly. There is also the emerging area of wearable technologies and the internet of things, which could play a major role in the management of patients with PD in the future.[13]

At AIIMS, an NGO, a patient-centric organization called Parkinsonism and Related disorders Awareness Network (PRAN, meaning life in Hindi) is working towards increasing awareness and knowledge of Parkinson's disease among general public, people with PD and their caregivers. It was established and registered 20 years ago and enrols all the patients of PD and their caregivers. It holds camps in residential colonies and public lectures are held for the knowledge and treatment options available for persons with PD. This way even the caregivers are provided knowledge and they are more sympathetic towards the people with PD. Under the aegis of PRAN, a one day program is held at AIIMS on first Sunday of April every year (the month dedicated to PD), in which people with PD and their caregivers are provided information about PD and its treatment options, result of any new research, exercises and relaxation and yoga exercises. They are given opportunity to share their experiences and ask any questions that they may have. This is followed by a sumptuous lunch. This program is so very successful and popular, that people with PD and their caregivers plan their visit months in advance and come from far off places to attend it and also take advice regarding their problems during their visit (Figure 1 and 2).

Dr Ravi Yadav did a study on the participants of the annual PRAN program (Knowledge, attitude and practice of Persons with PD and their care givers, Chapter 27). He observed that people who had attended the program earlier or were attending the Movement disorders clinic at AIIMS had very good knowledge about Parkinson's disease and were able to answer on an average 9 out of 10 multiple choice questions on PD.

To conclude, we believe "knowledge is strength". To be able to better manage any chronic disease, it is important that the persons who bear the brunt of disease are well informed and are able to take decisions about their treatment have to better quality of life. And this indeed is our motto: "To improve quality of life irrespective of disease or its severity."

References

1. Tengland PA. Empowerment: a conceptual discussion. Health Care Anal. 2008;16(2):77-96
2. Watling CJ, Brown JB. Education research: communication skills for neurology residents: structured teaching and reflective practice. Neurology. 2007;69(22);E20-6.
3. Lemmon ME, Strowd RE 3rd. Right Brain: Breaking bad news: Communication education for neurology trainees. Neurology. 2016;87(24);e285-e287.
4. Baile WF, Buckman R, Lenzi R, Glober G, Beale EA, Kudelka AP. SPIKES-A six-step protocol for delivering bad news: application to the patient with cancer. Oncologist. 2000;5(4);302-11.
5. Tol A, Alhani F, Shojaeazadeh D, Sharifirad G, Moazam N. An empowering approach to promote the quality of life and self-management among type 2 diabetic patients. J Educ Health Promot. 2015;4;13.
6. Hibbard J, Collins P, Mahoney E, Baker L: The development and testing of a measure assessing clinician beliefs about patient self-management. Health Expect. 2009;13;65–72.
7. Haywood K, Marshall S, Fitzpatrick R: Patient participation in the consultation process: a structured review of intervention strategies. Pat Educ Couns .2006;63;12–23.
8. Small N, Bower P, Chew-Graham CA, Whalley D, Protheroe J. Patient empowerment in long-term conditions: development and preliminary testing of a new measure. BMC Health Serv Res. 2013;13;263.
9. Koerts J, König M, Tucha L, Tucha O. Working capacity of patients with Parkinson's disease - A systematic review. Parkinsonism Relat Disord. 2016;27;9-24.
10. Mohammadzadeh Z, Davoodi S, Ghazisaeidi M. Online Social Networks - Opportunities for Empowering Cancer Patients. Asian Pac J Cancer Prev. 2016;17(3);933-6.
11. Wu K, Politis M, O'Sullivan SS, Lawrence AD, Warsi S, Lees A, Piccini P. Problematic Internet use in Parkinson's disease. Parkinsonism Relat Disord. 2014;20(5);482-7.
12. Larner AJ. Medical hazards of the internet: gambling in Parkinson's disease. Mov Disord. 2006;21(10);1789.
13. Pasluosta CF, Gassner H, Winkler J, Klucken J, Eskofier BM. An Emerging Era in the Management of Parkinson's Disease: Wearable Technologies and the Internet of Things. IEEE J Biomed Health Inform. 2015;19(6);1873-81.

Chapter 27

Knowledge, attitude and perceptions in Parkinson's Disease

Ravi Yadav, Anush Rangarajan

Introduction

Parkinson's disease (PD) is second most common degenerative disorder globally. The increase in life expectancy in developing economies all over the world will lead to further increase in the prevalence of this disorder. India being the second most populous country in the world will have one of the highest numbers of PD population. The value of the awareness about the disease cannot be undermined because of several factors that contribute to the final outcome of the disease. The most important among them is the early diagnosis and timely medical treatment leading to better outcomes. This, however, varies in different regions of the world depending on the literacy rates, the activity of the awareness and education programs in the society as well as the local celebrities taking up the cause for mass education.

In the Indian scenario, the healthcare system already has challenges like tropical infections (dengue, malaria, tuberculosis, encephalitis) and lifestyle related diseases (diabetes mellitus, hypertension, coronary artery disease and stroke). The rise of the geriatric population is posing greater challenges due to the rise in dementia (vascular, degenerative) and PD.

The role of the current neurophysicians, general physicians and psychiatrists in managing all of these disorders is increasing. There is an increase in the literature in last many years from the different parts of the world publishing the findings from these countries.

In this chapter, we will review the current literature from India and the rest of the world, would provide suggestions for improving the knowledge, attitude and perceptions about PD.

DOI: 10.1201/9781003505488-27

Indian studies

Despite advances in the treatment in the form of drugs and neurosurgical procedures, the benefit of these is still not being passed to the vast majority of the patients in the developing countries. Very little information is available about the knowledge of PD and patients or their caregivers in India.

Yadav et al. conducted a questionnaire-based interview among the subjects collected on the annual PD day in 2006 at the All India Institute of Medical Sciences, New Delhi. [1] This was done to explore the level of knowledge among the patients and caregivers attending PD and movement disorders clinic at a tertiary care centre in north India.[1] Out of 200 questionnaires that were distributed, 172 subjects responded. Of these, there were 103 (59.8%) patients with PD and 69 (40.11%) caregivers. The mean age of patients was around 55 years. Ninety-nine patients out of the 103 (96.1%) and 57 caregivers out of 69 (82.6%) had previously attended PD education program. High scores (>90%) were obtained to questions on body parts affected, pathology in PD, main drug treatment, epidemiology and effect of exercise. Low scores were seen in questions on surgery in PD and biochemical abnormality in PD. No significant difference was noted in correct answers among patients and caregivers on the duration of PD. [1] This study emphasized the need for adequate patient education and the role of specialty clinics in imparting knowledge.[1]

More information about PD in print and electronic media is needed, and establishment of dedicated movement disorder clinics and centres of excellence with special education programs would go a long way in increasing awareness and knowledge of PD in India.

Asian Studies

Studies with a similar design (Bhidayshree et al.), but with a larger questionnaire and with control population also shows that there is lack of knowledge about PD in the general population.[2] This study was conducted in 108 PD patients. [2] The questionnaire covered three domains of PD, including diagnosis, therapeutic options and disease course with a total knowledge score ranging from 0 to 26. Subjects with dementia were excluded. Patient misconceptions were reflected in the mean (SD) total, diagnosis therapeutic options and disease course scores. Significant knowledge gaps were identified amongst PD patients in all three aspects of the questionnaire. [2]

International Studies

Many studies have addressed this aspect and the summary of these studies have been shown in Table 1. Also important studies on genetics in PD [8], in Tanzania[15] and European data have been publeshed[17].

Newer perspectives and options to improve knowledge

Internet and social media

With the availability and easy access to internet, smartphones and social media, Indian population has greater reach to the health-related information. This opportunity should be better utilized to spread the knowledge of PD and remove ignorance in remote areas of our country. These programs are already being used by public for other health issues. Social media similarly is also being used spread the videos showing improvements with dance therapy and exercise in patients with PD.

Table 1: Findings of various studies on the topic of knowledge, attitudes and perceptions in PD

	Study Topic	Author/ Year /Country	Sample Size	Objectives	Conclusions
1	Knowledge, attitudes and perceptions of Parkinson's disease: A cross-sectional survey of Asian patients	Jitkritsadakul et al/2017/ Thailand	108(34 working and 74 Retired)	To evaluate general understanding of PD and identify knowledge gaps amongst PD patients using a validated PD knowledge questionnaire	Significant Knowledge Gaps were identified. People relatively Had better knowledge about disease diagnosis than treatment and disease course. Female patients, Working Patients and Younger patients had Better Knowledge than old retired males. Knowledge about these gaps would help the physicians to make timely interventions[2]
2	Public Knowledge and Attitude toward Essential Tremor: A Questionnaire Survey.	Shalaby et al./ 2016/ USA	250(100 pts attending Vascular Clinic,100 pts attending General Neurology OPD,50 PD patients)	To evaluate the awareness of Public(Non-ET patients) on Essential Tremors	Public knowledge of the existence and features of ET is overall poor. Greater awareness is important for the ET community.[3]
3	Meanings of feeling well among women with Parkinson's disease.	Olsson et al/2015/ Sweden	9 women with PD	To describe the meaning of feeling well in women with PD	Women with PD were feeling well when they did not have bodily pains/fewer motor symptoms of PD/when they had good support in family.[5]
4	Knowledge of Parkinson's Disease in a Multiethnic Urban Asian Setting	Tan et al/2015/Malaysia	1258 people of general public	Aimed to explore the level of public knowledge regarding PD in a large multiethnic urban Asian cohort, and (as a secondary aim) in a smaller cohort of PD patients and caregivers	Important gaps in knowledge were evident, which could present a barrier to early diagnosis and appropriate treatment of PD.[4]
5	Time intervals in diagnosing Parkinson's disease: The patients' views	Plouvier et al/2015 Australia	Review of 52 essays written by PD patients	To explore patients' views on their pathway to the diagnosis of Parkinson's disease	The pathway to diagnosing PD is process influenced by patient-, health care provider- and disease-related factors.[6]
6	Hospitalization in Parkinson's disease: a survey of UK neurologists, geriatricians and Parkinson's disease nurse specialists	Skelly et al/2015/ UK	93 health care workers in 65 hospitals	To document current service provision in the UK and to explore clinicians' views on standards of care and potential service improvements	Care for Parkinson's hospitalized patients is not highly rated by UK Parkinson's clinicians and interventions to improve care should be studied.[7]
7	Identifying gaps in knowledge about Parkinson disease among medical professionals in Thailand	Bhidayasiri et al/2014/ Thailand	114	To Identify gaps in knowledge about Parkinson disease among medical professionals in Thailand	There were significant knowledge gaps about Parkinson disease among practitioners to correct misconceptions about PD, authors propose that education and training on PD should be extensively covered in medical schools and continue as CME in both content and training methodologies throughout the careers of doctors.[9]

	Study Topic	Author/ Year /Country	Sample Size	Objectives	Conclusions
8	Knowledge and attitudes about Parkinson's disease among a diverse group of older adults	Pan et al/2014/ USA	229	The purpose of this mixed-methods study was to understand knowledge and attitudes about PD among a racially/ethnically diverse group of community members	There are several conclusions and future directions . First, there was a low level of knowledge about PD in all communities. Second, multiple barriers to care were identified: 1) fear and stigma due to PD across all racial/ethnic groups; 2)To care among Chinese-Americans due to language barriers and African-Americans due to mistrust; 3) lower cues to action among Chinese-Americans and African-Americans when compared to whites. Significant improvement is needed in improving knowledge and attitude.[11]
9	The experiences of Taiwanese older individuals at different stages of Parkinson disease	Taiwan	15(9men and 6 women)	To asses the experiences of Taiwanese older individuals at different stages of Parkinson disease and plan for appropriate interventions	Conclusions (1) more information on the disease in stage 1 is needed, (2) more guidance on regular medication usage and establish a support system in stage 2, (3) stress the importance of a safe living environment and improve participation in support groups in stage 3, and (4) counselling regarding life.[21]
10	Knowledge and attitudes towards genetic testing in those affected with Parkinson's disease	Scuffham et al/2014/ Australia	3000	The aim of this study was to assess the knowledge of genetics and the attitudes toward genetic testing of patients affected with PD. A secondary aim was to identify the determinants of attitudes towards genetic testing.	This study provides the basis to guide the development of specific genetic counseling protocols for PD, and other neuro-genetic diseases where informed decision-making around genetic testing may be required.[13]
11	Impact of the Parkinson's disease medication protocol program on nurses' knowledge and management of Parkinson's disease medicines in acute and aged care settings	Chenoweth et al/2013/ Australia	-	To determine the impact of a Parkinson's medicine education program on nurses' knowledge and practices in two settings where people with Parkinson's disease are cared for: hospitals and residential aged care facilities	These study results agree with the studies which identifies that without focussed clinical education nurses don thave sufficient knowledge to manage Parkinson'streatment and avoid poor outcome arising from delays, errors and omissions, nor do they know how to provide safe and effective care for persons with Parkinson's.[14]
12	Genetic testing and Parkinson disease: assessment of patient knowledge, attitudes, and interest	Falcone et al/2011/ USA	168	To assess patient knowledge, attitudes, and interest towards genetic testing and counseling	Patients had high level of interest but low level of knowledge about genetics of Parkinson disease. Study shows the need for genetics education and counseling in the PD population.[16]

	Study Topic	Author/ Year /Country	Sample Size	Objectives	Conclusions
13	Perceptions of cause and control in people with Parkinson's disease	Eccles et al/2010/UK	11	This study aims to investigate how people with Parkinson's disease (PD) perceived both the cause and their control of the disease	Different methods were used by participants so that control could be maintained, at least to some extent. However, a better social and healthcare system is needed to support patients.[22]
14	Patient perception of dyskinesia in Parkinson's disease.	Hung et al/2010/ Canada	259	To evaluate the perception of patients with Parkinson's disease (PD) regarding dyskinesia	Patients with PD with dyskinesia are less likely to be concerned about dyskinesia and more likely to choose dyskinesia over slowness and rigidity than patients without dyskinesia.[23]
15	Lay persons' beliefs and knowledge about Parkinson's disease: prevalence and socio-demographic correlates.	Werner et al/2009/ Israel	632	To assess the prevalence and socio-demographic correlates of lay persons' beliefs and knowledge about PD.	Study suggest the need to expand this line of research and to encourage the development of programs aimed at disseminating knowledge about PD.[19]
16	Caregiver education in Parkinson's disease: formative evaluation of a standardized program in seven European countries.	A'Campo et al/2010/ Esto-nia, Finland, Germany, Italy, The Netherlands, Spain and United Kingdom	PD patients (151) and caregivers (137)	To formatively evaluate a standardized psychosocial education program for patients with Parkinson's disease (PD) and their caregivers	The study found the program feasible to carry out in the various countries. This exploratory study led to improvements in the program and recommendations for further research.[24]
17	Knowledge of risk factors for falling reported by patients with Parkinson disease	Sadowski et al/2007/ Canada	28	To evaluate awareness of the risk factors for falling among a group of community dwelling patients with Parkinson disease (PD) using the Falls Risk Awareness Questionnaire	Persons with PD are at substantial risk of falling, yet many appear to be unaware of common risk factors, especially medication use.[20]
18	Comparing knowledge and attitudes towards genetic testing for Parkinson's disease in an American and Asian population	Tan et al/2009/Asia and America	515	Investigating the relationship between knowledge of and attitude towards predictive genetic testing of PD in two independent centers in America and Asia involving PD patients and caregivers.	The relationship between the level of genetic knowledge and attitude towards potential risks and benefits of predictive genetic testing in PD was different in study groups. These observations are important to develop gentic counselling protocols.[25]

Involvement of celebrities and role models of the society

There is also a need to involve more people who have celebrity status to spread awareness about PD. The example of Michel J Fox is known to all. He has been spearheading the task of research, spreading awareness and also motivating people to donate for the cause.

Future trends and policies

There is a need to have multipronged approach to spread the awareness, knowledge and true perceptions about PD in Indian society.

1. Television, print, electronic and social media have imoprtant role in spreading awareness
2. By organizing regular programs on PD day at all the healthcare institutions that are involved in care of the patients with PD
3. By involving the important personalities of our country to speak about it at various fora

Conclusion

To conclude, globally the level of knowledge of PD among the general public is very low. Systematic efforts should be made to allay the misperceptions about the aetiology and treatment of PD. There is urgent need to make a collective effort through the international and national societies for this noble cause. This would go a long way in management of the patients with PD in both rural as well as urban healthcare setting.

References

1. Yadav R, Shukla G, Goyal V, Singh S, Behari M. Knowledge of Parkinson's disease among patients and caregivers attending movement disorder clinic at a tertiary care centre in north India. Ann Indian Acad Neurol. 2012;15(4);294-6. doi: 10.4103/0972-2327.104339. PubMed PMID: 23349596; PubMed Central PMCID: PMC3548369.
2. Jitkritsadakul O, Boonrod N, Bhidayasiri R. Knowledge, attitudes and perceptions of Parkinson's disease: A cross-sectional survey of Asian patients. J Neurol Sci. 2017;374;69-74. doi: 10.1016/j.jns.2016.12.063. Epub 2016 Dec 29. PubMed PMID: 28104234.
3. Shalaby S, Indes J, Keung B, Gottschalk CH, Machado D, Patel A, Robakis D, Louis ED. Public Knowledge and Attitude toward Essential Tremor: A Questionnaire Survey. Front Neurol. 2016;7;60. doi: 10.3389/fneur.2016.00060. eCollection 2016 Apr 22. PubMed PMID: 27148160; PubMed Central PMCID: PMC4840213.
4. Tan AH, Tan CT, Marras C, Loh KW, Wye Ho NW, Lim QH, Tan PW, Lim CC, Cheong YW, Kong ST, Schee JP, Tan KH, Soo SK, Vanderschaaf C, Lai Heong Lew S, Mahamad UA, Goh KJ, Yong HS, Lim SY. Knowledge of Parkinson's Disease in a Multiethnic Urban Asian Setting. J Parkinsons Dis. 2015;5(4);865-79. doi: 10.3233/JPD-150594. PubMed PMID: 26444089.
5. Olsson M, Nilsson C. Meanings of feeling well among women with Parkinson's disease. Int J Qual Stud Health Well-being. 2015;10;28730. doi: 10.3402/qhw.v10.28730. eCollection 2015 Oct 20. PubMed PMID: 26489404; PubMed Central PMCID: PMC4613899.

6. Plouvier AO, Olde Hartman TC, Boots LP, Bloem BR, van Weel C, Lagro-Janssen AL. Time intervals in diagnosing Parkinson's disease: The patients' views. Patient Educ Couns. 2015;98(6);777-82. doi: 10.1016/j.pec.2015.02.010. Epub 2015 Feb 25. PubMed PMID: 25770049.
7. Skelly R, Brown L, Fakis A, Walker R. Hospitalization in Parkinson's disease: a survey of UK neurologists, geriatricians and Parkinson's disease nurse specialists. Parkinsonism Relat Disord. 2015;21(3);277-81. doi: 10.1016/j.parkreldis.2014.12.029. Epub 2015 Jan 6. PubMed PMID: 25585995.
8. Gupte M, Alcalay RN, Mejia-Santana H, Raymond D, Saunders-Pullman R, Roos E, Orbe-Reily M, Tang MX, Mirelman A, Ozelius L, Orr-Urtreger A, Clark L, Giladi N, Bressman S, Marder K. Interest in genetic testing in Ashkenazi Jewish Parkinson's disease patients and their unaffected relatives. J Genet Couns. 2015;24(2);238-46. doi: 10.1007/s10897-014-9756-x. Epub 2014 Aug 17. PubMed PMID: 25127731; PubMed Central PMCID: PMC4331260.
9. Bhidayasiri R, Brenden N, Viwattanakulvanid P, Jitkritsadakul O, Tabucanon R, Jagota P, Hojer H. Identifying gaps in knowledge about Parkinson disease among medical professionals in Thailand. Neurology. 2014;82(24);2238-40.doi:0.1212/WNL.0000000000000515. PubMed PMID: 24960833.
10. Hamberg K, Hariz GM. The decision-making process leading to deep brain stimulation in men and women with parkinson's disease - an interview study. BMC Neurol. 2014;14;89. doi: 10.1186/1471-2377-14-89. PubMed PMID: 24761767; PubMed Central PMCID: PMC4000892.
11. Pan S, Stutzbach J, Reichwein S, Lee BK, Dahodwala N. Knowledge and attitudes about Parkinson's disease among a diverse group of older adults. J Cross Cult Gerontol. 2014;29(3);339-52. doi: 10.1007/s10823-014-9233-x. PubMed PMID: 24752756; PubMed Central PMCID: PMC4134410.
12. Li J, Chen D, Song W, Chen K, Cao B, Huang R, Yang R, Shang H. Survey on general knowledge on Parkinson's disease in patients with Parkinson's disease and current clinical practice for Parkinson's disease among general neurologists from Southwest China. Clin Neurol Neurosurg. 2014;118;16-20. doi: 10.1016/j.clineuro.2013.12.009. Epub 2013 Dec 30. PubMed PMID: 24529223.
13. Scuffham TM, McInerny-Leo A, Ng SK, Mellick G. Knowledge and attitudes towards genetic testing in those affected with Parkinson's disease. J Community Genet. 2014;5(2);167-77. doi: 10.1007/s12687-013-0168-7. Epub 2013 Sep 10. PubMed PMID: 24018619; PubMed Central PMCID: PMC3955457.
14. Chenoweth L, Sheriff J, McAnally L, Tait F. Impact of the Parkinson's disease medication protocol program on nurses' knowledge and management of Parkinson's disease medicines in acute and aged care settings. Nurse Educ Today. 2013;33(5);458-64. doi: 10.1016/j.nedt.2012.04.022. Epub 2012 May 22. PubMed PMID: 22626862.
15. Mshana G, Dotchin CL, Walker RW. 'We call it the shaking illness': perceptions and experiences of Parkinson's disease in rural northern Tanzania. BMC Public Health. 2011;11;219. doi: 10.1186/1471-2458-11-219. PubMed PMID: 21477284; PubMed Central PMCID: PMC3088907.
16. Falcone DC, Wood EM, Xie SX, Siderowf A, Van Deerlin VM. Genetic testing and Parkinson disease: assessment of patient knowledge, attitudes, and interest. J Genet Couns. 2011;20(4);384-95. doi: 10.1007/s10897-011-9362-0. Epub 2011 Apr 8. PubMed PMID: 21476119; PubMed Central PMCID: PMC3280083.

17. Chou KL, Zamudio J, Schmidt P, Price CC, Parashos SA, Bloem BR, Lyons KE, Christine CW, Pahwa R, Bodis-Wollner I, Oertel WH, Suchowersky O, Aminoff MJ, Malaty IA, Friedman JH, Okun MS. Hospitalization in Parkinson disease: a survey of National Parkinson Foundation Centers. Parkinsonism Relat Disord. 2011;17(6);440-5. doi: 10.1016/j.parkreldis.2011.03.002. Epub 2011 Apr 1. PubMed PMID: 21458353; PubMed Central PMCID: PMC3895941.
18. Valldeoriola F, Coronell C, Pont C, Buongiorno MT, Cámara A, Gaig C, Compta Y; ADHESON Study Group. Socio-demographic and clinical factors influencing the adherence to treatment in Parkinson's disease: the ADHESON study. Eur J Neurol. 2011;18(7);980-7. doi: 10.1111/j.1468-1331.2010.03320.x. Epub 2010 Dec 29. PubMed PMID: 21199185.
19. Werner P, Korczyn AD. Lay persons' beliefs and knowledge about Parkinson's disease: prevalence and socio-demographic correlates. Parkinsonism Relat Disord. 2010;16(6);415-7. doi: 10.1016/j.parkreldis.2009.12.008. Epub 2010 Jan 12. PubMed PMID: 20061175.
20. Sadowski CA, Jones CA, Gordon B, Feeny DH. Knowledge of risk factors for falling reported by patients with Parkinson disease. J Neurosci Nurs. 2007;39(6);336-41. PubMed PMID: 18186418.
21. Liao YC, Wu YR, Tsao LI, Lin HR. The experiences of Taiwanese older individuals at different stages of Parkinson disease. J Neurosci Nurs. 2013 Dec;45(6):370-7. doi: 10.1097/JNN.0b013e3182a3cd5c. PubMed PMID: 24217147.
22. Eccles FJ, Murray C, Simpson J. Perceptions of cause and control in people with Parkinson's disease. Disabil Rehabil. 2011;33(15-16):1409-20. doi: 10.3109/09638288.2010.533241. Epub 2010 Nov 22. PubMed PMID: 21091132.
23. Hung SW, Adeli GM, Arenovich T, Fox SH, Lang AE. Patient perception of dyskinesia in Parkinson's disease. J Neurol Neurosurg Psychiatry. 2010 Oct;81(10):1112-5. doi: 10.1136/jnnp.2009.173286. Epub 2010 Jul 28. PubMed PMID: 20667858.
24. A'Campo LE, Spliethoff-Kamminga NG, Macht M; EduPark Consortium, Roos RA. Caregiver education in Parkinson's disease: formative evaluation of a standardized program in seven European countries. Qual Life Res. 2010 Feb;19(1):55-64. doi: 10.1007/s11136-009-9559-y. Epub 2009 Nov 28. PubMed PMID: 19946755; PubMed Central PMCID: PMC2804793.
25. Tan EK, Lee J, Hunter C, Shinawi L, Fook-Chong S, Jankovic J. Comparing knowledge and attitudes towards genetic testing in Parkinson's disease in an American and Asian population. J Neurol Sci. 2007 Jan 31;252(2):113-20. Epub 2006 Dec 14. PubMed PMID: 17174340.

INDEX

C

D

E

J

K

L

M

N

O

P

Q

R

S

T

U

V

W

X

Y

Z

For Product Safety Concerns and Information please contact our
EU representative GPSR@taylorandfrancis.com Taylor & Francis
Verlag GmbH, Kaufingerstraße 24, 80331 München, Germany